CAMBRIDGE LIBRARY COLLECTION

Books of enduring scholarly value

History of Oceania

This series focuses on Australia, New Zealand and the Pacific region from the arrival of European seafarers and missionaries to the early twentieth century. Contemporary accounts document the gradual development of the European settlements from penal colonies and whaling stations to thriving communities of farmers, miners and traders with fully-fledged administrative and legal systems. Particularly noteworthy are the descriptions of the indigenous peoples of the various islands, their customs, and their differing interactions with the European settlers.

Polynesian Researches during a Residence of Nearly Six Years in the South Sea Islands

From humble origins, and trained by the London Missionary Society in theology, printing and rudimentary medicine, William Ellis (1794–1872) sailed for the Society Islands in 1816. He found himself at the cusp of major cultural change as Western influences affected the indigenous Polynesians. During his time there, Ellis became a skilled linguist and able chronicler of the traditional yet rapidly shifting way of life. He succeeded in capturing vivid stories of a leisured people who, without written language, had developed a rich oral tradition, social structure and belief system. Published in 1829, this two-volume collection proved to be an important reference work, notably for its natural history; it soon accompanied Darwin aboard the *Beagle*. In Volume 2, Ellis moves between Huahine and Raiatea, giving further background on the existing customs and polytheistic rituals, contrasted with the introduction of Western religion, dress, schools, housing, medicine and law.

Cambridge University Press has long been a pioneer in the reissuing of out-of-print titles from its own backlist, producing digital reprints of books that are still sought after by scholars and students but could not be reprinted economically using traditional technology. The Cambridge Library Collection extends this activity to a wider range of books which are still of importance to researchers and professionals, either for the source material they contain, or as landmarks in the history of their academic discipline.

Drawing from the world-renowned collections in the Cambridge University Library and other partner libraries, and guided by the advice of experts in each subject area, Cambridge University Press is using state-of-the-art scanning machines in its own Printing House to capture the content of each book selected for inclusion. The files are processed to give a consistently clear, crisp image, and the books finished to the high quality standard for which the Press is recognised around the world. The latest print-on-demand technology ensures that the books will remain available indefinitely, and that orders for single or multiple copies can quickly be supplied.

The Cambridge Library Collection brings back to life books of enduring scholarly value (including out-of-copyright works originally issued by other publishers) across a wide range of disciplines in the humanities and social sciences and in science and technology.

Polynesian Researches
during a Residence of Nearly Six Years in the South Sea Islands

Volume 2

William Ellis

CAMBRIDGE
UNIVERSITY PRESS

University Printing House, Cambridge, CB2 8BS, United Kingdom

Published in the United States of America by Cambridge University Press, New York

Cambridge University Press is part of the University of Cambridge.
It furthers the University's mission by disseminating knowledge in the pursuit of education, learning and research at the highest international levels of excellence.

www.cambridge.org
Information on this title: www.cambridge.org/9781108065399

This edition first published 1829
This digitally printed version 2014

ISBN 978-1-108-06539-9 Paperback

IDOLS.

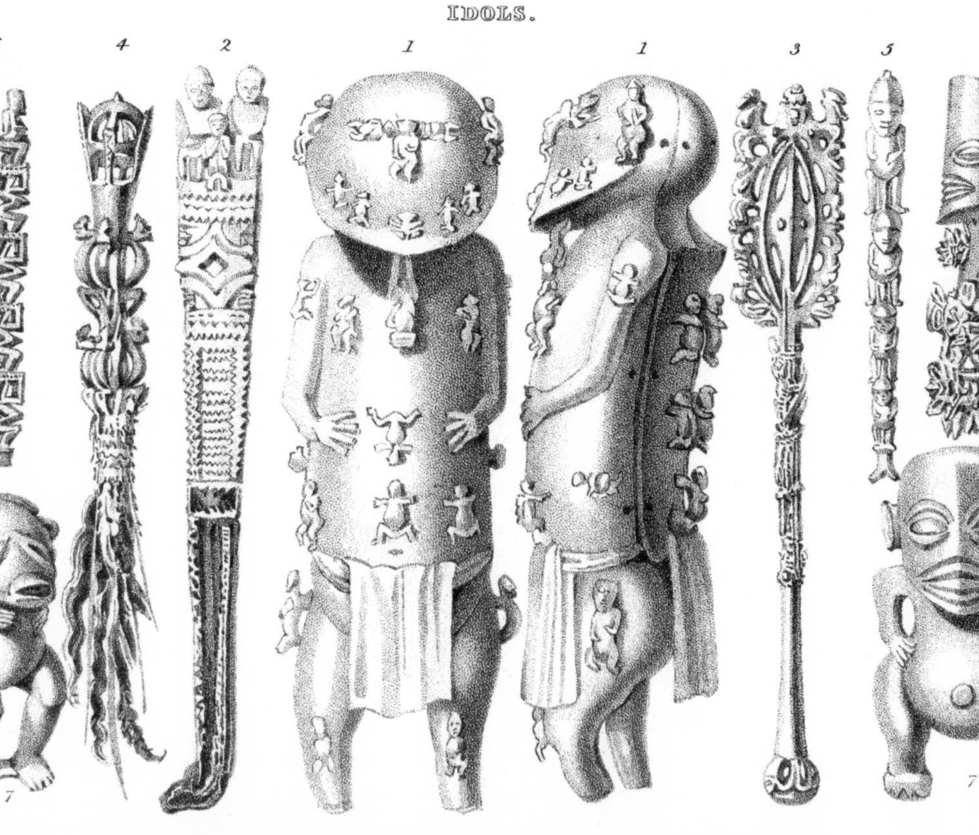

Worshipped by the Inhabitants of the South Sea Islands.

POLYNESIAN RESEARCHES,

DURING

A RESIDENCE OF NEARLY SIX YEARS

IN THE

SOUTH SEA ISLANDS;

INCLUDING

DESCRIPTIONS OF THE NATURAL HISTORY AND SCENERY OF THE ISLANDS—WITH REMARKS ON THE HISTORY, MYTHOLOGY. TRADITIONS, GOVERNMENT, ARTS, MANNERS, AND CUSTOMS OF THE INHABITANTS.

BY

WILLIAM ELLIS,

MISSIONARY TO THE SOCIETY AND SANDWICH ISLANDS, AND AUTHOR OF THE "TOUR OF HAWAII."

"In so vast a field, there will be room to acquire fresh knowledge for centuries to come, coasts to survey, countries to explore, inhabitants to describe, and perhaps to render more happy." COOKE.

IN TWO VOLUMES.

VOL. II.

LONDON:

FISHER, SON, & JACKSON, NEWGATE STREET,

M,DCCC,XXIX.

CONTENTS OF VOL. II.

CHAP. I.

CHAP. II.

CHAP. III.

CHAP. IV.

CHAP. V.

CHAP. VI.

CHAP. VII.

CHAP. VIII.

CHAP. IX.

CHAP. XVIII.

CHAP. XIX.

PLATES IN VOL. II.

WOOD ENGRAVINGS.

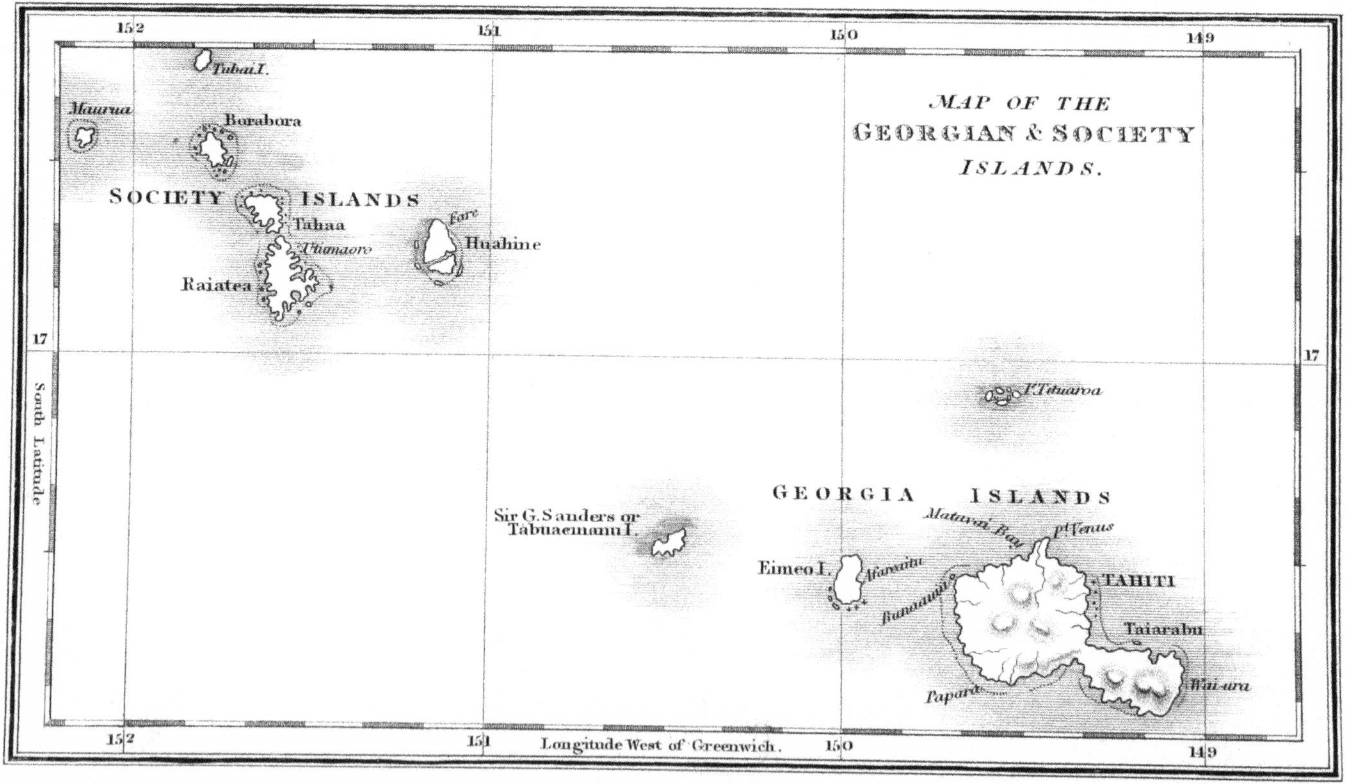

MAP OF THE
GEORGIAN & SOCIETY
ISLANDS.

152
151
150
149
17
South Latitude
Longitude West of Greenwich.

Tubai I.
Maurua
Borabora
SOCIETY ISLANDS
Tahaa
Raiatea
Fare
Huahine
Sir G. Sanders or Tabuaemanu I.
GEORGIA ISLANDS
Eimeo I.
Matavai Bay
Pt. Venus
TAHITI
Taiarabu
Wai-ura
Papara
Tetuaroa

London; Published by H. Fisher Son & P. Jackson 1829.

Drawn & Engraved by J. & C. Walker.

POLYNESIAN RESEARCHES.

CHAP. I.

Voyage to Raiatea—Appearance of the coral reefs—Breaking of the surf—Islets near the passage to the harbours—Landing at Tipaemau—Description of the islands—Arrival at Vaóaara—Singular reception—Native salutations—Improvement of the settlement—Traditionary connexion of Raiatea with the origin of the people—General account of the South Sea Islanders—Physical character, stature, colour, expression, &c.—Mental capacity, and habits—Aptness to receive instruction—Moral character—Hospitality—Extensive and affecting moral degradation—Its enervating influence—Longevity—Comparative numbers of the inhabitants—Indications and causes of depopulation—Beneficial tendency of Christianity.

DURING the first years of our establishment in Huahine, frequent voyages were necessary; and, early in 1819, circumstances rendered it expedient that we should revisit Raiatea. As we expected to be absent for several weeks, Mrs. Barff and Mrs. Ellis accompanied us; Mr. Orsmond was returning to his station, and we embarked in his boat, although it was scarcely large enough to contain our party and half a dozen native rowers. The morning on which we sailed was fine; the sea gently rippled with the freshening breeze, which was fair and steady, without being violent. Our voyage was pleasant; and soon after two in the afternoon of the same day, we entered an opening in the reef, a few miles to the northward of that leading to Opoa. This entrance is called by the inhabitants *Tipae mau*, True, or permanent, landing (place.)

The coral reef, around the eastern shores of Raiatea and Tahaa, often exhibits one of the most sublime and beautiful marine spectacles that it is possible to behold. It is generally a mile, or a mile and a half, and occasionally two miles, from the shore. The surface of the water within the reef is placid and transparent; while that without, if there be the slightest breeze, is considerably agitated; and, being unsheltered from the wind, is generally raised in high and foaming waves.

The trade-wind, blowing constantly towards the shore, drives the waves with violence upon the reef, which is from five, to twenty or thirty yards wide. The long rolling billows of the Pacific, extending sometimes, in one unbroken line, a mile or a mile and a half along the reef, arrested by this natural barrier, often rise ten, twelve, or fourteen feet above its surface; and then, bending over it their white foaming tops, form a graceful liquid arch, glittering in the rays of a tropical sun, as if studded with brilliants. But, before the eyes of the spectator can follow the splendid aqueous gallery which they appear to have reared, with loud and hollow roar they fall in magnificent desolation, and spread the gigantic fabric in froth and spray upon the horizontal and gently broken surface of the coral.

In each of the islands, and opposite the large valleys, through which a stream of water falls into the ocean, there is usually a break, or opening, in the line of reef that surrounds the shore—a most wise and benevolent provision for the ingress and egress of vessels, as well as a singular phenomenon in the natural history of these marine ramparts to the islands. Whether the current of fresh water, constantly flowing from the rivers to the ocean, prevents the tiny architects from building their concentric walls in one continued line, or whether in the fresh water itself there is any quality inimical to the growth or increase of coral, is not easy to determine; but it is a remarkable fact, that few openings occur in the reefs which surround the South Sea Islands, excepting opposite those parts of the shore from which streams of fresh water flow into the sea. Reefs of varied, but generally circumscribed extent, are frequently observed within the large outer barrier, and near the shore, or mouth of the river; but they are formed in shallow places, and the coral is of a different and more slender kind, than that of which the larger reef, rising from the depths of the ocean, is usually composed. There is no coral in the lagoons of the large islands.

The openings in the reefs around Sir Charles Sander's Island, Maurua, and other low islands, are small and intricate, and sometimes altogether wanting, probably because the land, composing these islands, collects but a scanty portion of water; and, if any, only small and frequently interrupted streams flow into the sea. The openings in the reefs around the larger islands, not only afford direct access to the indentations in the coast, and the mouths of the valleys, which form the best harbours,—but secure to shipping a supply of fresh water, in equal, if not greater abundance, than it could be procured in any other part of the island. The circumstance, also, of the rivers near the harbours flowing into the sea, affords the greatest facility in procuring fresh water, which is so valuable to seamen.

The openings in the reef, on the eastern side of Raiatea, are not only serviceable to navigation, but highly ornamental, adding greatly to the beauty of the surrounding scenery. At the *Ava Moa*, or Sacred Entrance leading to Opoa, there is a small island, on which a few cocoa-nut trees are growing. At Tipaemau there are two, one on each side of the opening, rising from the extremity of the line of reef. The little islets, elevated three or four feet above the water, are clothed with shrubs and verdure, and adorned with a number of lofty cocoa-nut trees. At Te-Avapiti, several miles to the northward of Tipaemau, and opposite the Missionary settlement—where, as its name indicates, are two openings—there are also two beautiful, green, and woody islands, on which the lowly hut of the fisherman, or of the voyager waiting for a favourable wind, may be often seen. Two large and very charming islands adorn the entrance at Tomahahotu, leading to the island of Tahaa. The largest of these is not more than half a mile in circumference, but both are covered with fresh and evergreen shrubs and trees.

Detached from the large islands, and viewed in connexion with the ocean rolling through the channel on the one side, or the foaming billows dashing, and roaring, and breaking over the reef on the other, they appear like emerald gems of the ocean, contrasting their solitude and verdant beauty with the agitated element sporting in grandeur around. They are useful, as well as ornamental. The tall cocoa-nuts that grow on their surface, can be seen many miles distant; and the native mariner is thereby enabled to steer directly towards the spot where he knows he shall find a passage to the shore. The constant current passing the opening, probably deposited on the ends of the reef fragments of coral, sea weeds, and drift-wood, which in time rose above the surface of the water. Seeds borne thither by the waves, or wafted by the winds, found a soil on which they could germinate—decaying vegetation increased the mould—and by this process it is most likely these beautiful little fairy-looking islands were formed on the ends of the reefs at the entrance to the different harbours.

We landed on one at Tipaemau, partook of some refreshment under the shade the shrubbery afforded, while our boat's crew climbed the trees, and afterwards made an agreeable repast on the nuts which they gathered. We planted, as memorials of our visit, the seeds of some large ripe oranges, which we had brought with us; then launched our boat, and prosecuted our voyage within the reef, towards the other side of the island, where the Missionary settlement was then established. This part of our voyage, for twelve or fourteen miles, was most delightful. The beauty of the wooded or rocky shores now appeared more rich and varied than before; the stillness of the smooth waters around was only occasionally disturbed by the passage of a light, nautilus-like canoe, with its little sail of white native cloth, or the rapid flight of a shoal of flying-fish, which, when the dashing of our oars or the progress of our boat intercepted their course or awakened their alarm, sprang from their native element, and darted along, three or four feet above the water

Ioretea, the Ulitea of Captain Cook, or, as it is now more frequently called by the natives, Raiatea, is the largest of the Society Islands. Its form is somewhat triangular, and its circumference about fifty miles. The mountains are more stupendous and lofty than those of Huahine, and in some parts equally broken and picturesque. The northern and western sides are singularly romantic; several pyramidal and cônical mountains rising above the elevated and broken range, that stretches along in a direction nearly parallel with the coast, and from one to three miles distant from the beach. Though the shore is generally a gradual and waving ascent from the water's edge to the mountain, it is frequently rocky and broken. At Mahapoto, about half way between Opoa, the site of their principal temple, the ancient residence of the reigning family, and Utumaoro at the north-east angle of the island, there is a deep indentation in the coast. The rocks rise nearly perpendicular in some places on both sides, and the smooth surface of the ocean extends a mile and a half, or two miles, towards the mountains. The shores of this sequestered bay are covered with sand, shells, and broken coral. At the openings of several of the little glens which surround it, the cottages of the natives are seen peeping through the luxuriant foliage of the pandanus, or the purau; while the cultivated plantations in various parts extend from the margin of the sea to the foot of the mountains. The rivers that roll along their rocky courses from the head of the ravines to the ocean below—and the distant mountains, that rise in the interior—combine to form, though on a limited scale, the most rich, romantic, and beautiful landscapes. The islands in general are well supplied with water. The mountains are sufficiently elevated to intercept the clouds that are wafted by the trade-winds over the surface of the Pacific; being clothed with verdure to their very summits, while they attract the moisture, they also prevent its evaporation. Most of the rivers or streams rise in the mountainous parts, and though, from the peculiar structure of these parts, and the circumscribed extent of the islands, the distance from their source to their union with the sea is comparatively short; yet the body of water is often considerable, and the uneven ground through which they have cut their way, the rocky projections that frequently divide the streams, and the falls that occur between the interior and the shore, cause the rivers to impart a charming freshness, vivacity, and splendour to the surrounding scenery.

Next to Tahiti, Raiatea perhaps is better supplied with rivers, or streams of excellent water, than any other island of the group. Its lowland is extensive, and the valleys, capable of the highest cultivation, are not only spacious, but conveniently situated for affording to the inhabitants intercourse with other parts of the island. On the north-west is a small but very secure harbour, called Hamaniino. Most of the ships formerly visiting Raiatea anchored in this by no means capacious, but convenient and sequestered, harbour. Such vessels usually entered the reefs that surround the two islands, either at the opening called Teavapiti, a little to the southward of Utumaora, or at that denominated Tomahahotu opposite the south end of the island of

Tahaa. They then proceeded within the reefs along the channel between the islands, to the harbour.

Water and wood were at all times procured with facility from the adjacent shore; and supplies of stock, poultry, and vegetables might generally be obtained by barter with the inhabitants. The mountains of the interior sheltered the bay from the strong eastern and southerly winds; and the wide opening in the reef, opposite the mouth of the valley forming the head of the bay, favoured the departure of vessels with the ordinary winds. A small and partially wooded island on the north side of the opening in the reefs opposite the harbour, distinctly points out the passage, and is very serviceable to ships going to sea. A few miles beyond the harbour of Hamaniino, Vaóaara is situated, which was the former Missionary station, the residence of the chiefs, and principal part of the population. There are two open bays on the east side of the island, Opoa and Utumaoro. They were occasionally visited by shipping; and the latter has, since the removal of the Missionary station, become the general place of anchorage. But although they are secured from heavy waves by the reefs of coral that stretch along the eastern shore, they are exposed to the prevailing winds, excepting so far as they are sheltered by the islands at the entrance from the sea.

There are no lakes in Raiatea or Tahaa, but both islands are encircled in one reef, which is in some parts attached to their shores, and in others rises to the water's edge, at the distance of two or three miles from it. The water within the reefs, is as smooth as the surface of a lake in a gentleman's pleasure-ground, though often from eighteen to thirty fathoms deep. The coral reefs form natural and beautiful breakwaters, preserving the lowland and the yielding soil of the adjacent shore from the force and encroachment of the heavy billows of the ocean. Numbers of singular and verdant little islands are remarkably useful, as they are most frequently found at those points where the openings into the harbours are formed. They are, therefore, excellent sea-marks, and furnish convenient temporary residences for the fishermen, who resort to them during the season for taking the operu, *scomber scomber* of Linneus, and other fish, periodically visiting their shores. Here they dry and repair their nets while watching the approach of the shoals, and find them remarkably advantageous in prosecuting the most important of their fisheries.

The sun had nearly set when we reached the settlement. As we approached the shore, crowds of the natives, who had recognized some of our party, came off to meet us, wading into the sea above their waist, in order to welcome our arrival. While gazing on the motley group that surrounded our boat, or thronged the adjacent shore, and exchanging our salutations with those nearest us, before we were aware of their design, upwards of twenty stout men actually lifted our boat out of the water, and raised it on their shoulders, carrying us, thus elevated in the air, amid the shouts of the bearers, and the acclamations of the multitude on the shore, first to the beach, and then to the large court-yard in front of the king's house, where, after experiencing no small apprehension from this unusual mode of conveyance, we were set down safe and dry upon the pavement. Here we experienced a very hearty welcome from the chiefs and people. Their salutations were cordial, though unaccompanied by the ceremonies that were formerly regarded as indispensable. Considering the islanders as an uncivilized people, they seem to have been the most ceremonious of any with whom we are acquainted. This peculiarity appears to have accompanied them to the temples, to have distinguished the homage and the service they rendered to their gods, to have marked their affairs of state, and the carriage of the people towards their rulers, to have pervaded the whole of their social intercourse, to have been mingled with their most ordinary avocations, and even their rude and diversified amusements. Their salutations were often exceedingly ceremonious. When a chieftain from another island, or from any distant part, arrived, he seldom proceeded at once to the shore, but usually landed, in the first instance, on some of the small islands near. The king often attended in person, to welcome his guest, or, if unable to do this himself, sent one of his principal chiefs.

When the canoes of the visitor approached the shore, the chiefs assembled on the beach. Long orations were pronounced by both parties before the guests stepped on the soil: as soon as they were landed, a kind of circle was formed by the people; the king or chiefs on the one side, and the strangers on the other; the latter brought their marotai, or offering, to the king and the gods, and accompanied its presentation with an address, expressive of the friendship existing between them: the priest, or orators of the king, then brought the presents, or manufaiti, bird of recognition. Two young plantain-trees were first presented, one for *te atua,* the god; the other for *te hoa,* the friend. A plantain-tree and a pig were brought for the king, a similar offering for the god; this was followed by a plantain and a pig, for the *toe moe,* the sleeping hatchet. A plantain-tree and a bough were then brought for the taura, the cord or bond of union, and then a plantain and a pig for the friend.

In some of their ceremonies, a plantain-tree was substituted for a man, and in the first plantain-trees offered in this ceremony to the god and the friend, they might perhaps be so regarded. Considerable ceremony attended the reception of a company of Areois. When they approached a village or district, the inhabitants came out of their doors, and, greeting them, shouted Manava, Manava, long before they reached the place. They usually answered, Teie "Here," and so proceeded to the rendezvous appointed, where the marotai was presented to the king, and a similar offering to the god.

Our mode of saluting by merely shaking hands, they consider remarkably cold and formal. They usually fell upon each others necks, and tauahi, or embraced each other, and saluted by touching or rubbing noses. This appears to be the common mode of welcoming a friend, practised by all the inhabitants of the Pacific. It also prevails among the natives of Madagascar. During my visit to New Zealand, I was several times greeted in this manner by chiefs, whose tataued countenances, and ferocious appearance, were but little adapted to cherish any predisposition to so close a contact. This method of saluting is called by the New Zealanders Ho-gni, Honi by the Sandwich Islanders, and Hoi by the Tahitians. In connexion with this, the custom of cutting themselves with sharks' teeth, and indulging in loud wailing, was a very singular method of receiving a friend, or testifying gladness at his arrival; it was, however, very general when the Europeans first arrived.

In the court-yard of the king we were met by our friends Messrs. Williams and Threlkeld, who, considering the short time they had been among the people, had been the means of producing an astonishing change, not only in their habits and appearance, but even in the natural face of the district. A carpenter's shop had been erected, the forge was daily worked by the natives, neat cottages were rising in several directions, and a large place of worship was building. The wilderness around was cleared to a considerable extent; the inhabitants of other parts were repairing to Vaoaara, and erecting their habitations, that they might enjoy the advantage of instruction. A flourishing school was in daily operation, and a large and attentive congregation met for public worship in the native chapel every Sabbath-day. In the society of our friends we spent a fortnight very pleasantly, and having adjusted our public arrangements, returned to Huahine, in the Haweis, in which Messrs. Barff, Williams, and myself, proceeded to Tahiti.

The island of Raiatea is not only the most important in the leeward group, from its central situation and its geographical extent, but on account of its identity, in tradition, with the origin of the people, and their preservation in the general deluge. It has been celebrated as the cradle of their mythology, the seat of their oracle, and the abode of those priests whose predictions for many generations regulated the expectations of the nation. It is also intimately connected with the most important matters in the traditionary history and ancient religion of the people.

The inhabitants of the Georgian, Society, and adjacent isles, comprehended, according to the ideas they entertained prior to the arrival of foreign vessels, the whole of the human race. In the island of Raiatea, their traditions informed them, their species originated, and hither after death they repaired. In connexion with my first voyages to the spot, which many have been accustomed to consider as the birth-place of mankind, and the region to which their disembodied spirits were supposed to resort, it may be proper to introduce the facts and observations, in reference to their origin, physical, intellectual, and moral character, which have resulted from subsequent visits to the island, and more extensive acquaintance with the people.

The inhabitants of the South Sea Islands are generally above the middle stature; but their limbs are less muscular and firm than those of the Sandwich Islanders, whom in many respects they resemble. They are, at the same time, more robust than the Marquesans, who are the most light and agile of the inhabitants of Eastern Polynesia. In size and physical power they are inferior to the New Zealanders, and probably resemble in person the Friendly Islanders, as much as any others in the Pacific; exhibiting, however, neither the gravity of the latter, nor the vivacity of the Marquesans. Their limbs are well formed, and although where corpulency prevails there is a degree of sluggishness in their actions, they are generally active in their movements, graceful and stately in their gait, and perfectly unembarrassed in their address. Those who reside in the interior, or frequently visit the mountainous parts of the islands, form an exception to this remark. The constant use of the naked feet in climbing the steep sides of the rocks, or the narrow defiles of the ravines, probably induces them to turn their toes inwards, which renders their gait exceedingly awkward.

Among the many models of perfection in the human figure that appear in the islands, (presenting to the eye of the stranger all that is beautiful in symmetry and graceful in action,) instances of deformity are now frequently seen, arising from a loathsome disease, of foreign origin, affecting the features of the face, and muscular parts of the body. There is another disease, which forms such a curvature of the upper part of the spine, as to produce what is termed a humped or broken back. The disease which produces this distortion of shape, and deformity of appearance, is declared by the natives, to have been unknown to their ancestors; and, according to the accounts some of them give it, was the result of a disease left by the crew of Vancouver's ship. It does not prevail in any of the other islands; and although such numbers are now affected with it, there is reason to believe, that, except the many disfigurements produced by the elephantiasis, which appears to have prevailed from their earliest antiquity, a deformed person was seldom seen.

The countenance of the South Sea Islander is open and prepossessing, though the features are bold, and sometimes prominent. The facial angle is frequently as perpendicular as in the European structure, excepting where the frontal and the occipital bones of the skull were pressed together in infancy. This was frequently done by the mothers, with the male children, when they were designed for warriors. The forehead is sometimes low, but frequently high and finely formed; the eye-brows are dark and well defined, occasionally arched, but more generally straight; the eyes seldom large, but bright and full, and of a jet black colour; the cheek-bones by no means high; the nose either rectilinear or aquiline, often accompanied with a fulness about the nostrils; it is seldom flat, notwithstanding it was formerly the practice of the mothers and nurses to press the nostrils of the female children, a flat and broad nose being by many regarded as more ornamental than otherwise. The mouth in general is well formed, though the lips are sometimes large, yet never so much so as to resemble those of the African. The teeth are always entire, excepting in extreme old age, and, though rather large in some, are remarkably white, and seldom either discoloured or decayed. The ears are large, and the chin retreating or projecting, most generally inclining to the latter. The form of the face is either round or oval, and but very seldom exhibits any resemblance to the angular form of the Tartar visage, while their profile frequently bears a most striking resemblance to that of the European countenance. Their hair is of a shining black or dark brown colour; straight, but not lank and wiry like that of the American Indian, nor, excepting in a few solitary instances, woolly like the New Guinea or New Holland negroes. Frequently it is soft and curly, though seldom so fine as that of the civilized nations inhabiting the temperate zones.

There is a considerable difference between the stature of the male and female sex here, as well as in other parts of the world, yet not so great as that which often prevails in Europe. The females, though generally more delicate in form and smaller in size than the men, are, taken altogether, stronger and larger than the females of England, and are sometimes remarkably tall and stout. A roundness and fulness of figure, without extending to corpulency, distinguishes the people in general, particularly the females.

It is a singular fact in the physiology of the inhabitants of this part of the world, that the chiefs, and persons of hereditary rank and influence in the islands, are, almost without exception, as much superior to the peasantry or common people, in stateliness, dignified deportment, and physical strength, as they are in rank and circumstances; although they are not elected to their station on account of their personal endowments, but derive their rank and elevation from their ancestry. This is the case with most of the groups of the Pacific, but peculiarly so in Tahiti and the adjacent isles. The father of the late king was six feet four inches high; Pomare was six feet two. The present king of Raiatea is equally tall. Mahine, the king of Huahine, but for the effects of age, would appear little inferior. Their limbs are generally well formed, and the whole figure proportioned to their height; which renders the difference between the rulers and their subjects so striking, that some have supposed they were a distinct race, the descendants of a superior people, who at a remote period had conquered the aborigines, and perpetuated their supremacy. It does not, however, appear necessary, in accounting for the fact, to resort to such a supposition; different treatment in infancy, superior food, and distinct habits of life, are quite sufficient.

The prevailing colour of the natives is an olive, a bronze, or a reddish brown—equally removed from the jet-black of the African and the Asiatic, the yellow of the Malay, and the red or copper-colour of the aboriginal American, frequently presenting a kind of medium between the two latter colours. Considerable variety, nevertheless, prevails in the complexion of the population of the same island, and as great a diversity among the inhabitants of different islands. The natives of the Paliser or Pearl Islands, a short distance to the eastward of Tahiti, are darker than the inhabitants of the Georgian group. It is not, however, a blacker hue that their skin presents, but a darker red or brown. The natives of Maniaa, or Mangeea, one of the Harvey cluster, and some of the inhabitants of Rurutu, and the neighbourhood to the south of Tahiti, designated by Malte Brun, "the Austral Islands," and the majority of the reigning family in Raiatea, are not darker than the inhabitants of some parts of southern Europe.

At the time of their birth, the complexion of Tahitian infants is but little if any darker than that of European children, and the skin only assumes the bronze or brown hue as they grow up under repeated or constant exposure to the sun. Those parts of the body that are most covered, even with their loose draperies of native cloth, are, through every period of life, much lighter coloured than those that are exposed; and, notwithstanding the dark tint with which the climate appears to dye their skin, the ruddy bloom of health and vigour, or the sudden blush, is often seen mantling the youthful countenance under the light brown tinge, which, like a thin veil, but partially conceals its glowing hue. The females who are much employed in beating cloth, making mats, or other occupations followed under shelter, are usually fairer than the rest; while the fishermen, who are most exposed to the sun, are invariably the darkest portion of the population.

Darkness of colour was generally considered an indication of strength; and fairness of complexion, the contrary. Hence, the men were not solicitous either to cover their persons, or avoid the sun's rays, from any apprehension of the effect it would produce on the skin. When they searched the field of battle for the bones of the slain, to use them in the manufacture of chisels, gimlets, or fish-hooks, they always selected those whose skins were dark, as they supposed their bones were strongest. When I have seen the natives looking at a very dark man, I have sometimes heard them say, "*Taata ra e, te ereere! ivi maitai tona:*" The man, how dark! good bones are his. A fair complexion was not an object of admiration or desire. They never considered the fairest European countenance seen among them, handsomer than their own; and sometimes, when a fine, tall, well-formed, and personable man has landed from a ship, they have remarked as he passed along, "A fine man that, if he were but a native." They formerly supposed the white colour of the European's skin to be the effect of illness, and hence beheld it with pity. This opinion probably originated from the effects of a disease with which they are occasionally afflicted—a kind of leprosy, which turns the skin of the parts affected, white. This impression, however, is now most probably removed by the lengthened intercourse they have had with foreigners, and the residence of European families among them.

The mental capacity of the Society Islanders has been hitherto much more partially developed than their physical character. They are remarkably curious and inquisitive, and, compared with other Polynesian nations, may be said to possess considerable ingenuity, together with mechanical invention and imitation. Totally unacquainted with the use of letters, their minds could not be improved by any regular or continued culture; yet the distinguishing features of their civil polity—the imposing nature, the numerous observances, and diversified ramifications of their mythology—the legends of their gods—the historical songs of their bards—the beautiful, figurative, and impassioned eloquence sometimes displayed in their national assemblies—and, above all, the copiousness, variety, precision, and purity of their language, with their extensive use of numbers—warrant the conclusion, that they possess no contemptible mental capabilities.

This conclusion has been abundantly confirmed since the establishment of schools, and the introduction of letters. Not only have the children and young persons learned to read, write, cipher, and commit their lessons to memory with a facility and quickness not exceeded by individuals of the same age in any civilized country; but the education of adults, and even persons advanced in years—which in England with every advantage is so difficult an undertaking, that nothing but the use of the best means and the most untiring application ever accomplished it—has been effected here with comparative ease. Multitudes, who were upwards of thirty or forty years of age when they commenced with the alphabet, have, in the course of twelve months, learned to read distinctly in the New Testament, large portions, and even whole books of which, some of them have in a short period committed to memory.

They acquired the first rules of arithmetic with equal facility, and have readily received the different kinds of instruction hitherto furnished, as fast as their teachers could prepare lessons in the native language. It is probable that not less than ten thousand persons have learned to read the sacred Scriptures, and that nearly an equal number are either capable of writing, or are under instruction. In the several stations and branch stations, many thousands are still receiving daily instruction in the first principles of human knowledge and Divine truth.

The following extract from the journal of a Tahitian, now a native Missionary in the Sandwich group, is not only most interesting from the intelligence it conveys, but creditable to the writer's talents. It was published in the American Missionary Herald, and refers to the young princess of the Sandwich Islands, the only sister of the late and present king.

"Nahienaena, in knowledge and words, is a woman of matured understanding. All the fathers and mothers of this land are ignorant and left-handed; they become children in the presence of Nahienaena, and she is their mother and teacher. Her own men, women, and children, those composing her household (or domestic establishment,) listen to the good word of God from her lips. She also instructs Hoapiri and wife in good things. She teaches them night and day. She is constantly speaking to her steward, and to all her household. Very numerous are the words which she speaks, to encourage, and to strengthen them in the good way.

"The young princess has always been pleasant in conversation. Her words are good words. She takes pleasure in conversation, like a woman of mature years. She orders her speech with great wisdom and discretion, always making a just distinction between good and evil. She manifests much discernment in speaking to others the word of God, and the word of love. It was by the maliciousness of the people, old and young, that she was formerly led astray. She was then ignorant of the devices of the wicked. They have given her no rest; but have presented every argument before her that this world could present, to win her over to them.

"Nahienaena desires now to make herself very low. She does not wish to be exalted by men. She desires to cast off entirely the rehearsing of names; for her rejoicing is not now in names and titles. This is what she desires, and longs to have rehearsed—'Jesus alone; let him be lifted up; let him be exalted; let all rejoice in him; let our hearts sing praise to him.' This is the language of her inmost soul."

On a public occasion, in the island of Raiatea, during the year 1825, a number of the inhabitants were conversing on the wisdom of God; which, it was observed, though so long unperceived by them, was strikingly exhibited in every object they beheld. In confirmation of this, a venerable and gray-headed man, who had formerly been a sorcerer, or priest of the evil spirit, stretched forth his hand, and looking at the limbs of his body, said, "Here the wisdom of God is displayed. I have *hinges* from my toes to my finger-ends. This finger has its hinges, and bends at my desire—this arm, on its hinge, is extended at my will—by means of these hinges, my legs bear me where I wish; and my mouth, by its hinge, masticates my food. Does not all this display the wisdom of God?"

The above will shew, more clearly than any declaration I can make, that the inhabitants of these distant isles, though shut out for ages from intercourse with every other part of the world, and deprived of every channel of knowledge, are, notwithstanding, by no means inferior in intellect or capacity to the more favoured inhabitants of other parts of the globe. These statements also warrant the anticipation that they will attain an elevation equal to that of the most cultivated and enlarged intellect, whenever they shall secure the requisite advantages.

They certainly appear to possess an aptness for learning, and a quickness in pursuit of it, which is highly encouraging, although in some degree counteracted by the volatile disposition, and fugitive habits, of their early life, under the influence of which their mental character was formed.

The moral character of the South Sea Islanders, though more fully developed than their intellectual capacity, often presents the most striking contradictions. Their hospitality has, ever since their discovery, been proverbial, and cannot be exceeded. It is practised alike by all ranks, and is regulated only by the means of the individual by whom it is exercised. A poor man feels himself called upon, when a friend from a distance visits his dwelling, to provide an entertainment for him, though he should thereby expend every article of food he possessed; and would generally divide his fish or his bread-fruit with any one, even a stranger, who should be in need, or who should ask him for it.

I am willing to afford them every possible degree of credit for the exercise of this truly amiable disposition; yet, when it is considered that a guest is not entertained day after day at his friend's table, but that after one large collection of food has been presented, the visitor must provide for himself, while the host frequently takes but little further concern about him—we are induced to think that the force of custom is as powerful in its influence on his mind, as that of hospitality. In connexion with this, it should be recollected, that for every such entertainment, the individual expects to be reimbursed in kind, whenever he may visit the abode of his guest. Their ancient laws of government, also, imperiously required the poor industrious landholder, or farmer, to bring forth the produce of his garden or his field for the use of the chiefs, or the wandering and licentious Areois, whenever they might halt at his residence; and more individuals have been banished, or selected as sacrifices, for withholding what these daring ramblers required, than perhaps for all other crimes. To withhold food from the king or chiefs, when they might enter a district, was considered a crime next to resisting the royal authority, or declaring war against the king; and this has in a great degree rendered the people so ready to provide an entertainment for those by whom they may be visited.

Next to their hospitality, their cheerfulness and

good nature strikes a stranger. They are seldom melancholy or reserved, always willing to enter into conversation, and ready to be pleased, and to attempt to please their associates. They are, generally speaking, careful not to give offence to each other: but though, since the introduction of Christianity, families dwell together, and find an increasing interest in social intercourse, yet they do not realize that high satisfaction experienced by members of families more advanced in civilization. There are, however, few domestic broils; and were fifty natives taken promiscuously from any town or village, to be placed in a neighbourhood or house—where *they* would disagree once, fifty Englishmen, selected in the same way, and placed under similar circumstances, would quarrel perhaps twenty times. They do not appear to delight in provoking one another, but are far more accustomed to jesting, mirth, and humour, than irritating or reproachful language.

Their jests and raillery were not always confined to individuals, but extended to neighbourhoods, or the population of whole islands. The inhabitants of one of the Leeward Islands, Tahaa, I believe, even to the present time furnish matter for mirthful jest to the natives of the other islands of the group, from the circumstance of one of their people, the first time she saw a foreigner who wore boots, exclaiming, with astonishment, that the individual had *iron legs*. It is also said, that among the first scissors possessed by the Huahineans, one pair became exceedingly dull, and the simple-hearted people, not knowing how to remedy this defect, tried several experiments, and at length *baked the scissors* in a native oven, for the purpose of sharpening them. Hence the people of Huahine are often spoken of in jest by the Tahitians, as the *feia eu paoti*, or people that baked the scissors. The Tahitians themselves were in their turn subjects of raillery, from some of their number who resided at a distance from the sea, attempting, on one occasion, to kill a turtle by pinching its throat, or strangling it, when the neck was drawn into the shell, on which they were surprised to find they could make no impression with their fingers. The Huahineans, therefore, in their turn, spoke of the Tahitians as the *feia uumi honu*, the people that strangled the turtle.

Their humour and their jests were, however, but rarely what might be termed innocent sallies of wit, and were in general low and immoral to a disgusting degree. Their common conversation, when engaged in their ordinary avocations, was often such as the ear could not listen to without pollution, presenting images, and conveying sentiments, whose most fleeting passage through the mind left contamination. Awfully dark, indeed, was their moral character, and notwithstanding the apparent mildness of their disposition, and the cheerful vivacity of their conversation, no portion of the human race was ever perhaps sunk lower in brutal licentiousness and moral degradation, than this isolated people. The veil of oblivion must be spread over this part of their character, of which the appalling picture, drawn by the pen of inspiration in the hand of the apostle, in the first chapter of the epistle to the Romans, revolting and humiliating as it is, affords but too faithful a portraiture.

The depraved moral habits of the South Sea Islanders undoubtedly weaken their mental energies, and enervate their physical powers; and although remarkably strong men are now and then met with among them, they seem to be more distinguished by activity, and capability of endurance, than muscular strength. They engage in some kinds of work with great spirit for a time, but they soon tire. Regular, steady habits of labour are only acquired by long practice. When a boat manned with English seamen, and a canoe with natives, have started together from the shore—at their first setting out, the natives would soon leave the boat behind, but, as they became weary, they would relax their vigour; while the seamen, pulling on steadily, would not only overtake them, but, if the voyage occupied three or four hours, would invariably reach their destination first.

The natives take a much larger quantity of refreshment than European labourers, but their food is less solid and nutritive. They have, however, the power of enduring fatigue and hunger in a greater degree than those by whom they are visited. A native will sometimes travel, in the course of a day, thirty or forty miles, frequently over mountain and ravine, without taking any refreshment, except the juice from a piece of sugarcane, and apparently experience but little inconvenience from his excursion. The facility with which they perform their journeys is undoubtedly the result of habit, as many are accustomed to traverse the mountains, and climb the rocky precipices, even from their childhood.

The longevity of the islanders does not appear to have been, in former times, inferior to that of the inhabitants of more temperate climates. It is, however, exceedingly difficult to ascertain the age of individuals in a community destitute of all records; and although many persons are to be met with, whose wrinkled skin, decrepit form, silver hair, impaired sight, toothless jaws, and tremulous voice, afford every indication of extreme age; these alone would be fallacious data, as climate, food, and habits of life might have prematurely induced them. Our inferences are therefore drawn from facts connected with comparatively recent events in their history, the dates of which are well known. When the Missionaries arrived in the Duff, there were natives on the island who could recollect the visit of Captain Wallis: he was there in 1767. There are, in both the Sandwich and Society Islands, individuals who can recollect Captain Cook's visit, which is fifty years ago; there are also two now in the islands, that were taken away in the Bounty, forty years since; and these individuals do not look more aged, nor even so far advanced in years, as others that may be seen. The opinion of those Missionaries who have been longest in the islands is, that many reach the age of seventy years, or upwards. There is, therefore, every reason to believe, that the period of human life, in the South Sea Islands, is not shorter than in other parts of the world, unless when it is rendered so by the inordinate use of ardent spirits, and the influence of diseases, prevailing among the lower classes, from which they were originally exempt, and the ravages of which they are unable to palliate or remove.

The mode of living, especially among the farmers, their simple diet, and the absence of all stimulants, their early hours of retiring to rest, and rising in the morning with or before the break of day, their freedom from irritating or distressing cares, and sedentary habits, which so often, in artificial or civilized society, destroy health, appear favourable to the longevity of this portion of the inhabitants, and present a striking contrast to the dissipated and licentious habits of the Areois, the dancers, and similar classes.

It is impossible for any one who has visited these shores, or traversed any one of the districts, to entertain the slightest doubt that the number of inhabitants in the South Sea Islands was formerly much greater than at present. What their number, in any remote period of their history, may have been, it is not easy to ascertain: Captain Cook estimated those residing in Tahiti at 200,000. The grounds, however, on which he formed his conclusions were certainly fallacious. The population was at all times so fugitive and uncertain, as to the proportion it bore to any section of geographical surface, that no correct inference, as to the amount of the whole, could be drawn from the numbers seen in one part. Captain Wilson's calculation, in 1797, made the population of Tahiti only about 16,000; and, not many years afterwards, the Missionaries declared it as their opinion, that this island did not contain more than 8000 souls; and I cannot think that, within the last thirty years, it has ever contained fewer inhabitants.

The present number of natives is about 10,000. That of Eimeo and Tetuaroa probably 2,000. The Leeward Islands perhaps contain nearly an equal number. The Austral Islands have about 5,000 inhabitants; 4,000 of whom reside in the islands of Rapa and Raivavai. Rarotogna, or Rarotoa, has a population of nearly 7,000; and the whole of the Harvey

Islands contain not less than ten or eleven thousand. Connected with these may be considered the Paumotu, or Pearl Islands, of whose population it is difficult to form any correct estimate, as there are no means of ascertaining their numbers, excepting from the reports of the natives, and the observations of masters of vessels, who generally make a very short stay among them. Anaa, or Prince of Wales's Island, is said to be inhabited by several thousands, and as the islands are numerous, though small, it is to be presumed that their population does not amount to less than ten thousand. From these statements it will appear, that the population of the Georgian and Society Islands, together with the adjacent clusters, with which the natives maintain constant intercourse, and to which Christianity has been conveyed by native or European teachers, comprises between forty-eight and fifty thousand persons. In this number, the Marquesas, to which native teachers have gone, and which one of the Missionaries has recently visited, are not included. Their population is probably about thirty thousand.

With respect to the Society and neighbouring islands, although no ancient monuments are found indicating that they were ever inhabited by a race much further advanced in civilization than those found on their shores by Wallis, Cook, and Bougainville; yet that race has evidently, at no very remote period, been much more numerous than it was when discovered by Europeans. In the bottom of every valley, even to the recesses in the mountains, on the sides of the inferior hills, and on the brows of almost every promontory, in each of the islands, monuments of former generations are still met with in great abundance. Stone pavements of their dwellings and court-yards, foundations of houses, and ruins of family temples, are numerous. Occasionally they are found in exposed situations, but generally amidst thickets of brushwood or groves of trees, some of which are of the largest growth. All these relics are of the same kind as those observed among the natives at the time of their discovery, evidently proving that they belong to the same race, though to a more populous æra of their history. The stone tools occasionally found near these vestiges of antiquity demonstrate the same lamentable fact.

The present generations are deeply sensible of the depopulation that has taken place, even within the recollection of those most advanced in years, and have felt acutely in prospect of the annihilation that appeared inevitable. Their priests formerly denounced the destruction of the nation, as the greatest punishment the gods could inflict, and the following was one of the predictions: *E tupu te fau, e toro te farero, e mou te taata:* "The fau (*hibiscus*) shall grow, the *farero* (coral) shall spread or stretch out its branches, but man shall cease."—The fau is one of the most spreading trees, and is of quickest growth; it soon over-runs uncultivated lands; while the branching coral, *farero,* is perhaps more rapid in its formation than any of the corallines that close up the openings in the reefs, and, wherever it is shallow, rise to the water's surface, so as to prevent the passage of the canoe, and destroy the resort of the fish. This was denounced as the punishment that would follow disobedience to the junctions or requisitions of the priest, delivered in the name, and under the authority, of the gods. Tati, however, remarked to Mr. Davies, that it was the observing, not the neglecting of the directions of the priest, that had nearly produced its actual accomplishment.

At the time when the nation renounced idolatry, the population was so much reduced, that many of the more observant natives thought the denunciation of the prophet was about to be literally fulfilled. Tati, the chief of Papara, talking with Mr. Davies on this subject, in 1815, said, with great emphasis, that "if God had not sent his word at the time he did, wars, infant-murder, human sacrifices, &c. would have made an end of the small remnant of the nation." A similar declaration was pathetically made by Pomare soon after, when some visitors from England waited upon him at his residence. He addressed them to the following effect: "You have come to see us under circumstances very different from those under which your countrymen formerly visited our ancestors. They came in the æra of men, when the islands were inhabited, but you are come to behold just the remnant of the people." I have often heard the chiefs speak of themselves and the natives as only a small *toea*, remainder, left after the extermination of Satani, or the evil spirit; comparing themselves to a firebrand unconsumed among the mouldering embers of a recent conflagration. These figures, and others equally affecting and impressive, were but too appropriate, as emblems of the actual state to which they were at that time reduced. Under the depopulating influence of vicious habits—the dreadful devastation of diseases that followed, and the early destruction of health—the prevalence of infanticide—the frequency of war—the barbarous principles upon which it was prosecuted, and the increase of human sacrifices, it does not appear possible that they could have existed, as a nation, for many generations longer.

An inquiry naturally presents itself in connexion with this subject, viz.—To what cause is this recent change in the circumstances of the people to be attributed? It is self-evident, that if these habits had always prevailed among the Tahitians, they must long since have been annihilated. Society must, at some time, have been more favourable, not only to the preservation, but to the increase of population, or the inhabitants could never have been so numerous as they undoubtedly were a century or a century and a half ago. There is no question but that depopulation had taken place to a considerable extent prior to their discovery by Captain Wallis, and it is not easy to discover the causes which first led to it. Infanticide and human sacrifices, together with their wars, appear to have occasioned the diminution of the inhabitants before the period alluded to. Whether war was more frequent immediately preceding their discovery, than it had been in earlier ages, we have not the means of knowing, nor have we been able to ascertain, with any great accuracy, how long the Areoi society had existed, or child-murder was practised. There is reason to believe that infanticide is not of recent origin, and the antiquity of the Areoi fraternity, according to tradition, is equal to that of the first inhabitants.

Human sacrifices, we are informed by the natives, are comparatively of modern institution, and were not admitted until a few generations antecedent to the discovery of the islands. They were first offered at

Raiatea, in the national marae at Opoa, having been demanded by the priest in the name of the god, who had communicated the requisition to his servant in a dream. Human sacrifices were presented at Raiatea and the Leeward Islands for some time before they were introduced among the offerings to the deities of Tahiti; but soon after they began to be employed, they were offered with great frequency, and in appalling numbers: but of this, an account will hereafter be given.

The depopulation that has taken place during the last two or three generations, viz. since their discovery, may be easily accounted for. In addition to a disease, which, as a desolating scourge, spread, unpalliated and unrestrained, its unsightly and fatal influence among the people, two others are reported to have been carried thither—one by the crew of Vancouver in 1790; and the other by means of the Britannia, an English whaler, in 1807. Both these disorders spread through the islands; the former almost as fatal as the plague, the latter affecting nearly every individual throughout all the islands of the group. The maladies originally prevailing among them, appear, compared with those by which they are now afflicted, to have been few in number and mild in character.

Next to these diseases, the introduction of fire-arms, although their use in war has not perhaps rendered their engagements more cruel and murderous than when they fought hand to hand with club and spear—has most undoubtedly cherished, in those who possessed them, a desire for war, as a means of enlarging their territory, and augmenting their power. Pomare's dominion would never have been so extensive and so absolute, but for the aid he derived, in the early part of his reign, from the mutineers of the Bounty, who attended him to battle with arms which they had previously learned to use with an effect, which his opponents could not resist. Subsequently, the hostile chieftains, having procured fire-arms, and succeeding in attaching to their interest European deserters from their ships, considered themselves, if not invincible, at least equal to their enemies, and sought every opportunity for engaging in the horrid work of accelerating the depopulation of their country. Destruction was the avowed design with which they commenced every war, and the principle of extermination rendered all their hostilities fatal to the vanquished party.

Another cause most influential in the diminution of the Tahitian race, has been the introduction of the art of distillation, and the extensive use of ardent spirits. They had, before they were visited by our ships, a kind of intoxicating beverage called *ava*, but the deleterious effects resulting from its use were confined to a comparatively small portion of the inhabitants. The growth of the plant from which it was procured was slow; its culture required care; it was usually tabued for the chiefs; and the common people were as strictly prohibited from appropriating it to their own use, as the peasantry are in reference to the game of England. Its effects also were rather sedative, than narcotic or inebriating.

But after the Tahitians had been taught by foreign seamen, and natives of the Sandwich Islands, to distil spirits from indigenous roots, and rum had been carried to the islands in abundance as an article of barter, intoxication became almost universal; and all the demoralization, crimes, and misery, that follow in its train, were added to the multiplied sorrows and wasting scourges of the people. It nurtured indolence, and spread discord through their families, increased the abominations of the Areoi society, and the unnatural crime of infanticide. Before going to the temple to offer a human sacrifice to their gods, the priests have been known to intoxicate themselves, in order that they might be insensible to any unpleasant feelings this horrid work might excite.

These causes operating upon a people, whose simple habits of diet rendered their constitutions remarkably susceptible of violent impressions, are, to a reflecting mind, quite sufficient to account for the rapid depopulation of the islands within the last fifty or sixty years.

The philanthropist, however, will rejoice to know, that although sixteen years ago the nation appeared on the verge of extinction, it is now, under the renovating and genial principles of true religion, and the morality with which this is inseparably connected, rapidly increasing. When the people in general embraced Christianity, we recommended that a correct account of the births and deaths occurring in each of the islands should be kept. From the operation of the causes above enumerated, for some years even after the crimes in which they originated had ceased, the number of deaths exceeded that of births. About the years 1819 and 1820 they were nearly equal, and since that period population has been rapidly increasing.

It was not till the account of deaths and births was presented, that we had an adequate idea of the affecting depopulation that had been going on; and if, for several years after infanticide, inebriation, human sacrifices, and war, were discontinued, the number of deaths exceeded that of the births; how appalling must that excess have been, when all these destructive causes were in full operation! There is now, however, every ground to indulge the expectation that the population will become greater than it has been in any former period of their history; and it is satisfactory, in connexion with this anticipation, to know—that extent of soil capable of cultivation, and other rescourses, are adequate to the maintenance of a population tenfold increased above its present numbers.

CHAP. II.

Origin of the inhabitants of the South Sea Islands—Traditions—Legend of Taaoroa and Hina—Resemblance to Jewish history—Coincidences in language, mythology, &c. with the language, &c. of the Hindoos and Malays, Madagasse, and South Americans—Difficulty of reaching the islands from the west—Account of different native voyages—Geographical extent over which the Polynesian race and language prevail—Account of the introduction of animals—Predictions of their ancient prophets relating to the arrival of ships—Traditions of the deluge, corresponding with the accounts in sacred and profane writings.

THE origin of the inhabitants of the South Sea Islands, in common with other parts of Polynesia, is a subject perhaps of more interest and curiosity, than of importance and practical utility. The vast extent of geographical surface covered by the race of which they form an integral portion, the analogy in character, the identity in language, &c., the remote distance at which the different tribes are placed from each other, and the isolated spots and solitary clusters which they occupy in the vast expanse of surrounding water, render the source whence they were derived, one of the mysteries connected with the history of our species.

To a Missionary, the business of whose life is with the people among whom he is stationed, every thing relating to their history is, at least, interesting; and the origin of the islanders has often engaged our attention, and formed the subject of our inquiries. The early history of a people destitute of all records, and remote from nations in whose annals contemporaneous events would be preserved, is necessarily involved in obscurity. The greater part of the traditions of this people are adapted to perplex rather than facilitate the investigation.

A very generally received Tahitian tradition is, that the first human pair were made by Taaroa, the principal deity formerly acknowledged by the nation. On more than one occasion, I have listened to the details of the people respecting his work of creation. They say, that after Taaroa had formed the world, he created man out of *araea*, red earth, which was also the food of man until bread-fruit was made. In connexion with this, some relate that Taaroa one day called for the man by name. When he came, he caused him to fall asleep, and that, while he slept, he took out one of his *ivi*, or bones, and with it made a woman, whom he gave to the man as his wife, and that they became the progenitors of mankind. This always appeared to me a mere recital of the Mosaic account of creation, which they had heard from some European, and I never placed any reliance on it, although they have repeatedly told me it was a tradition among them before any foreigner arrived. Some have also stated that the woman's name was Ivi, which would be by them pronounced as if written *Eve*. Ivi is an aboriginal word, and not only signifies a bone, but also a widow, and a victim slain in war. Notwithstanding the assertion of the natives, I am disposed to think that *Ivi*, or Eve, is the only aboriginal part of the story, as far as it respects the mother of the human race. Should more careful and minute inquiry confirm the truth of their declaration, and prove that this account was in existence among them prior to their intercourse with Europeans, it will be the most remarkable and valuable oral tradition of the origin of the human race yet known.

Another extensive and popular tradition referred the origin of the people to Opoa, in the island of Raiatea, where the *tiis*, or spirits, formerly resided, who assumed of themselves, or received from the gods, human bodies, and became the progenitors of mankind. The name of one was Tii Maaraauta; *Tii*, branching or extending towards the land, or the interior; and of the other, Tii Maaraatai, *Tii*, branching or spreading towards the sea. It is supposed that prior to the period of Tii Maaraauta's existence, the islands were only resorted to by the gods or spiritual beings, but that these two, endowed with powers of procreation, produced the human species. They first resided at Opoa, whence they peopled the island of Raiatea, and subsequently spread themselves over the whole cluster. Others state, that Tii was not a spirit, but a human being, the first man made by the gods; that his wife was sometimes called Tii, and sometimes Hina; that when they died, their spirits were supposed to survive the dissolution of the body, and were still called by the same name, and hence the term *tii* was first applied to the spirits of the departed, a signification which it retained till idolatry was abolished.

In the Ladrone Islands, departed chiefs, or the spirits of such, are called *aritis*, and to them prayers were addressed. The *tiis* of Tahiti were also considered a kind of inferior deities, to whom, on several occasions, prayers were offered. The resemblance of this term to the dæmon or *dii* of the ancients, is singular, and might favour the conjecture that both were derived from the same source.

The origin of the islands, as well as their inhabitants, was generally attributed to Taaroa, or the joint agency of Taaroa and Hina, and although one of their traditions states that all the islands were formerly united in one *fenua nui*, or large continent, which the gods in anger destroyed, scattering in the ocean the fragments, of which Tahiti is one of the largest; yet others ascribe their formation to Taaroa, who is said to have laboured so hard in the work of creation, that the profuse perspiration induced thereby, filled up the hollows, and formed the sea; accounting, by this circumstance, for its transparency and saltness. Others attribute the origin of the world, the elements, the heavenly bodies, and the human species, to the procreative powers of their deities; and, according to their account, one of the descendants of Taaroa, and the son of the sun and moon, and, in reference to his descent, the Manco Capac of their mythology, embracing the sand on the sea shore, begat a son, who was called Tii, and a daughter, who was called Opiira. These two, according to their tradition, were the father and mother of mankind.

But the most circumstantial tradition, relative to the origin of mankind, is one for which, as well as for much valuable information on the mythology and worship of the idols of the South Sea Islanders, I am indebted to the researches of my esteemed friend and coadjutor, Mr. Barff. According to this legend, man was the fifth order of intelligent beings created by Taaroa and Hina, (of whom an account will hereafter be given,) and was called the *Rahu taata i te ao ia Tii*, "The class, or order of the world, of, or by, Tii." Hina is reported to have said to Taaroa, "What shall be done, how shall man be obtained? Behold, classed or fixed are gods of the *po*, or state of night, and there are no men." Taaroa is said to have answered, "Go on the shore to the interior, to your brother." Hina answered, "I have been inland, and he is not." Taaroa then said, "Go to the sea, perhaps he is on the sea; or if on the land, he will be on the land." Hina said, "Who is at sea?" The god answered, "Tiimaaraatai." Who is Tiimaaraatia? is he a man?" "He is a man, and your brother," answered the god; "Go to the sea, and seek him." When the goddess had departed, Taaroa ruminated within himself as to the means by which man should be formed, and went to the land, where he assumed the appearance and substance which should constitute man. Hina returning from her unsuccessful search for Tiimaaraatai at sea, met him, but not knowing him, said, "Who are you?" "I am Tiimaaraatai," he replied. "Where have you been?" said the goddess: "I have sought you here, and you were not; I went to the sea, to look for Tiimaaraatai, and he was not. "I have been here in my house, or abode," answered Tiimaaraatai," and behold you have arrived, my sister, come to me." Hina said, "So it is, you are my brother; let us live together." They became man and wife; and the son that Hina afterwards bore, they called Tii. He was the first-born of mankind. Afterwards Hina had a daughter, who was called Hinaereeremonoi; she became the wife of Tii, and bore to him a son, who was called Taata, the general name (with slight modification) for *man* throughout the Pacific. Hina, the daughter and wife of Taaroa, the grandmother of Taata, being transformed into a beautiful young woman, became the wife of *Taata* or Man, bore him a son and a daughter, called

Ouru and Fana, who were the progenitors of the human race.

Another tradition stated, that the first inhabitants of the South Sea Islands originally came from a country in the direction of the setting sun, to which they say several names were given, though none of them are remembered by the present inhabitants.

Their traditions are numerous, though it is difficult to obtain a correct recital of them from any of the present inhabitants; and there is but little reason to suppose they can impart any valuable information as to the country whence the inhabitants originally came. Some additional evidence, small indeed in quantity, but rather more conclusive, may be gathered from the traditions of the mythology, customs, and language preserved among the Tahitians, and inhabitants of other isles of the Pacific, when they are compared with those prevailing in different parts of the world. One of their accounts of creation, that in which Taaroa is stated to have made the first man with earth or sand, and the very circumstantial tradition they have of the deluge, if they do not, as some have supposed, (when taken in connexion with many customs, and analogies in language,) warrant the inference that the Polynesians have an Hebrew origin; they shew that the nation, whence they emigrated, was acquainted with some of the leading facts recorded in the Mosaic history of the primitive ages of mankind. Others appear to have a striking resemblance to several conspicuous features of the more modern Hindoo, or Braminical mythology. The account of the creation given in Sir W. Jones's translation of the Institutes of Menu, accords in no small degree with the Tahitian legends of the production of the world, including waters, &c., by the procreative power of their god. The Bramitical account is, that "He (i. e. the divine Being) having willed to produce various beings from his own Divine substance, first, with a thought, created the waters, and placed in them a productive seed. That seed became an egg, bright as gold, blazing like the luminary with a thousand beams, and in that egg he was born himself, in the form of Brama, the great forefather of all spirits. The waters were called *nara*, because they were the production of *narau*, the spirit of God; and since they were his first *ayana*, or place of motion, he is thence named *Narayana*, or moving in the waters. In the egg the great power sat inactive a whole year (of the creator;) at the close of which, by his thought alone, he caused the egg to divide itself. From its two divisions he formed the heavens (above) and the earth (beneath)" &c. It is impossible to avoid noticing the identity of this account, contained in one of the ancient writings of the Bramins, with the ruder version of the same legend in the tradition prevailing in the Sandwich Islands, that the islands were produced by a bird, a frequent emblem of deity, a medium through which the gods often communicated with men; who laid an egg upon the waters, which afterwards burst of itself, and produced the islands; especially, if with this we connect the appendages Tahitian tradition furnishes, that at first the heavens joined the earth, and were only separated by the *teva*, an insignificant plant, *draconitum pollyphillum*, till their god, *Ruu*, lifted up the heavens from the earth. The same event is recorded in one of their songs, in the following line:

Na Ruu i to te rai.
Ruu did elevate or raise the heavens.

Meru, or mount Meru, the abode of the gods, the heaven of the Hindoos, is also the paradise of some classes of the South Sea Islanders, the dwelling-place of departed kings, and others who have been deified.

The institutes of Menu* also forbade a Bramin to eat with his wife, or to be present when she ate; and in this injunction may have originated the former universal practice among these islands, of the man and his wife eating their meat separately. *Varuna* and *Vahni* are among the gods of the Hindoos; the latter, among the eight guardian deities of the world, appears to have been the Neptune of the Bramins, as we learn from the following lines in Sir W. Jones's beautiful translation of the hymn to Indra: "Green Varuna, whom foaming waves obey:" and also, "Vahni flaming like the lamp of day." Both the terms in the South Sea language for spirit, or spiritual being, bear a strong resemblance to these names; the one being *varua*, in which the *n* only is omitted; and in many words, as they are used among the other islanders, some of their consonants are omitted by the Tahitians. *Vaiti* is also another apparently more ancient term for spirit used by them, which somewhat resembles the *Vahni* of the Hindoos. Bishop Heber, the most recent writer on the usages and appearance of the Hindoos, informs us, in his admirable journal, that many things which he saw among the inhabitants of India reminded him of the plates in Cook's voyages.

The points of resemblance between the Polynesians and the Malayan inhabitants of Java, Sumatra, and Borneo, and the Ladrone, Caroline, and Philippine Islands, are still greater. Among the Battas of Sumatra, men

* Menu was the Noah of the Hindoos; and Miru, pronounced Meru, was the first king of the Sandwich Islands.

and women eat separately, cannibalism prevails, and they are much addicted to gaming. War is determined, and its results predicted, by observing the entrails of the animals offered in sacrifice; these all prevail in the isles of the Pacific.

The principal portion of the marriage ceremony, in some of these islands, consists in the bridegroom throwing a piece of cloth over the bride, or the friends throwing it over both. This is also practised among the Tahitians. The bodies of the dead are kept by the inhabitants of the Caroline Islands, in a manner resembling the tupapaus of Tahiti; and, in the Ladrones, they feast round the tomb, and offer food, &c. to the departed. This practice also prevailed extensively in the South Sea Islands.

In the former also, according to the accounts of the Jesuit Missionaries, a licentious society existed, called by the people *Uritoy*, strikingly analogous, in all its distinguishing features, to that institution in the South Seas called the Areoi society. Their implements of war are alike. Dr. Buchanan states, that in Pulo Panang he saw a chief of the Malay tribe, who had a staff, the head of which was ornamented with a bushy lock of human hair, which the chief had cut from the head of his enemy when he lay dead at his feet. This exactly accords with the conduct of the Marquesans; many of whose clubs, and even walking-sticks, I have seen decorated with locks of human hair taken from those slain in battle.

Between the canoes and the language, of these islands and the southern groups, there is a more close resemblance. Their language has a remarkably close affinity with that of the eastern Polynesia. There are also many points of resemblance in language, manners, and customs, between the South Sea Islanders and the inhabitants of Madagascar in the west; the inhabitants of the Aleutian and Kurile islands, in the north, which stretch along the mouth of Behring's straits, and form the chain which connects the old and the new worlds; and also between the Polynesians and the inhabitants of Mexico, and some parts of South America. The general cast of feature, and frequent shade of complexion—the practice of tatauing, which prevails among the Aleutians, and some of the tribes of America—the process of embalming the dead bodies of their chiefs, and preserving them uninterred—the game of chess among the Araucanians—the word for God being *tew* or *tev*—the exposure of their children—their games—their mode of dressing the hair, ornamenting it with feathers—the numerous words in their language resembling those of Tahiti, &c.; their dress, especially the *poncho*, and even the legend of the origin of the Incas, bear no small resemblance to that of Tii, who was also descended from the sun.

The points of resemblance are not so many as in the Asiatic continent and islands; but that probably arises from the circumstance of the great facilities furnished by the Hindoo records, and the absence of all original records relating to the history, mythology, manners, language, &c. of the aborigines of South America. Were we better acquainted with the history and institutions of the first inhabitants of the new world, more numerous points of resemblance would probably be discovered.

Other coincidences, of a more dubious character,

occur in the eastern, western, and intermediate or oceanic tribes; among which might be mentioned the account given by Sir John Mandeville. He is stated to have commenced his travels early in the fourteenth century. In a country near the river Indus, he met with the fountain of youth, the water of which being odoriferous, tasted of all manner of spices; and of this, whoever drank for a few days upon a fasting stomach, was quickly cured of every internal disorder with which he might be afflicted. To this description he added, it was certain those who lived near, and drank frequently of it, had a wonderful appearance of youth through their whole lives, and that he himself drank of it three or four times, and imagined his health was better afterwards. The expedition which led to the discovery of Florida was undertaken not so much from a desire to explore unknown countries, as to find an equally celebrated fountain, described in a tradition prevailing among the inhabitants of Puerto Rico, as existing in Binini, one of the Lucayo Islands. It was said to possess such restorative powers as to renew the youth and vigour of every person who bathed in its waters. It was in search of this fountain, which was the chief object of their expedition, that Ponce de Leon ranged through the Lucayo Islands, and ultimately reached the shores of Florida.* Although it may throw no light on the origin of the South Sea Islanders, nor furnish any evidence of their former connexion with the inhabitants either of India or America, the coinci-

* In reference to this enterprise, Robertson remarks: "That a tale so fabulous should gain credit among the uninstructed Indians, is not surprising; that it should make any impression on an enlightened people, appears, in the present age, altogether incredible. The fact, however, is certain.

dence is striking between these fabulous traditions, and those so circumstantially detailed by the natives of some of the islands of the Pacific, especially in the Hawaiian account of the voyage of Kamapiikai, to the land where the inhabitants enjoyed perpetual health and youthful beauty, where the *wai ora* (life-giving fountain) removed every internal malady, and every external deformity or paralyzed decrepitude, from all those who were plunged beneath its salutary waters. A tabular view of a number of words in the Malayan, Asiatic, or the Madagasse, the American, and the Polynesian languages, would probably shew, that at some remote period, either the inhabitants of these distant parts of the world maintained frequent intercourse with each other, or that colonies from some one of them, originally peopled, in part or altogether, the others. The striking analogy between the numerals and other parts of the language, and several of the customs, of the aborigines of Madagascar, and those of the Malays who inhabit the Asiatic islands, many thousands of miles distant in one direction, and of the Polynesians more remote in another, shews that they were originally one people, or that they had emigrated from the same source. Many words in the language, and several of the traditions, customs, &c. of the Americans, so strongly resemble those of Asia, as to warrant the inference that they originally came from that part of the world. Whether some of the tribes who originally passed from Asia, along the Kurile or Aleutian Islands, across Behring's straits, to America, left part of their number, who were the progenitors of the present race inhabiting those islands; and that they, at some subsequent period, either attempting to follow the tide of emigration to the east, or steering to the south, were by the north-east trade-winds driven to the Sandwich Islands, whence they proceeded to the southern groups; or whether those who had traversed the north-west coast of America, sailed either from California or Mexico across the Pacific under the favouring influence of the regular easterly winds, peopled Easter Island, and continued under the steady easterly or trade-winds advancing westward till they met the tide of emigration flowing from the larger groups or islands, in which the Malays form the majority of the population—it is not now easy to determine. But a variety of facts connected with the past and present circumstances of the inhabitants of these countries, authorize the conclusion, that, either part of the present inhabitants of the South Sea Islands came originally from America, or that tribes of the Polynesians have, at some remote period, found their way to the continent.

The origin of the inhabitants of the Pacific is involved in great mystery, and the evidences are certainly strongest in favour of their derivation from the Malayan tribes inhabiting the Asiatic Islands; but, allowing this to be their source, the means by which they have arrived at the remote and isolated stations they now occupy, are still inexplicable. If they were peopled from the Malayan Islands, they must have possessed better vessels, and more accurate knowledge of navigation, than they now exhibit, to have made their way against the constant trade-winds prevailing within the tropics, and blowing regularly, with but transient and uncertain interruptions, from east to west. The nations at present inhabiting the islands of the Pacific, have undoubtedly been more extensively spread than they now are. In the most remote and solitary islands occasionally discovered in recent years, such as Pitcairn's, on which the mutineers of the Bounty settled, and on Fanning's Island near Christmas Island, midway between the Society and Sandwich Islands, although now desolate, relics of former inhabitants have been found. Pavements of floors, foundations of houses, and stone entrances, have been discovered; and stone adzes or hatchets have have been found at some distance from the surface, exactly resembling those in use among the people of the North and South Pacific at the time of their discovery. These facts prove that the nations now inhabiting these and other islands have been, in former times, more widely extended than they are at present. The monuments or vestiges of former population found in these islands are all exceedingly rude, and therefore warrant the inference that the people to whom they belonged were rude and uncivilized, and must have emigrated from a nation but little removed from a state of barbarism—a nation less civilized than those must have been, who could have constructed vessels, and traversed this ocean six or seven thousand miles against the regularly prevailing winds, which must have been the fact, if we conclude they were peopled only by the Malays.

On the other hand, it is easy to imagine how they could have proceeded from the east. The winds would favour their passage, and the incipient stages of civilization in which they were found, would resemble the condition of the aborigines of America, far more than that of the Asiatics. There are many well-authenticated accounts of long voyages performed in native vessels by the inhabitants of both the North and South Pacific. In 1696, two canoes were driven from Ancarso to one of the Philippine islands, a distance of 800 miles. In 1720, two canoes were drifted from a remote distance to one of the Marian islands. Captain Cook found in the island of Wateo inhabitants of Tahiti, who had been drifted by contrary winds in a canoe, from some islands to the eastward, unknown to the natives. Several parties have, within the last few years, reached the Tahitian shores from islands to the eastward, of which the Society Islanders had never before heard. In 1820, a canoe arrived at Maurua, about twenty miles west of Borabora, which had come from Rurutu, one of the Austral Islands. This vessel had been at sea between a fortnight and three weeks, and, considering its route, must have sailed seven or eight hundred miles. A more recent instance occurred in 1824: a boat belonging to Mr. Williams of Raiatea, left that island with a westerly wind for Tahiti. The wind changed after the boat was out of sight of land. They were driven to the island of Atui, a distance of nearly 800 miles in a south-westerly direction, where they were discovered several months afterwards. Another boat, belonging to Mr. Barff of Huahine, was passing between that island and Tahiti about the same time, and has never since been heard of. The traditions of the inhabitants of Rarotogna, one of the Harvey Islands, preserve the most satisfactory accounts, not only of single parties, at different periods for many generations back, having arrived there from the Society Islands, but also derive the origin of the population from the island of Raiatea. Their traditions according with those of the Raiateans on the leading points, afford the strongest evidence of these islands having been peopled from those to the eastward.

If we suppose the population of the South Sea Islands to have proceeded from east to west, these events illustrate the means by which it may have been accomplished; for it is a fact, that every such voyage related in the accounts of voyagers, or preserved in the traditions of the natives, has invariably been from east to west, directly opposite to that in which it must have been, had the population been altogether derived from the Malayan archipelago.

From whatever source, however, they have originated, the extent of geographical surface over which they have spread themselves, the variety, purity, and copiousness of their language, the ancient character of some of the best traditions, as of the deluge &c. justify the supposition of their remote antiquity. Yet their ignorance of letters, of the use of iron till a short time prior to their discovery, and the rude character of all their implements, and of the monuments of their ancestry, seem opposed to the idea of their having been derived, as supposed by some eminent modern geographers, from an ancient, powerful nation, which cultivated maritime habits, but which has been frittered down into detached local communities unknown to each other.

The accounts the natives give of the introduction of the animals found on the islands by the first European visitors, are most of them as fabulous as those relating to their own origin. Some, indeed, say that pigs and dogs were brought from the west by the first inhabitants; but others refer their origin to man. One of their traditions states, that after Taaroa had made the world and mankind, he created the quadrupeds of the earth, the fowls of the air, and the fishes of the sea; but one of their most indelicate accounts states, that in ancient times a man died, and after death his body was destroyed by worms, which ultimately grew into swine—and were the first known in the islands. We never observed among them any traces of the Asiatic doctrine of the transmigration of souls; although they believed that hogs had souls, and that there was a distinct place, called Ofetuna, whither they supposed the souls of the pigs repaired after their death. This idea some carried so far as to suppose, that, not only had animals souls, but to imagine that even flowers and plants were organized beings, also possessing souls. Another singular practice in reference to their pigs, was that of giving them some distinct, though often arbitrary name; so that each pig had his own proper name, by which he was called, as well as the several members of the family. This difference, however, prevailed—a man frequently changed his name, but the name of the pig, once received, was usually retained until his death.

The island of Raiatea has not only been distinguished among the surrounding cluster by its identity with the traditions of the past, but also by its being the source of predictions respecting the future. There are some which regarded the destiny of the people, but the most remarkable (because, according to the interpretation of the natives themselves, they have received a partial fulfilment) were those referring to the strange events that should occur. Among the native prophets of former times, there appear to have been several of the name of *Maui*. One of the most celebrated of this name resided at Raiatea, and on one occasion, when supposed to be under the inspiration of the god, he predicted that in future ages a *vaa ama ore*, literally an "outriggerless canoe," would arrive in the islands from some foreign land. Accustomed to attach that appendage to their single canoes, whatever might be the size or quality, they considered an outrigger essential to their remaining upright upon the water, and consequently could not believe that a canoe without one would live at sea. The absence of this has ever appeared to the South Sea Islanders one of the greatest wonders connected with the visits of the first European vessels. At one of the Harvey Islands, where the natives had never seen a vessel until recently visited by a Missionary, when the boat was lowered down to the water, and pushed off by the rowers from the ship's side, the natives simultaneously and involuntarily exclaimed—"It will overturn and sink, it has no outrigger."

The chiefs and others, to whom Maui delivered his prophecy, were also convinced in their own minds, that a canoe would not swim without this necessary balance, and charged him with foretelling an impossibility. He persisted in his predictions, and, in order to remove their scepticism as to its practicability, launched his umete, or oval wooden dish, upon the surface of a pool of water near which he was sitting, and declared that in the same manner would the vessel swim that should arrive.

We have not been able to ascertain the period of their history during which this prediction was delivered. It was preserved among the people by oral tradition, until the arrival of Captain Wallis's and Cook's vessels. When the natives first saw these, they were astonished at their gigantic size, imposing appearance, and the tremendous engines on board. These appearances induced them first to suppose the ships were islands inhabited by a supernatural order of beings, at whose direction the lightnings flashed, the thunders roared, and the destroying demon slew, with instantaneous but invisible strokes, the most daring and valiant of their warriors. But when they afterwards went alongside, or ventured on board, and saw that they were floating fabrics of timber, borne on the surface of the waters, and propelled by the winds of heaven, they unanimously declared that the prediction of Maui was accomplished, and the canoes without outriggers had arrived. They were confirmed in this interpretation, when they saw the small boats belonging to the ships employed in passing to and fro between the vessel and the shore. These being simple in their structure, and approaching their own canoes in size, yet conveying in perfect safety those by whom they were manned, excited their astonishment, and confirmed their convictions that Maui was a prophet.

When a boat or a vessel has been sailing in or out of the harbour, I have often heard the natives, while gazing at the stately motion, exclaim, *Te vaa a Maui e! Ta vaa ama ore.* "Oh the canoe of Maui! the outriggerless canoe!" They have frequently asked us how he could have known such a vessel would arrive, since it was at that time considered by all besides as an impossibility. We have told them it was probable he had observed the steadiness with which his umete, or other hollow wooden vessels, floated on the water, and had thence inferred that at some future period they might behold larger vessels equally destitute of any exterior balancing power. They in general consider the use of boats and shipping among them as an accomplishment of his prediction.

The islanders also state, that there is another prediction, still to be fulfilled; and although it appears to them as great an improbability as this, yet the actual appearance of one, leads many to think that possibly they may witness the other. This remaining prediction also has reference to a canoe, and declares that after the arrival of the canoe without an outrigger, *e vaa taura ore*, a canoe or vessel without ropes or cordage, shall come among them. What idea Maui designed to convey by this declaration, it is perhaps not easy to ascertain; but the people say it is next to impossible that the masts should be sustained, the sails attached, or the vessel worked, without ropes or cordage. They say, however, that one prediction respecting the vessels has been accomplished, but that the other remains to be realized. I have often thought, when contemplating the little use of rigging on board our steam-vessels, that should a specimen of this modern invention ever reach the South Sea Islands, although the natives would not, perhaps, like the inhabitants of the banks of the Ganges, be ready to fall down and worship the wonderful exhibition of mechanical skill, they would be equally astonished at that power within itself by which it would be propelled, and would at once declare that the second prediction of Maui was accomplished, and the vessel without rigging or cordage had arrived.

They have other predictions, but less circumstantial or probable, yet I never could learn that they have ever been led, from the declarations of their wise men, to anticipate the arrival of any distinguished personage in their country. The expectation of some wise and great prince or ruler rising up among them, or coming from some distant region, which has prevailed among many nations, and is generally supposed to refer to the appearance of the Saviour, does not seem to have existed among them; unless we suppose the anticipated return of Rono to the Sandwich Islands, an *Avatar* of whom, the inhabitants supposed Captain Cook to be, refers to this event.

Traditions of the deluge, the most important event in reference to the external structure and appearance of our globe that has occurred since its creation, have been found to exist among the natives of the South Sea Islands, from the earliest periods of their history. Accounts, more or less according with the scripture narrative of this awful visitation of Divine justice upon the antediluvian world, have been discovered among most of the nations of the earth; and the striking analogy between those religiously preserved by the inhabitants of the islands of the Pacific, and the Mosaic account, would seem to indicate a degree of high antiquity belonging to this isolated people.

The principal facts are the same in the traditions prevailing among the inhabitants of the different groups, although they differ in several minor particulars. In one group the accounts stated, that in ancient times Taaroa, the principal god according to their mythology, the creator of the world, being angry with men on account of their disobedience to his will, overturned the world into the sea, when the earth sunk in the waters, excepting a few *aurus*, or projecting points, which remaining above its surface, constituted the present cluster of islands. The memorial preserved by the inhabitants of Eimeo, states, that after the inundation of the land, when the water subsided, a man landed from a canoe near Tiataepua, in their island, and erected an altar, or marae, in honour of his god.

The tradition, which prevails in the Leeward Islands, is intimately connected with the island of Raiatea. According to this, shortly after the first peopling of the world by the descendants of Taata, *Ruahatu,* the Neptune of the South Sea Islanders, was reposing among the coralline groves in the depths of the ocean, on a spot that, as his resort, was sacred. A fisherman, either through forgetfulness or disregard of the tabu, and sacredness of the place, paddled his canoe upon the forbidden waters, and lowered his hooks among the branching corals at the bottom. The hooks became entangled in the hair of the sleeping god. After remaining some time, the fisherman endeavoured to pull up his hooks, but was for a long period unable to move them. At length they were suddenly disentangled from whatever they had been attached to, and he began to draw them towards the surface. In an instant, however, the god, whom he had aroused from his slumbers, appeared at the surface of the water, and, after upbraiding him for his impiety, declared, that the land was criminal or convicted of guilt, and should be destroyed.

The affrighted fisherman prostrated himself before the god of the sea, confessed his sorrow for what he had done, and implored his forgiveness, beseeching him that the judgment denounced might be averted, or that he might escape. Ruahatu, moved by his penitence and importunity, directed him to return home for his wife and child, and then proceed to a small island called Toa-marama, which is situated within the reefs on the eastern side of Raiatea. Here he was promised security, amid the destruction of the surrounding islands. The man hastened to his residence, and proceeded with his wife and child to the place appointed. Some say he took with him a friend who was residing under his roof, with a dog, a pig, and a pair of fowls, so that the party consisted of four individuals, besides the only domesticated animals known in the islands.

They reached the refuge appointed, before the close of the day; and as the sun approached the horizon, the waters of the ocean began to rise, the inhabitants of the adjacent shore left their dwellings on the beach, and fled to the mountains. The waters continued to rise during the night, and the next morning the tops of the mountains only appeared, above the wide-spread surface of the sea. These were afterwards covered, and all the inhabitants of the land perished. The waters subsequently retired, the fisherman and his companions left their retreat, took up their abode on the main land, and became the progenitors of the present inhabitants.

Toamarama, the ark in which those individuals are stated to have been preserved, is a small and low coralline island, of exceedingly circumscribed extent, while its highest parts are not more than two feet above the level of the sea. Whether, on the occasion above referred to, it was raised by Ruahatu to a greater elevation than the summits of the lofty mountains on the adjacent shore, or whether the waters, when, according to their representations, they rose several thousand feet above their present level, formed a kind of cylindrical wall around Toamarama, the natives do not pretend to know, and usually decline discussing this circumstance. Their belief in the event was, however, unshaken; and whenever we have conversed with them on the subject, they have alluded to the *farero*, coral, shells, and other marine substances, occasionally found near the surface of the ground, on the tops of their highest mountains. These, they say, would never have been carried there by the people, and could not have originally existed in the situations in which they are now found, but must have been deposited there by the waters of the ocean, when the islands were inundated.—We do not consider these marine substances as evidences that the islands were overflowed at the deluge, but have generally been accustomed to attribute to the whole a formation, if not posterior, yet not of more than equal antiquity with that event. We have usually viewed the coral, shells, &c. which do not appear to be fossils, as indications of the submarine origin of the mountains, and have supposed they were deposited on the rocks, near the surface of which they are now found, when those rocks formed the bed of the ocean, and prior to those violent explosive convulsions by which they were raised to their present elevation, and formed the groups of islands now under consideration.

These are but mere speculative opinions, and however strong the indications of such an origin might appear to our own minds, we could not demonstrate that the different islands now existing had not formerly belonged to one large island. Neither could we shew that they were not the remains of a continent, originally stretching across the Pacific, and uniting Asia and America, which, having been overflowed by the waters of the deluge, might have disappeared after those disruptions had taken place, by which the fountains of the great deep were broken up. Such speculations would have been useless, and we should only have perplexed the minds of the people with our own opinions. In general, we endeavoured to direct them to the records of that great event preserved in the Scriptures; in the traditionary accounts of which, perpetuated, as they were likely to be, by the descendants of the family of Noah for many generations, their own traditions, with those of the Sandwich Islanders, and other neighbouring tribes, had probably originated. I have frequently conversed with the people on the subject, both in the northern and southern groups, but could never learn that they had any accounts of the windows of heaven having been opened, or the rain having descended. In the legend of Ruahatu, the Toamarama of Tahiti, and the Kai of Kahinarii in Hawaii, the inundation is ascribed to the rising of the waters of the sea. In each account, the anger of the god is considered as the cause of the inundation of the world, and the destruction of its inhabitants. The element employed in effecting it is the same as that mentioned in the Bible; and in the Tahitian tradition, the boat or canoe being used, as the means of safety to the favoured family, and the preservation of the only domestic animals found on the islands, appear corrupted fragments of the memorial of Noah, the ark, and its inmates. These, with other minor points of coincidence between the native traditions and the Mosaic account of the deluge are striking, and warrant the inference, that although the former are deficient in many particulars, and have much that is fabulous in their composition, they yet refer to the same event.

The memorial of an universal deluge, found among all nations existing in those communities, by which civilization, literature, science, and the arts, have been carried to the highest perfection, as well as among the most untutored and barbarous, preserved through all the migrations and vicissitudes of the human family, from the remote antiquity of its occurrence to the present time, is a most decisive evidence of the authenticity of revelation. The brief yet satisfactory testimony to this event, preserved in the oral traditions of a people secluded for ages from intercourse with other parts of the world, is adapted to furnish strong additional evidence that the scripture record is irrefragable. In several respects, the Polynesian account resembles not only the Mosaic, but those preserved by the earliest families of the postdiluvian world, and supports the presumption that their religious system has descended from the Arkite idolatry, the basis of the mythology of the gentile nations. The mundane egg is conspicuous in the cosmogony of some of the most ancient nations. One of the traditions of the Hawaiians states, that a bird deposited an egg (containing the world in embryo) upon the surface of the primeval waters. If the symbol of the egg be supposed to refer to the creation, and the bird is considered a corrupted memorial of the event recorded in the sacred writings, in which it is said, "The Spirit of God moved upon the face of the waters," the coincidence is striking. It is no less so, if it be referred to the ark, floating on the waters of the deluge. The sleep of Ruahatu accords with the slumber of Brama, which was the occasion of the crime that brought on the Hindoo deluge. The warning to flee, and the means of safety, resemble a tradition recorded by Kœmpfer, as existing among the Chinese. The canoe of the Polynesian Noah has its counterpart in the traditions of their antipodes, the Druids, whose memorial states the bursting of the waters of the lake Lleon, and the overwhelming of the face of all lands, and drowning all mankind excepting two individuals, who escaped in a naked vessel, (a vessel without sails,) by whom the island of Britain was re-peopled. The safety which the progenitors of the Peruvian race are said to have found in caves, or the summits of the mountains, when the waters overflowed the land, bears a resemblance to the Hawaiian; and that of the Mexican, in which Coxcox, or Tezpi and his wife, were preserved in a bark, corresponds with the Tahitian tradition. Other points of resemblance between the Polynesian account, and the memorial of the deluge preserved among the ancient nations, might be cited; but these are sufficient to shew the agreement in the testimony to the same event, preserved by the most distant tribes of the human family.

CHAP. III.

General state of society—Former modes of living—Proposed improvement in the native dwellings—Method of procuring lime from the coral-rock—First plastered houses in the South Sea Islands—Progress of improvement—Appearance of the settlement—Described by Captain Gambier—Sensations produced by the scenery, &c.—Irregularity of the buildings—Public road—Effect on the surrounding country—Duration of native habitations—Building for public worship—Division of public labour—Manner of fitting up the interior—Satisfaction of the people—Chapel in Raiatea—Native chandeliers—Evening services.

THE change which had taken place in Tahiti and Eimeo, in consequence of the abolition of idol worship, had been exceedingly gratifying, as it regarded the general conduct of the people, their professed belief in the truth of revelation, and their desire to regulate their lives by its injunctions; but the visible change which resulted from the establishment of the Missions in Huahine and Raiatea, was more striking, and did not fail to attract the notice, and command the approbation, of the most superficial observer.

We did not deem what is usually termed civilization essential to their receiving the forgiveness of sin, enjoying the favour of God, exercising faith in Christ, and being after death admitted to the heavenly state; yet we considered an improvement of their circumstances, and a change in their occupations, necessary to their consistent profession of Christianity, and the best means of counteracting that inveterate love of indolence to which from infancy they had been accustomed. Habits of application were also essential to the cultivation and enlargement of intellect, the increase of knowledge, and enjoyment in every department and every period of the present life. This was peculiarly desirable in reference to the rising generation, who were to be the future population, and who would arrive at years of maturity, under circumstances and principles as opposite as light and darkness to those under which their parents had been reared. Under these impressions, those who were stationed in the Leeward Islands, next to the attention they paid to religious instruction, directed their attention to the advancement of civilization among the people, and the improvement of their temporal condition. We had already persuaded them to extend the culture of the soil beyond the growth of the articles necessary for their support during the season when the bread-fruit yielded no supply, and to raise cotton and productions, which they might exchange for clothing, tools, &c. We now directed them to the improvement of their dwellings, which, generally speaking, were temporary sheds, or wide unpartitioned buildings, by no means favourable to domestic comfort or Christian decency.

When we landed at Fare in Huahine, I do not think there were more than ten or twelve houses in the whole district. Four, besides those we occupied, were of considerable size, belonging to the chiefs; the others were mere huts. In the latter, the inmates took their food, and rested upon their mats spread upon the floor, which, had it been simply of earth, would have been comparatively clean and comfortable. The temporary roof of thatch was often pervious to the rays of the sun, and the drops of the frequently descending shower. In these cabins, parents, children, dogs, and frequently pigs and fowls, passed the night, and the greater part of the day. The houses of the chiefs were better built, and more capacious. The roofs generally impervious, and the sides frequently enclosed with straight white poles of the hibiscus tree. Their interior, however, was but little adapted to promote domestic comfort. The earthen floor was usually covered with long grass. This, by being repeatedly trodden under foot, became dry, broken, and filled with dust, furnishing also a resort for vermin, which generally swarmed the floors in such numbers, as to become intolerable. In these houses the people took their meals, sitting in circles on the grass-spread floor. Here, the fresh water used in washing their hands, the cocoa-nut water which was their frequent beverage, and the sea-water in which they dipped their food, was often spilt. Moisture induced decay, and although over these parts of the floor they often spread a little fresh grass, yet many places in the native houses frequently resembled a stable, or a stable-yard, more than any thing else.

In the drier parts of the house, along each side, the inmates slept at night. However large the building might be, there were no partitions or skreens. Some of their houses were two hundred feet long, and on the floor, hundreds have, at times, lain down promiscuously to sleep. They slept on mats manufactured with palm-leaves, spread on the ground. These mats were generally rolled up like a sailor's hammock in the morning, and spread out at night. The chief and his wife usually slept at one end of the house, without the least partition between them and the other inmates of their dwelling. Instead of a single mat, three or four, or even ten, were sometimes spread one upon the other, to give elevation and softness; and this, with the finer texture of the mats, was the only difference between the bed of the chief, and that on which the meanest of his dependents slept. Instead of being spread on the floor, the mats were sometimes spread on a low bedstead, raised nine or twelve inches above the floor. The sides and bottom of this bedstead were made with the boards of the bread-fruit-tree. Next to the chief, the members of his own family spread their mats on the floor, and then the friends and attendants—the females nearest the chief, the men towards the opposite end of the building.

I have sometimes entered the large houses in Huahine, soon after our arrival there, and have seen, I think, forty, fifty, or sixty sleeping places of this kind, in one house, consisting of a mat spread on the ground, a wooden pillow or bolster, in the shape of a low stool, next the side or wall; and a large thick piece of cloth, like a counterpane or shawl, which they call *ahu taoto*, sleeping-cloth, and which is their only covering, lying in the middle of each mat. There was no division or skreen between the sleeping places, but the whole ranged along in parallel lines from one end of the house to the other. What the state of morals must necessarily be among such a community, it is unnecessary to shew; yet such were the modes of life that prevailed among many, even after they had renounced idolatry. Such we found society in Huahine, and such our friends in Raiatea found it there. One of the reasons which they gave why so many slept in a house, was, their constant apprehensions of evil spirits, which were supposed to wander about at night, and grasp or strangle those who were objects of their displeasure, and whom they might find alone. Great numbers passing the night under the same roof, removed this fear, and inspired a confidence of security from the attacks their idolatrous absurdities led them to expect.

The evils necessarily resulting from these habits were too palpable to allow us to delay attempting at least an alteration. We recommended each family to build distinct and comfortable cottages for themselves, and the chiefs to partition bed-rooms in their present dwellings, in which they must reside while building others; even in these we recommended them to reduce the number of their inmates, and to erect distinct sleeping rooms for those they retained.

We were happy to perceive on their part a willingness to follow our advice. The first native improvement was made by Mai, the chief of Borabora, residing at that time at Fare in Huahine. He directed his servants to clear out all the grass from the floor of the house he occupied; they then levelled the earth, procured lime, and plastered it over nearly an inch thick with mortar; this hardened, and formed an excellent, solid, durable, and clean floor. With this material we had made the floors of our own temporary dwellings, in which we had erected slight partitions of hibiscus poles, covered with thick native cloth, to separate the different apartments from each other. In this also we soon perceived the chiefs promptly following our example. At the same time we commenced the erection of permanent places of residence for ourselves, and spared no pains to induce the people to do the same. Our first effort was to build a lime-kiln, on which we bestowed considerable labour, though it did not ultimately answer. The natives prepared their lime by burning it in a large pit, in a manner resembling that in which they had prepared their ovens for opio. This was done with greater facility than they could burn it in the kiln they had built, though with less economy in fuel.

Specimens of fibrous limestone, and small fragments of calcareous rock, have been occasionally found in some of the islands, but not in quantity or kind to be available in the preparation of lime for building. Shells might be procured in tolerable abundance; but the white coral rock, of which the extensive reefs surrounding these islands are composed, and which appears inexhaustible, is used in the manufacture of lime.

The natives dive into the sea, sometimes several fathoms deep, in order to procure the solid or sponge-shaped coral, which for this purpose is better than the forked or branching kinds. They also prefer that which is attached to the main reef, and growing, or, as they sometimes call it, live coral, to that which is broken off and hardened or dead. The large fragments or blocks of coral, sometimes three or four feet in diameter, thus procured, are conveyed on rafts to the shore, where they are broken into small pieces. A capacious hole is then dug, wherein fuel in immense logs is piled up till it assumes the appearance of a mound four or five feet high. On the outside of this, the pieces of coral are placed, twelve or eighteen inches thick. The pile is then kindled, the fuel consumed, and the lime, thus burnt, sinks down into the pit. They are generally so impatient to see whether it is well burnt, that they throw water upon it often before the fire is extinct;

and if they find it crumble and become pulverized, they cover it over with cocoa-nut leaves, and use it as occasion requires.

The coral rock makes excellent lime, not perhaps so strong as that made from rock-limestone, but fine, beautifully white, and durable. It may be obtained in any quantity, but the labour of procuring the fuel necessary for preparing it on the present plan, is exceedingly irksome. Could they be induced to erect kilns, and burn it after the European manner, it might be furnished with great facility, and the fact of their being able to prepare with little trouble, lime from the coral rock, would encourage them in building comfortable houses.

Our friends in Raiatea were perhaps more urgent than ourselves, in their recommendation of improved dwellings. On our first visit to Raiatea, in January 1819, the servants of Tamatoa, the king of that island, were plastering a house for his residence: it was nearly finished; the outside was completed, and they were at work within. A day or two after our return to Huahine, we were delighted to see one in the district of Fare actually finished. It was smaller than Tamatoa's, and differently shaped, his being oval, and this being nearly square, with high gable-ends. It belonged to an ingenious and industrious young man, whose name was Navenavehia, and who, although an inferior chief in Huahine, had accompanied Mahine to Eimeo, where he had resided in the family of Mr. George Bicknell, by whom he had been taught in some degree the use of tools, and the art of burning lime. It is not easy, nor is it material, to determine which of these two houses was finished first. They were certainly both in hand at the same time, and the periods of their completion were probably not very remote from each other. A new order of architecture was thus introduced to the nation, and the names of Tamatoa, king of Raiatea, and of Navenavehia, the more humble chief in Huahine, ought not to be forgotten, in connexion with the introduction of a style of building which has since prevailed so extensively among the people, greatly augmenting their social and domestic comforts, changing the appearance of their villages, and improving the beautiful scenery of their islands.

These two houses were not only the first in the Leeward group, but they were the first of the kind ever erected, for their own abode, by any of the natives of the South Sea Islands.

The success of these individuals encouraged others, although we found great difficulty in persuading them to persevere in the heavy labour this improvement required, particularly as they were now actively employed in the erection of a spacious chapel, and the frames of our dwellings. It was no easy task for them to build houses of this kind; there were no regular carpenters and masons. Every man had, in the first place, to go to the woods or the mountains, and cut down trees for timber, trim them into posts, &c. and remove them to the spot where his house was to be built, then to erect the frame, with the doorway and windows. This being done, he must again repair to the woods for long branches of hibiscus for rafters, with which he framed the roof.

The leaves of the pandanus were next gathered, and soaked, and sewed on reeds, with which the roof was thatched. This formerly would have completed his dwelling, but he now had to collect, with great labour, a large pile of firewood, to dig a pit, to dive into the sea for coral rock, to burn it, to mix it with sand so as to form mortar, wattle the walls and partitions of his house, and plaster them with lime. He then had to ascend the mountains again, to cut down trees, which he must either split or saw into boards for flooring his apartments, manufacturing doors, windows, shutters, &c. This was certainly a great addition of labour; and hence many occupy their cottages as soon as they have finished the roof, the walls, and the door—levelling the ground for the floor, and spreading grass over it—occupying one part, while they board or plaster the other.

In this state we found Navenavehia's house, when we paid him our first visit. We recommended him to persevere in completing it, and, in order to encourage him, promised him a few nails to make doors, and whatever else was wanting. He assured us of his intention to board the floor, and partition off their bed-room; but said, he thought they might as well live in it while he was doing this, and therefore had occupied it as soon as the walls were dry.

The settlements in the Leeward Islands now began to assume an entirely new aspect. Multitudes flocked from the different districts, to attend the means of instruction in the school, and on the Sabbath. The erection of a house upon the improved plan, regulating its size by the rank or means of the family for whom it was designed, became a kind of test of sincerity in professions of desire to be instructed; for to embrace Christianity, with the precepts which it inculcated, nothing could be more at variance than the habits of indolence and unsightly filthiness of their former habitations.

Activity was now the order of the day. Frames of buildings were seen rising with astonishing rapidity, in every part of the district; and houses of every size, from the lowly snug little cottage with a single door and window in front, to the large two-storied dwelling of the king or the chief. Buildings, in every stage of their progress, might be seen in a walk through the settlement: sometimes only a heap of spars and timber lay on the spot where the house was to be raised, but at other places the principal posts of the house were erected, others were thatched, and some partially or entirely enclosed with the beautiful white coral-lime plaster. Axes, hatchets, planes, chisels, gimlets, and saws, were, next to their books, the articles in greatest demand and highest esteem.

No small portion of our time was occupied in directing and encouraging them in their labours. We had, however, occasion to regret, that we were sometimes at as great a loss as the people themselves. They usually formed the walls of their dwellings, either by mortising upright posts into large trees laid on the earth, or planting the posts in the ground about three feet apart. The spaces between the posts, excepting those for doors or windows, were filled with a kind of hurdle-work, or wattling of small rods or sticks, of the tough casuarina. This they plastered with the mortar composed of coral-lime and sand, forming a plain surface, and covering also the posts on the outside, but leaving them projecting within.

The next object was to make the doors and window-shutters; thus far they had been able to proceed in the erection of their dwellings without nails; but to make doors and shutters without these, brought them at first to a stand. We were glad to furnish the chiefs and others with these most valuable articles, so far as our stock would allow, but it was useless to think of supplying the wants of the entire population; we only regretted that we could not have more ready access to our friends in England, many of whom, we had no doubt, would readily have supplied them with an article easily procured in abundance there, but which was here exceedingly scarce. Nails are still among the most valuable manufactures they can receive. Their invention and perseverance at length overcame the difficulty, and they constructed their doors by fastening together three upright boards, about six feet long, by means of three narrow pieces placed across, one at each end, the other in the middle. These latter were fastened to the long boards by strong wooden pegs. What the pegs wanted in strength, they determined to supply by numbers, and I think I have seen upwards of fifty or sixty hard pegs driven through one of these cross-pieces into the boards forming the door. In order to prevent their dropping out when the wood shrunk by the heat, they drove small wedges into the ends of the pegs, which frequently kept them secure. In the same manner they fastened most of their floors to the sleepers underneath, using, however, large pegs resembling the treenails in a ship's plank, more than the nails in a house-floor.

When the door was made, it was necessary to hang it; but only a few of the most highly favoured were, for many years, able to procure iron hinges. Some substituted tough pieces of fish-skin, pieces of the skin of other animals, or leather procured from the ships; but these soon broke, and many of the natives set to work to make wooden hinges. They were generally large, and, when attached to a light thin door, looked remarkably clumsy: but they were made with great industry and care, and the joints very neatly fitted. A man would sometimes be a fortnight in making a single pair of hinges. After all, they were easily broken, and made a most unpleasant noise every time the door was opened or shut.

In our walks through the native settlements, we were often amused at the state in which we found the houses occupied by their proprietors. Some appeared with only the walls on the outside plastered, others with both sides plastered; some having their doors and window-shutters fixed, others with a low fence only across the door-way; some with grass spread over the whole floor, while others had a portion boarded sufficiently large to contain their sleeping-mats at night. A few, whose dwellings were completely finished, inhabited them with all the conscious satisfaction attending the enjoyment of what had cost them long and persevering labour. All confessed that the new kind of houses were better than the old: that when the weather was warm, they could have as much air as was agreeable; and when the night was cold and the wind high, or the rain drifting, they had not, as formerly, to rise and move their beds, or secure their clothing from wet, but could sleep on, sheltered from the influence of the elements without.

This was the state of the settlement in Huahine when visited by Captain Gambier, of H. M. ship Dauntless, Captain Elliot, and other naval officers, whom I had the pleasure of meeting there. The account of the settlement given by the former, and the emotions excited in his own mind by his visit, are so interesting, that I think it would be almost unjust to deprive the readers of these pages of the satisfaction his description is adapted to afford. In reference to Tahiti, and the change generally, Captain Gambier observes, "The testimony is a strong one; as I had never felt any interest in the labours of Missionaries, I was not only not prepossessed in favour of them, but I was in a measure suspicious of their reports. It will appear as clear as light to the spiritual mind, that the account of their state, and the gratification experienced in the contemplation of it, was altogether of a temporal nature; that the progress made towards civilization and earthly happiness, in consequence of the moral influence of Christianity, was the cause of that delight. The hand of a superintending Providence is generally acknowledged, it is true, but it is so only with respect to the temporal state. So true it is, that the mind itself, untaught by the Divine Spirit, knows nothing of the awful and overwhelming importance of the eternal interests of the soul over the things of this short-lived scene." In reference to Huahine, and the station now described, though not more forward than others in the same group, Captain Gambier observes: "At about ten o'clock on the morning of the 20th of January, 1822, the ship being hove-to outside the reef, a party of us proceeded towards the village of Fare. After passing the reef of coral which forms the harbour, astonishment and delight kept us silent for some moments, and was succeeded by a burst of unqualified approbation at the scene before us. We were in an excellent harbour, upon whose shores industry and comfort were plainly perceptible; for, in every direction, white cottages, precisely English, were seen peeping from amongst the rich foliage, which every where clothes the lowland in these islands. Upon various little elevations, beyond these, were others, which gave extent and animation to the whole. The point on the left in going in, is low, and covered with wood, with several cottages along the shore.* On the right, the high land of the interior slopes down with gentle gradual descent, and terminates in an elevated point, which juts out into the harbour, forming two little bays. The principal and largest is to the left, viewing them from seaward; in this, and extending up the valley, the village is situated. The other, which is small, has only a few houses—but so quiet, so retired, that it seems the abode of peace and perfect content. Industry flourishes here. The chiefs take a pride in building their own houses, which are now all after the European manner; and think meanly of themselves, if they do not excel the lower classes in the arts necessary for the construction. Their wives also surpass their inferiors in making cloth. The queen and her daughter-in-law, dressed in the English fashion, received us in their neat little cottage.† The furniture of her house was all made on the island, and by the natives, with a little instruction originally from the Missionaries. It consisted of sofas with backs and arms, with (cinet) bottoms, really very well constructed; tables and bedsteads by the same artificers. There were curtains to the windows, made of their white cloth, with dark leaves stained upon it for a border, which gave a cheerful and comfortable air to the rooms. The

This part of Fare Harbour is represented in the plate, vol. i. p. 414.

See No. 2. in the plate of "Eastern part of Fare Harbour," page 79.

bed-roms were up stairs, and were perfectly clean and neat. These comforts they prize exceedingly; and such is the desire for them, that a great many cottages, after the same plan, are rising up every where in the village.

"The sound of industry was music to my ears. Hammers, saws, and adzes were heard in every direction. Houses in frame met the eye in all parts, in different stages of forwardness. Many boats, after our manner, were building, and lime burning for cement and white-washing.

"Upon walking through the village, we were very much pleased to see that a nice, dry, elevated foot-path or causeway ran through it, which must add to their comfort in wet weather, when going to prayers in their European dresses. As we stopped occasionally to speak to some of the natives standing near their huts, we had frequent opportunities of observing the value they set upon the comforts of our English style of cottage, and other things introduced among them of late. They said they were ashamed to invite us into their huts, but that their other house was building, and then they would be happy to see us there.

"Afterwards I walked out to the point forming the division between the two bays. When I had reached it, I sat down to enjoy the sensations created by the lovely scene before me. I cannot describe it; but it possessed charms independent of the beautiful scenery and rich vegetation. The blessings of Christianity were diffused amongst the fine people who inhabited it; a taste for industrious employment had taken deep root; a praiseworthy emulation to excel in the arts which contribute to their welfare and comfort, had seized upon all, and,

Drawn by Captn. Robt. Elliot R.N. 4 Engraved by W. Le Petit

EASTERN PART OF FA-RE HARBOUR, IN HUAHINE.

Fisher, Son & Co. London, 1829.

in consequence, civilization was advancing with wonderfully rapid strides."

The point referred to by Captain Gambier, is situated at a short distance to the right of the view of Fare, as given in the annexed plate engraved from a sketch taken on the spot by Captain Elliot. It is a delightful spot, and affords an extensive view of the unruffled waters of the bay, and the infant settlement rising on its shores. The figures along the bottom refer to the following buildings: No. 1. The chapel; 2. The residence of Mahine, the chief of the island,—this was the first house with an upper room which the natives erected. No. 3. is placed beneath the schools. 4. Marks the site where our dwelling stood, and that of my coadjutor, Mr. Barff; both these were erected at some distance from the shore, and stood on an elevation at the foot of the mountains forming the boundary of the valley.

Although we always urged the completion of their houses as soon as they could, we were often highly interested in visiting their partially finished dwellings. There is something peculiarly pleasing in watching the process which periodically changes the face of the natural world. The swelling bud—the opening blossom—the expanding leaves—the tiny fruit-formations, as they regularly pass under the eye of the observer, are not less interesting than the bough bending with full-ripe fruit. The process which effects the changes marking the progress from birth to maturity in the animal creation, is not less curious; and at this time we beheld a work advancing which was rapidly transforming the character and habits of a nation, and materially altering even the aspect of the habitable portions of their country. This gave a peculiar interest to the nondescript sort of dwelling, half native hut, and half European cottage, which many of the people at this time inhabited. They marked the steps, and developed the process, by which they were rising from the rude and cheerless degradation of the one, to the elevation and enjoyment of the other. These sensations were often heightened by our beholding in the neighbourhood of these half-finished houses, the lonely and comfortless hut they had abandoned, and the neatly finished cottage in which the inmates enjoyed a degree of comfort, that, to use their own powerful expression, made them sometimes ready to doubt whether they were the same people who had been contented to inhabit their former dwellings, surrounded by pigs and dogs, and swarms of vermin, while the wind blew over them, and the rain beat upon them.

The greater number of houses, already erected, contain only two or three rooms on one floor, but several of the chiefs have built spacious, and, considering the materials with which they are constructed, substantial habitations, with two stories, and a number of rooms in each, having also some of the windows glazed. Mahine, the king of Huahine, was, we believe, the first native of the South Sea Islands, who finished a house with upper rooms. When done, it was quite a curiosity, or occasion of wonder, among the natives of the Leeward Islands, and multitudes came on purpose to see it. It was built with care, and, considering it as a specimen of native workmanship, was highly creditable to their industry, perseverance, and ingenuity. Many of the natives, especially those who have been native house-builders, are tolerably good carpenters, and handle tools with facility. They have also been taught to saw trees into a number of boards, instead of splitting them into two planks, which was their former practice.

The stone in the northern parts of the island is a kind of compact ancient lava, and, though rather hard, is, we think, adapted for buildings. We were desirous to induce some of the chiefs to attempt the erection of a stone house; but they had no proper tools for preparing the stone, and the labour was also greater than in their present state of civilization they were disposed to undertake. It is not, however, improbable that stone buildings will ultimately supersede the neat, yet, compared with those erected of less perishable materials, temporary dwellings they are now occupying. The coral rock is also more durable than the plaster; and although soft, and easily hewn when first taken out of the sea, it afterwards assumes a degree of hardness which resists the weather for a long series of years. A chapel has been built with this material in the island of Eimeo, and will probably last longer than any yet erected.

When we arrived in Eimeo, Messrs. Hayward and Bicknell were residing in boarded dwellings with chambers, and Mr. Nott, in a house, the walls of which were neatly plastered. The earth in some parts of the islands would probably answer for bricks; and the Missionaries formerly made one or two attempts to prepare them for ovens, &c. but did not succeed. Individuals professing to understand making bricks have once or twice offered to teach the natives; but much as we have wished to possess permanent brick houses for ourselves, or to recommend the natives to prepare such, we are convinced that the labour would be too great, and the failures in burning them too frequent, to allow at present of their being made with advantage,—yet we hope they will follow the plastered cottage, as that now occupies the place of the native hut.

The timber principally employed in their buildings, is the wood of the bread-fruit; and although they are careful of this valuable tree, it is necessary frequently to urge the duty of planting, in order to ensure a future supply not only of timber but of food, as the large trees are now comparatively few, and the population is evidently increasing.

In the commencement of a new settlement, or the establishment of a town, like that rising around us at the head of Fare harbour, we were desirous that it should assume something like a regular form, as it regarded the public buildings and habitations of the chiefs and people. We repeatedly advised the chiefs and others to build their houses and form their public roads in straight lines, and to leave regular and equal distances between the roads and the houses, and also between their respective dwellings. Our endeavours, however, were unavailing. They could perceive nothing that was either desirable or advantageous in a straight road, or regularity in the site, and uniformity in the size or shape, of their dwellings. Every one, therefore, followed his own inclinations. The size of the building was regulated by the number in the family, the rank or the means of its proprietor, and the shape by his fancy. It was oblong or square, with high gable, or circular ends covered with thatch, so that the building resembled an oval more than any other shape.

The situations selected were either parts of their own ground, or such places as accorded with their taste and habits. Those who were frequently upon the waters, and enjoyed the gentle sea-breezes, or wished to excel their neighbours, built a massy pier or causeway in the sea, and, raising it four or five feet above high-water mark, covered it with smooth flat stones, and then erected their houses upon the spot they had thus recovered from the sea, by which it was on three sides surrounded. The labour required for effecting this, prevented any but chiefs from building in such situations. Others, actually building upon the sand, erected their dwelling upon the upper edge of the beach, within four or five yards of the rising tide.

The public road, from six to twelve feet wide, which led through the district, extending in a line parallel with the coast, presented all its curvatures. Some of the natives built their houses facing the sea; others, turning their fronts towards the mountain, reared them within five or six feet of the road; while several, of a more retiring disposition, built in the centre of their plantations, or under the embowering shade of a grove of bread-fruit trees, enclosing them within the fence that surrounded their dwelling. Some of the leading chiefs, in order to enjoy a more extensive prospect, and to breathe a purer atmosphere, left the humidity and shade of the lowland and the valley, and built their houses on the sides of the verdant hills that rise immediately behind the bay, and form the connecting link between the rocks around the beach and the high mountains of the interior.

A settlement thus formed could never possess any approximation to uniformity; and although we had endeavoured to persuade the people to render it more regular, yet it often seemed as if the variety in size and shape among the buildings, and the irregularity of their situation, was in perfect keeping with the wild, untrained luxuriant loveliness, and romantic appearance, of the rocks, the hills, the mountains, the valleys, and every natural object by which the rising settlement was surrounded. The chiefs vied with each other in the size, elevation, or conveniences of their houses: some being, like Pohuetea's and Terlitaria's, built upon a pier in the sea; others preparing to attach verandas, by which they could remain cool under a meridian sun; others erected rude covered balconies, in which they might enjoy a more extended prospect, be shaded from the sun, and breathe purer air. The rustic palm-leaf thatch, and beautifully white plastered walls, of all the buildings, whether standing on the sea-beach, on the mountain's side, embowered under the bread-fruit and cocoa-nut grove, or situated in the midst of their plantations, with a walk strewed with fragments of coral and shells leading from the road to the door, appeared in delightful contrast with the thick dark foliage of the trees, the perpetual luxuriance of vegetation, and the variegated blossoms of the native flowers.

The duration of the buildings was in general according to the nature of the thatch; the same house frequently received two or three new roofs, and if the frame was well put together, and the timber seasoned, a plastered cottage would probably last ten or fifteen years. Many, however, from the rude and hurried manner in which they were built, became dilapidated in a much shorter period.

While individuals and families were thus engaged in the erection of their domestic habitations, the people of the island were occupied in raising a spacious and substantial chapel. They commenced it in the beginning of 1819, and completed it early in the following year. It was one hundred feet long, and sixty wide.

The sides were fourteen or sixteen feet high, and the centre not less than thirty. The walls were plastered within and without. The roof was covered with pandanus leaves, the windows closed with sliding shutters, and the doors hung with iron hinges of native workmanship. Altogether, the building was finished in a manner highly creditable to their public spirit, skill, and persevering industry. All classes cheerfully united in the work, and the king of the island—assisted by his only son, a youth about seventeen years of age—might be seen every day directing and encouraging those employed in the different parts of the building, or working themselves with the plane or the chisel, in the midst of their chiefs and subjects.

The interior of the roof was remarkable for the neatness of its appearance, and the ingenuity of its structure. The long rafters, formed with slender cocoanut, casuarina, or hibiscus trees, were perfectly straight, and polished at the upper end. The lower extremities were ornamented with finely-woven variegated matting, or curiously braided cord, stained with brilliant red or black and yellow native colours, ingeniously wound round the polished wood, exhibiting a singularly neat and chequered appearance. The ornament on the rafter terminated in a graceful fringe or bunch of tassels.

The pulpit, situated at a short distance from the northern end, was hexagonal, and supported by six pillars of the beautiful wood of the pua, *beslaria laurifolia* of Parkinson, which resembles, in its grain and colour, the finest satin-wood. The pannels were of rich yellow bread-fruit, and the frame of mero, *thespesia populnea,* a beautiful fine-grained, dark, chestnut-coloured wood. The stairs, reading-desk, and communion table, were all of deep umber-coloured bread-fruit; and the whole, as a specimen of workmanship, was such as the native carpenters were not ashamed of. The floor was boarded with thick sawn planks, or split trees; and, although it exhibited great variety of timber and skill, was by no means contemptible.

According to ancient usage in the erection of public buildings, the work had been divided among the different chiefs of the islands; these had apportioned their respective allotments among their peasantry or dependants, and thus each party had distinct portions of the wall, the roof, and the floor. The numbers employed rendered these allotments but small, seldom more than three or six feet in length, devolving on one or two families. This, when finished, they considered their own part of the chapel; and near the part of the wall they had built, and the side of the roof they had thatched, they usually fitted up their sittings. The principal chiefs, however, fixed their seats around the pulpit, that they might have every facility of hearing.

Uniformity was as deficient in the sittings of the chapel, as in the houses of the town, each family fitting up their own according to their inclination or ability. For a considerable extent around the pulpit, the seats were in the form of low boarded pews neatly finished. Behind them appeared a kind of open, or trellis-work line of pews, which were followed by several rows of benches with backs; and, still more remote from the pulpit, what might be called free or unappropriated sittings, were solid benches or forms, without any support for the back or arms.

The colour and the kind of wood, used in the interior,

was as diversified as the forms in which it was employed; it was, nevertheless, only when empty, that its irregularity and grotesque variety appeared. When well filled with respectably dressed and attentive worshippers, as it generally was on the Sabbath, the difference in the material or structure of the places they occupied, was not easily noticed.

A remarkably ingenious and durable low fence, called by the natives *aumoa*, was erected round it, and the area within the enclosure was covered with small fragments of white branching coral, called *anaana*, and found on the northern shores of the bay.

In the month of April, 1820, it was finished, and on the 3d of May I had the pleasure of opening it for Divine service.

A distressing epidemic had raged for some time among the people, and still confined many to their habitations, yet there were not fewer than fifteen hundred present. Many of them were arrayed in light European dresses, and all evidently appeared to feel a high degree of satisfaction in assembling for the public adoration of the Almighty in a building, in many respects an object of astonishment through the island, and which their own toil and perseverance had enabled them to finish.

Individuals in England, who have materially contributed by personal exertions or pecuniary aid to the erection or enlargement of a church or chapel, have, when the object of their solicitude and their toil has been accomplished, experienced emotions of satisfaction during the subsequent opportunities they have had of rendering divine homage there; but the satisfaction of the Tahitians, though the same in kind, I am disposed to believe is stronger in degree, when standing on the floor, the trees constituting which, they cut down in the forest—when skreened from the wind by that portion of the wall their own hands reared—and covered by that section of the roof which they had thatched.

While the inhabitants of Huahine were thus laudably engaged in providing the means of increasing their domestic enjoyments, and accommodating the assemblies for public worship; their neighbours in the adjacent island of Raiatea were not behind them in the rapidity of their improvement. They had erected a number of dwelling-houses, and a building for divine service, larger than that at Huahine, but inferior in elevation and breadth; being forty-two feet wide, and at the sides about ten feet high. It was finished a week or two earlier than the chapel in Huahine, and was opened on the 11th of April in the same year; when upwards of 2400 inhabitants of that and the adjacent islands assembled within its walls.

To the natives of Raiatea, this work of their own hands appeared a wonderful specimen of architecture; and the manner in which its interior was finished perfectly astonished them, and appeared no less surprising to the natives of the other islands. It was not only furnished with a pulpit, a desk, a boarded floor throughout, constructed of the tough planks of the reva, or (*galaxa sparta,*) and filled with pews and seats, but, by the invention and ingenuity of the Missionaries, it was subsequently furnished with a rustic set of chandeliers.

By this contrivance it could be lighted up for an evening congregation, while we were under the necessity of concluding all our public services before the sun departed. These chandeliers, as they may perhaps with propriety be called, were not indeed of curious workmanship or dazzling brilliancy, in polished metal or cut-glass, but of far more common materials, and rude simplicity of structure. The frame was of light tough wood, and the lamps, instead of being coloured and transparent, were opaque cocoa-nut shells. They were, however, the only inventions of the kind the natives had ever seen; and on the night when the chapel was first illuminated by their aid, as they came in one after another, and saw the glare of such a number of lights suspended from the roof in a manner that they could not at first understand, they involuntarily stopped to gaze as they entered the door, and few proceeded to their seats without an exclamation of admiration or surprise. Their astonishment was probably greater than would be experienced by an English peasant from a retired village, on beholding, for the first time, a spacious public building splendidly lighted up with gas.

Although we were pleased with the effect produced on the minds of the natives, and a thousand delightful associations reviving in our bosoms the first time we mingled with a crowded *evening* congregation, we did not recommend our people to follow the example their ingenious neighbours had set them. It appeared more desirable, in the partially organized state of society then prevailing in the islands, to conclude all our public meetings by daylight, rather than call the people from home after sunset.

CHAP. IV.

Schools erected in Huahine—Historical facts connected with the site of the former building—Account of Mai, (Omai)—His visit to England with Captain Furneux—Society to which he was introduced—Objects of his attention—Granville Sharp—His return with Captain Cook—Settlement in Huahine—His subsequent conduct—Present proprietors of the Beritani in Huahine—House for hidden prayer—Cowper's lines on Omai—Royal Mission Chapel in Tahiti—Its dimensions, furniture, and appearance—Motives of the king in its erection—Description of native chapels—Need of clocks and bells—Means resorted to for supplying their deficiency—Attendance on public worship—Habits of cleanliness—Manner of wearing the hair—Process of shaving—Artificial flowers—Native toilet.

As soon as the new building in Huahine was finished, and appropriated to the sacred use for which it had been reared, the original chapel was converted into a school, and was scarcely sufficient to accommodate the increasing number of scholars.

Two new places, upon the same plan as the chapel, and built with similar materials, were afterwards erected, one for the boys' school, and the other for the girls'; these, when finished, greatly facilitated the instruction of the people—the accommodation they afforded, encouraging those to attend who had before been deterred.

The spot on which the old chapel and subsequent school had been erected, was connected with an important event in the modern history, not only of Huahine, but the several adjacent clusters of islands. In September, 1773, when Captains Cook and Furneux left Huahine, the latter was accompanied by a native, who had intimated his desire to proceed in the ship on a visit to Britain. He was a Raiatean; who, after a defeat which his countrymen had sustained in an engagement with the daring and warlike natives of Borabora, had taken shelter in Huahine. His inducement to undertake a voyage, of the incidents and exposures of which he could form no idea, does not appear to have resulted so much from a wish to gratify a restless and ardent curiosity, as from the desire to obtain the means of avenging his country, and regaining the hereditary possessions of his family, which were now occupied by the victors.

The name of this individual was *Mai,* usually called *Omai,* from the circumstance of the *o* being prefixed in the native language to nouns in the nominative case. Mai is the name of the present king of Borabora, though I am not certain of his having descended from the same family. The Mai who accompanied Captain Furneux does not appear to have been connected by birth or rank with the regal or sacerdotal class, although, among other accounts circulated respecting him while in England, it was stated that he was a priest of the sun, an office and title unknown in his native islands. He represented himself as a *hoa,* friend or attendant, on the king. In person he was tall and thin, easy and engaging in his manners, and polite in his address; but in symmetry of form, expression of countenance, general outline of feature, and darkness of complexion, inferior to the majority of his countrymen. His conversation was said to be lively and facetious. He reached England when the interest of Captain Cook's first voyage, and the deep impression produced by his discoveries, were still vivid and universal, and anticipation was raised to the highest pitch, in reference to the developments expected from his second visit to that distant part of the world. Mai being the first native of the islands of the South Sea, brought to England, produced an excitement as unprecedented, in connexion with an untutored islander, as it was powerful and extensive, even in the most polished circles of society. Mai, on his arrival in London, was considered a sort of prodigy; he was introduced to the most fashionable parties, conducted to the splendid entertainments of the highest classes, and presented at the British court amidst a brilliant assemblage of all that was illustrious in rank, and dignified in station. The Tahitians in general are good imitators of others; this talent he possessed in an eminent degree, and adopted that polite, elegant, and unembarrassed address, whereby the class with which he associated has ever been distinguished. Naturally quick in his perceptions, and lively in his conversation, although the structure and idiom of his own language effectually prevented his speaking English with ease or fluency, he was soon able to make himself understood; and the embarrassment he occasionally felt, in giving utterance to his thoughts, perhaps added to the interest of those who were watching the effect which every object in a world so new to him must naturally occasion.

Every place of public amusement, and every exhibition adapted to administer pleasure, was repeatedly visited; and the multiplicity of spectacles thus presented in rapid succession, kept his mind in a state of perpetual excitement and surprise. The impression made by one object, was obliterated by the exhibition of some new wonder, which prevented his paying particular regard to any. This constant variety deprived him of all useful knowledge, and excluded his attention to the important subjects that demanded his notice while residing in the metropolis of Britain. A most favourable opportunity was afforded for his acquiring that knowledge of our agriculture, arts, and manufactures, our civil and religious institutions, which would have enabled him to introduce the most salutary improvements among his countrymen. Thus he might have become a father to his nation; and his visit to England might have been rendered a blessing to its latest generations. But, as Forster, who accompanied him on his return, laments, "no friendly Mentor ever attempted to cherish and to gratify this wish, much less to improve his moral character, to teach him our exalted ideas of virtue, and the sublime principles of revealed religion." To the censure thus passed upon those, under whose care he spent the period of his residence in England, one exception at least must be made, and that in favour of a name that will ever be dear to every friend of humanity. Granville Sharp became acquainted with Mai, taught him the first principles of writing, and, so far as his knowledge of our language allowed, endeavoured to pour the light of divine truth into his ignorant and untutored mind. He made such progress in the use of letters, that on his voyage to the South Seas, while staying at the Cape of Good Hope, he wrote a letter to his friend Dr. Solander.

During the two years he spent in this country, he was inoculated for the small-pox, from which he happily recovered; and, loaded with presents profusely furnished by his friends, he embarked for his native island at Plymouth, in the summer of 1776. He accompanied Captain Cook in the visits he made to New Zealand, the Friendly Islands, and Tahiti, and, after an absence of rather more than four years, returned to Huahine on the twelfth of October, 1777.

In this island Captain Cook judged it most prudent to establish his fellow-voyager, and consequently solicited for him a grant of land from the chiefs. It was readily furnished, and a spot marked out, measuring about two hundred yards, along the sea-shore, and extending from the beach to the mountain. Here a garden was enclosed, and many valuable seeds and roots, which had been brought from England or the Cape of Good Hope, were planted. The carpenters of the vessels erected for him a house in the European style, and on the 26th of October, the presents with which he had been so liberally supplied, were landed, and he took possession of his dwelling. In addition to the seeds and plants, a breed of horses, goats, and other useful animals, were landed; but the greater part of the presents was comparatively useless, and many were bartered to the sailors for hatchets or iron tools. It does not appear that there was any implement of husbandry, or useful tool, included in the catalogue of his presents, though he landed with a coat of mail, a suit of armour, musket, pistols, cartouch-box, cutlasses, powder, and ball! Besides these, however, he was furnished with a portable organ, an electrical machine, fire-works, and numerous trinkets.

The estimate Captain Cook formed of his character was correct: he appeared to have derived no permanent advantage from the voyage he had made, the attention he had received, or the civilized society with which he had been associated. He soon threw off his European dress, and adopted the costume, uncivilized manners, and indolent life, of his countrymen. Weakness and vanity, together with savage pride, appear to have been the most conspicuous traits of character he developed in subsequent life.

The horses, included among his presents, appear to have been regarded by Mai as mere objects of curiosity, and were occasionally ridden, in order to inspire terror or excite admiration in the minds of the inhabitants. His implements of war, and especially the fire-arms, rendered his aid and co-operation a desideratum with the king of the island, who, in order more effectually to secure the advantage of his influence and arms, gave him one of his daughters in marriage, and honoured him with the name of *Paari*, (wise or instructed,) by which name he is now always spoken of among the natives; several of whom still remember him. He appears to have passed the remainder of his life in inglorious indolence or wanton crime, to have become the mere instrument of the caprice or cruelty of the king of the island, who not only availed himself of the effects of his fire-arms in periods of war, but frequently ordered him to shoot at a man at a certain distance, to see how far the musket would do execution; or to despatch with his pistol, in the presence of the king, the ill-fated objects of his deadly anger.

The majority of those whom I have heard speak of him, have mentioned his name with execration rather than respect; and though some of the chiefs speak of him as a man who had seen much of the world, and possessed, according to their ideas, an amazing mass of information, his memory is certainly but lightly esteemed by any of his countrymen. He does not, however, seem to have evinced, either on board the vessels in which he sailed, or among the company with which he mingled while in England, any latent malignity of character or cruelty of disposition, and might perhaps have returned with very different sentiments and principles, had he fallen into other hands during his visit here.

The spot where Mai's house stood is still called Beritani,* or Britain, by the inhabitants of Huahine. A shaddock tree, which the natives say was planted by Captain Cook himself while the vessels lay at anchor, is still growing on a spot which was once part of his garden. The animals, with the exception of the dogs and pigs, have all died; and in this instance, the benevolent intentions of the British government, in sending out horses, cattle, &c. proved abortive. The helmet, and some other parts of his armour, with several cutlasses, are still preserved, and displayed on the sides of the house now standing on the spot where Mai's dwelling was erected by Captain Cook. A few of the trinkets, such as a jack-in-a-box, a kind of serpent that darts out of a cylindrical case when the lid is removed, are preserved with care by one of the principal chiefs, who at the time of our arrival considered them great curiosities, and exhibited them, as a mark of his condescension, to particular favourites. What became of the organ and electrical machine, I never knew. Among the curiosities preserved by the young chief of Tahaa,

* See No. 1, in the Engraving, vol. I, p. 414.

there was an article that I was very glad to see; it was a large quarto English Bible, with numerous coloured engravings, which were the only objects of attraction with the natives. I was told it belonged to Paari, or Mai, and hope it was given him among the presents from England, although no mention whatever is made of a Bible, or any other book, among the various articles enumerated by those who conveyed him to his native shores.

Within the limits of the grant made to Captain Cook for his friend Mai, some of the Missionaries, who in 1809 took shelter in Huahine, after their expulsion from Tahiti in 1808, erected their temporary habitations. A few yards distant from the spot in which Mai's house stood, and immediately in front of the dark and glossy-leaved shaddock-tree planted by Captain Cook, the first building for the worship of Jehovah was erected; and on the same spot, the first school in Huahine was opened, in which the use of letters, and the principles of religion, were inculcated.

Nearly in front of the site of Mai's dwelling now stands the residence of Pohuetea and Teraimano, to whom by right of patrimony Beritani belongs. It was, when I was last there, in 1824, one of the most neat, substantial, and convenient modern houses in the settlement, containing two stories, and eight apartments. The district around, which when we arrived was altogether uncultivated, and overrun with brushwood growing in wild luxuriance, has been cleared; the garden has been again enclosed, and planted with all that is useful in the vegetable productions of tropical regions. It is cultivated by its proprietors, who, there is reason to hope, are decided Christians. They erected, within the precincts of their garden, a beautiful but rustic little summer-house or cottage, which they call a *fare bure huna*, or house for hidden prayer. I one day visited this garden, a few weeks after it had been enclosed and stocked with the most valuable indigenous plants of the islands. Towering above the plantains, papaws, &c. the shaddock planted by Captain Cook appeared, like an inhabitant of another country, in solitary exile; for though the climate is similar in point of temperature to that in which it is accustomed to thrive, its shoots are not long and vigorous, its leaves are not so clear, dark, and glossy as those of the other plants, and the fruit, though large and abundant, falls prematurely to the ground.

After wandering some time among the clustering sugarcane, rows of pine-apples, plantains, and bananas, I approached this house for private devotion. A narrow path covered with sand and anaana, or branches of coral, led to the entrance. An elegant hibiscus spread its branches over the cottage, and threw its embowering shade on its rude and lowly roof. A native palm-leaf mat covered the earthen floor,—a rustic seat, a table standing by a little open window, with a portion of the Scripture, and a hymn-book in the native language, constituted its only furniture. The stillness of every thing around, the secluded retirement of the spot, and the diversified objects of nature with which it was associated, seemed delightfully adapted to contemplation and devotion. The scene was one of diversified beauty, and the only sounds were those occasioned by the rustling among the sugarcanes, or the luxuriant and broad-leaved plantains, while the passing breezes swept gently through them.

I naturally inferred that the house was appropriated to purposes of secret devotion; and meeting its proprietor, I asked its use. He informed me that it was devoted to that object, and spoke with apparent satisfaction of the happiness he enjoyed in the retirement it afforded.

The erection of their dwelling, culture of their garden, building the house for hidden prayer, &c. (the labours of the present proprietors of Beritani,) are very different from the erection of a boarded house merely as a fortress, in which are deposited, as the most valued treasures of its inhabitant, arms and ammunition. It does not appear that Mai's house was designed as a model by which the natives were to be encouraged to build their own, but a place of security for the property, which he was recommended to enclose with a spacious native building: and the pursuits of its present occupants are in delightful contrast with the childish exhibition of fireworks, or the display of those trinkets, by which it was endeavoured to impress the minds of the natives with ideas of English superiority over untutored barbarians. The events which have since transpired were but little anticipated by the distinguished navigator, who conducted this simple-hearted native from one end of the globe to the other, spared no pains to promote his welfare and comfort, and who, although mistaken in the means he employed, undoubtedly aimed at the prosperity of the interesting people whom he had introduced to the notice of the civilized world.

Visiting almost daily the spot, and living in habits of intercourse with the successors of Mai, I have been often led to compare the views and circumstances of the present inhabitants of Beritani with those of the resident originally left there by its discoverer; and in connexion with the circumstances of Mai after his return to his native islands, the following beautiful and pathetic lines have often occurred to my mind; and though perused on the spot with sensations probably unfelt elsewhere, I have nevertheless supposed, that could the poet have foreseen what has since taken place, not only in this island, but throughout the group—or had he lived in the present day—he would never, in anticipation of their abandonment so soon after their discovery, have recorded such mournful anticipations.

"But far beyond the rest, and with most cause,
Thee, gentle savage,* whom no love of thee
Or thine, but curiosity perhaps,
Or else vain-glory, prompted us to draw
Forth from thy native bowers, to shew thee here
With what superior skill we can abuse
The gifts of Providence, and squander life.
The dream is past. And thou hast found again
Thy cocoas and bananas, palms and yams,
And homestall thatched with leaves. But hast thou found
Their former charms? And having seen our state,
Our palaces, our ladies, and our pomp
Of equipage, our gardens, and our sports,
And heard our music; are thy simple friends,
Thy simple fare, and all thy plain delights,
As dear to thee as once? And have thy joys
Lost nothing by comparison with ours?
Rude as thou art, (for we returned thee rude
And ignorant, except of outward show,)
I cannot think thee yet so dull of heart
And spiritless, as never to regret
Sweets tasted here, and left as soon as known.
Methinks I see thee straying on the beach,
And asking of the surge that bathes thy foot,

* Omai

If ever it has wash'd our distant shore.
Thus fancy paints thee, and though apt to err,
Perhaps errs little when she paints thee thus.
She tells me too, that duly ev'ry morn
Thou climb'st the mountain-top, with eager eye
Exploring far and wide the wat'ry waste
For sight of ship from England. Ev'ry speck
Seen in the dim horizon, turns thee pale
With conflict of contending hopes and fears,
But comes at last the dull and dusky eve,
And sends thee to thy cabin, well prepar'd
To dream all night of what the day denied.
Alas! expect it not. We found no bait
To tempt us in thy country. Doing good,
Disinterested good, is not our trade.
We travel far, 'tis true, but not for nought;
And must be brib'd to compass earth again
By other hopes, and richer fruits, than yours."

In the visit of Mai, the experiment, in reference to the effect of refinement, civilization, and philosophy upon the ignorant and uncivilized, was tried under circumstances the most favourable for producing sympathy in one party, and impression on the other:—the result was most affecting. The individual who had been brought from the ends of the earth, and shewn whatever England could furnish, adapted to impress his wondering mind, returned, and became as rude and indolent a barbarian as before. With one solitary exception, the humanizing and elevating principles of the Bible had never been presented to his notice, and he appeared to have derived no benefit from his voyage. Well might the poet lament his fate. But the ship Duff had not sailed, and the spirit of Missionary enterprise was not aroused in the British churches. Institutions, the ornament and the glory of our country, had not arisen. The schoolmaster was not abroad in the earth, and, proceeding onward with the tide of commerce that rolled round the world, the progress of discovery and science penetrating every remote, inhospitable section of our globe; the Bible and the Missionary had not been sent. Had Cowper witnessed these operations of Christian benevolence, how he would have cheered, with his own numbers, those who had gone out from Britain, and other lands, not only to civilize, but to attempt the moral renovation of the heathen world.

The regularly framed and plastered chapels in Huahine and Raiatea were the first of the kind in the Leeward or Windward Islands; they were not, however, the only large buildings erected for public worship. Pomare had, ever since our arrival, been engaged in preparing materials, and erecting a chapel, at Papaoa, by far the largest ever built in the islands; it had been opened twelve months before those in the Leeward Islands were finished.

This building, which is called the Royal Mission Chapel, is certainly, when we consider the imperfect skill of the artificers, the rude nature of their tools, the amazing quantity of materials used, and the manner in which its workmanship is completed, an astonishing structure. It is seven hundred and twelve feet in length, and fifty-four wide. Thirty-six massy cylindrical pillars of the bread-fruit tree sustain the centre of the roof, and two hundred and eighty smaller ones, of the same material, support the wall-plate along the sides, and around the circular ends, of the building. The sides or walls around are composed of planks of the bread-fruit tree, fixed perpendicularly in square sleepers. The whole, either smoothed with a carpenter's plane, or polished, according to the practice of the natives, by rubbing the timber with smooth coral and sand. One hundred and thirty-three windows or apertures, furnished with sliding shutters, admit both light and air, and twenty-nine doors afford ingress and egress to the congregation. The building was covered with the leaves of the pandanus, enclosed with a strong and neat, low *aumoa*, or boarded fence; and the area within the enclosure was filled with basaltic pebbles, or broken coral. The roof was too low, and the width and elevation of the building too disproportioned to its length, to allow of its appearing either stupendous or magnificent.

The interior of this spacious structure was at once singular and striking. The bottom was covered in the native fashion with long grass, and, with the exception of a small space around each pulpit, was filled with plain, but substantial forms or benches. The rafters were bound with curiously-braided cord, coloured in native dyes, or covered nearly to the top of the roof with finely-woven matting, made of the white bark of the purau, or *hibiscus*, and often presenting a chequered mixture of opposite colours, by no means unpleasing to the eye. The end of the matting usually hung down from the upper part of the rafter three, six, or nine feet, and terminated in a fine broad fringe or border.

The most singular circumstance, however, connected with the interior of the Royal Mission Chapel, is the number of pulpits. There are no fewer than three. They are nearly two hundred and sixty feet apart, but without any partition between. The east and west pulpits are about a hundred feet from the corresponding extremities of the chapel. They are substantially built, and though destitute of any thing very elegant in shape or execution, answer exceedingly well the purpose for which they were erected.

This immense building was opened for divine service on the 11th of May, 1819, when the encampment of the multitudes assembled stretched along the sea-beach, on both sides of the chapel, to the extent of four miles. On this occasion, three distinct sermons, from different texts, were preached at the same time, to three distinct congregations. Each audience, consisting of upwards of two thousand hearers, assembled round the respective pulpits within the same building. The king and principal chiefs appeared at the east, which, contrary to the order observed in their antipodes, is considered the court end. The whole number of hearers, according to the nearest calculation, was about seven thousand; and, notwithstanding this number assembled, a space remained between the different congregations.

I have occasionally preached in the Royal Mission chapel, but never when any other person besides was engaged; consequently, I cannot say what effect is produced on the ear by the delivery of more than one discourse at the same time. In the account the Missionaries give of its opening, they say, the pulpits being at so great a distance from each other, no confusion ensued from the speakers preaching at once in the same house. To an individual who could have stood at one end of the building, a little above the assembly, and directed his glance to the other—the three pulpits and preachers—the seven thousand hearers assembled around in all the variety, and form, and colour of their different costume—must have presented an imposing and interesting spectacle.

Although divested of every thing like stateliness or

grandeur, the first visit I paid to the chapel left a strong impression on my mind. I entered from the west; and the perspective of a vista, extending upwards of seven hundred feet, partially illuminated by the bright glow of strong noon-day light entering through the windows, which were opened at distant intervals, along the lengthened line of pillars that supported the rafters—the clean rustic appearance of the grass-spread floor—the uniformity of the simple and rude forms extending thoughout the whole building—the pulpits raised above them—heightened the effect of the perspective. Besides these, the singular, novel, light, waving, and not inelegant adornments of the roof, all combined to increase the effect. The reflections also associated with the purpose for which it had been erected, and the recent events in the history of the people, whose first national Christian temple we were visiting, awakened a train of solemn and grateful emotions. How it might be when the house was filled, I do not know; but when empty, the human voice could be distinctly heard from one end to the other, without any great effort on the part of those who at this distance called or answered.

A long aisle or passage, between the forms, extends from one end to the other. In walking along this aisle on my first visit, I was surprised to see a watercourse five or six feet wide, crossing, in an oblique direction, the floor of the chapel. On inquiry of the people who accompanied our party, they said it was a natural water-course from the mountains to the sea; and that, as they could not divert its channel so as to avoid the building without great additional labour, and constant apprehension of its returning, they had judged it best to make a grating at each side under the wall, and allow it to pass in its accustomed channel. As it was not during the rainy season that we were there, it was dry; the sides were walled, and the bottom neatly paved; but in the rainy season, when the water is constantly flowing through, its effect must be rather singular on the minds of those sitting near it during public worship.

One end of the building was used by the inhabitants for divine service every Sabbath; the other parts are only occupied at the annual meetings of the Tahitian Missionary Society, or on similar occasions, when large national assemblies are convened. In 1822, when I last visited it, the roof had already begun to decay. The labour of keeping so large a place in repair, would be very great; and the occasions for its use so seldom occur, that no repairs have been made since the king's death; and the exposure being constant, it will not probably last many years longer. The texture of the palm-leaves composing the thatch, is not such as to resist for any protracted period the intense heat of the climate; and the heavy rains accelerate its destruction.

It has appeared matter of surprise to many, that the natives should desire, or the Missionaries recommend, the erection of such large places of worship; and I have often been asked, how we came to build such immense houses. The Royal Chapel at Papaoa, however, is the only one of the kind in the islands. It originated entirely with the king, and in its erection the Missionaries took no part. The king, determined in his purpose, levied a requisition for materials and labour on the chiefs and people of Tahiti and Eimeo—by whose combined efforts it was ultimately finished. The Missionaries were far from approving of the scale on which Pomare was proceeding; and, on more than one occasion, some of them expressed their regret that so much time and property should be appropriated to the erection of a building, which would be of far less general utility than one of smaller dimensions. But the king was not thus to be diverted from his original design; and however injudicious the plan he pursued might be, the motives by which he was influenced were certainly commendable. He frequently observed, that the heaviest labour and the most spacious and enduring buildings ever erected, were in connexion with the worship of their former deities, illustrating his remarks by allusion to the national maraes at Atehuru, Tautira, and other parts; declaring, at the same time, his conviction that the religion of the Bible was so much superior to that under which they formerly lived, and the service of the true God so happy and beneficial in its influence, that they ought to erect a much better place for the homage of Jehovah, than had ever been reared for the dark mysteries and cruel sacrifices connected with the worship of their idols.

In this statement of his motives, we have every reason to believe the king was sincere, and we consequently felt less inclined to object. It is probable, also, that considering the Tahitians as a Christian people, he had some desire to emulate the conduct of Solomon in building a temple, as well as surpassing in knowledge the kings and chieftains of the islands. When, in the course of conversation, the building was mentioned, or he was asked why he reared one so large, he inquired whether Solomon was not a good king, and whether he did not erect a house for Jehovah superior to every building in Judea, or the surrounding countries.

Excepting its lengthened vista, and the singular appearance of the ornamented roof, there is nothing very prepossessing in the interior of the Royal Mission Chapel; and its length is so very disproportioned to its width and elevation, that the exterior is neither elegant nor imposing; and although it breaks the uniformity and loneliness of the landscape, it can hardly be said that its introduction has been an improvement. Pomare, however, appeared to experience great satisfaction in superintending its erection, and in marking its progress. He was present, surrounded by not fewer than seven thousand of his subjects, when it was for the first time appropriated to the sacred purpose for which it had been built, and his feelings on that occasion were, no doubt, of a superior and delightful kind—very different from those of his predecessors in the government of Tahiti, and especially of his father; who, when the Missionaries built their little chapel at Matavai, for which he had furnished the timber, sent a large fish, requesting it might be suspended in the temple of the God of Britain, that he might share his favour, and secure his aid, as well as that of the gods of Tahiti.

The first places of worship erected by the natives, after the subversion of idolatry, were comparatively small in size, and differed but little from the common native houses, excepting in the manner in which the interior was fitted up. This was generally done by fixing benches from one end to the other, and erecting a kind of desk or table equally distant from both extremities, and near one of the sides. These chapels were formerly numerous, and the inhabitants of each district had their own *fare bure*, or house of prayer, in which they were accustomed, even in the most remote part of the island, to assemble regularly twice on the Sabbath, and once during the week, for reading the scriptures and prayer. Such was the rapidity with which places for public worship were erected, that at the close of 1818, twelve months only after the battle of Narii, near Bunaauïa, there were sixty-six in the island of Tahiti alone.

Since the establishment of the stations in Huahine and the other islands, the number has been greatly diminished; the people in many parts have resorted to the Missionary settlement, particularly on the Sabbath; and the places formerly used as chapels have been converted into schools. Places now used for worship in the islands, although not so numerous as formerly, are much more convenient and substantial. The walls are either of plank or plaster, the floors are boarded, and the area within is fitted up with a pulpit, desk, and pews, or seats. Some have neat and commodious galleries; and in the island of Eimeo, on the site of the temple of which Patii was priest, a neat and substantial chapel has been built with white hewn coral.

I have not heard that glass windows have been introduced into the chapels of any of the stations. Cushions have not yet intruded into any of the pews, and only into one of the pulpits.

No native chapel is yet furnished with a public clock; and although it would be a valuable article, there is not such a thing in the South Sea Islands. The stations have also been hitherto but indifferently supplied with a far more useful appendage to their places of public worship than even a dial; namely, a bell. Whatever may be said of the inutility of bells in churches or chapels in civilized countries, where public clocks are numerous, and watches almost universal—the same objections will not apply to a people destitute of these, and having no means of denoting the hour of the day, except by mentioning the situation of the sun in the heavens. In the South Sea Islands they certainly are not a needless article, and we found it impossible to induce the people to attend the schools, or assemble for public worship at any regular or appointed season, without some such method of calling them together. For several years there was, in all the islands, only one small hand-bell, not so large as that ordinarily used by the belman in an English market-town.

As the number of stations increased, bells were sent from England, but they were either too small, badly made, or carelessly used, and were frequently broken a few days after their arrival. Various were the expedients resorted to for supplying the deficiency thus occasioned, and I have often been amused at beholding the singular substitutes employed. In the Sandwich Islands they sometimes, I think, used a bullock's horn; in others, a long tin horn resembling that used by a mail-coach guard; but, in general, a far more classic instrument, a beautiful marine shell, a species of *turbo*, or trumpet-shell, varying in size according to the power of the individual by whom it might be sounded. This, in fact, was the trumpet carried by the king's messenger; and I have often been delighted to see a tall and active man, or a lively and almost ruddy boy, with a light cloak or scarf thrown loosely over his shoulder, a wreath of flowers on his head, and a *maro* or girdle around his loins,—a shell, suspended by a braided cord, carelessly hanging on his arm—going round the village, stopping at intervals to sound his shell, and afterwards, perhaps, inviting the listening throng to hasten to the school, or to attend the place of worship. I procured a trumpet-shell actually used for these purposes in Oahu, during my residence, and consider it one of the most interesting curiosities which I was enabled to deposit in the Missionary Museum.

At Eimeo, a thick hoop of iron, resembling the tier of a small carriage-wheel, suspended by a rope of twisted bark, and struck with an iron bolt, was substituted for a bell. At Huahine, during the greater part of my residence there, we had a square bar of iron hanging, by a cord of purau bark, from a high cocoa-nut tree that grew near the chapel; and our only means of calling the inhabitants of the settlement together was, by appointing a person, at the proper hour, to strike it several minutes with a hard stone. It had been so long in use, that the bar of iron was considerably battered, and almost flattened by the blows.

The Missionaries at Raiatea procured what is called a pig of cast-iron ballast, a solid piece about three or four feet long, and six or nine inches square, with a hole through one end. Near the chapel they erected a low frame, consisting of two upright posts, and a cross-piece at the top, resembling a gallows, from the centre of which the pig of iron was suspended; and when used, struck with a stone. What the natives thought of it I do not know, but to those who were accustomed to associate with a gallows, and any object so attached to it, only ideas of an execution, or of a criminal hung in irons, its appearance was not adapted to awaken very gratifying feelings.

At Borabora, for a long time after Mr. Orsmond's settlement there, their only substitute for a bell was a broad carpenter's axe. The handle was taken out, a string of braided cinet passed through the eye, and when the inhabitants were to assemble, a native boy went through the settlement, holding it up by the string with one hand, and striking it with a stone which he held in the other. When I last saw the boy going his accustomed rounds, I perceived that, in consequence of frequent and continued use, the side he struck had actually become concave, while the opposite one exhibited a corresponding convexity.

But the most rude and simple expedient I ever beheld was at Raivavai, or High Island, where every implement of iron was as precious and as scarce as bells or clocks were at the other stations. At Raiatea, a sun-dial was erected, by which the natives, when the sun shone, were informed of the proper time for ringing their bell: at the other stations they usually applied to the Missionaries, by whose watches the meetings were regulated, but here they had neither dial nor watch: they therefore regulated their time of assembling in the school or the chapel by the situation of the sun. At the appointed time, the person whose office it was to call them together, went to the green spreading tree, from one of whose lower branches their rude unpolished bell was suspended. It was a rough flattish oval-shaped stone, about three feet long, and twelve or eighteen inches wide. A piece of rugged twisted bark was tied across it, and fastened to the tree. A number of small round stones lay underneath, with which, when it was necessary to call the people together, the large one was struck; I could not imagine its use, until, in answer to my inquiry, the native teacher said, "It is the bell with which we call the people to prayers." It appeared metallic to a great degree, as the sound produced by striking it was considerable; but not, I should think, such as could be heard at a distance. These circumstances appear trivial, but they serve to shew the expedients resorted to in a state so peculiar as that now prevailing in the South Sea Islands.

For school the bell is rung, the shell sounded, or the bar of iron beaten, only once; which is about a quarter of an hour before it commences. For public worship it is repeated a second time—once at a quarter before the commencement, and again immediately preceding the service; and indifferent as the means of giving public notice are, there is no cause to complain of delay or interruption, from the late attendance of the people. They are punctual in repairing to the house of prayer immediately after the first intimation has been given, and are usually all assembled before the period for the service to commence has arrived. Their ready and early attendance is a circumstance cheering to the minds of their teachers, who often receive a message, informing them, that though it may not be time to ring the second bell, the house is full, and the people are waiting. This is not only manifested with regard to their Sabbath-day services, but their lecture on Wednesday evening, and their monthly Missionary prayer-meetings. It is true, their occupations at home are seldom very urgent, and they have not much to neglect; it is nevertheless encouraging to notice, that they do not wish to avoid a place of worship, when a public service is held.

To the influence of climate, the habit of frequent bathing, so prevalent among the South Sea Islanders, is probably to be attributed. This salutary custom is followed alike by all classes, without regard to sex or age. The infant immediately after its birth is with its mother taken to the sea; and the last effort often made by the aged and decrepit, is to crawl or totter to the water, and enjoy its refreshing influence. Their loose light mode of dressing, and the abundance of cool, clear, and secluded streams meandering through almost every valley in the islands, probably favour the frequency of the practice, and its grateful effects render it one of their greatest luxuries.

Contrary to the practice of those who are accustomed to resort to the sea-side for the purpose of bathing in salt-water, the natives of these islands, without exception, prefer on every account to bathe in the mountain streams. It is a principal remedy in many of their diseases; yet doubtless often aggravates what they design to alleviate. It is, however, a means of great benefit: for this, as well as every other purpose, they prefer the fresh water; and even those whose avocations lead them to frequent the sea for fishing, although they may have plunged beneath the wave fifty times in the day, yet invariably repair to the nearest stream to bathe, before they return to their houses. They say the sea-water produces an irritation which is peculiarly unpleasant. Children not more than three or four years, of age are often seen playing in groups along the margin of the sea, without the least apprehension of danger, and they as frequently resort for amusement to the rivers. It is probable that the people in general bathe less now than they were accustomed to do formerly, yet there are none, perhaps, who omit bathing once, and many who visit the river twice, in the course of the day. The universality and frequency of this custom is highly conducive to health, and produces a degree of personal cleanliness seldom met with among an uncivilized race.

Although some of their practices are offensive to every feeling of delicacy and propriety, yet they are certainly a remarkably cleanly people. This regards not only their repeated ablutions, but their care to remove every thing unsightly from their persons. No hair was allowed on their limbs; formerly it was plucked out by the roots, or shaved with a shell or a shark's tooth; and those who do not wear the European dress, are still very particular in removing the hair from their legs and arms. This is usually done with a knife, the razors they have among them being reserved for removing the beard.

The adults formerly wore their hair in a variety of forms; the heads of their children they always shaved with a shark's tooth. This operation was frequently repeated during their juvenile years. The females generally cut their hair short, but the men wore theirs in every diversity of form—sometimes half the head almost shaved, the hair being cut short, and the other half covered with long hair—sometimes the crown cut, and the edges left the original length. Frequently, it was plaited in a broad kind of tail behind, or wound up in a knot on the crown of the head, or in two smaller ones above each ear. Since the introduction of Christianity it has been worn remarkably neat: the men's hair is usually short, the females the same, excepting in the front, though some wear it long, curled in front, and bound up on the crown.

Nothing at first sight produces a stronger impression on the most careless observer, in the difference between the inhabitants of an island where paganism prevails, and those of one where Christianity has been introduced, than the appearance of their hair. I have often seen one who was an idolater, or who had but recently embraced Christianity, and whose hair was uncut and his beard unshaven, standing in a group of Christians, and I have been struck with the contrast.

Sometimes the men plucked the beard out by the roots, shaved it off with a shark's tooth, or removed it with the edges of two shells, acting like the blades of a pair of scissors, by cutting against each other; while others allowed the beard to grow, sometimes twisting and braiding it together. These fashions, however, have all disappeared, and the beard is generally shaved at least once a week, and by the chiefs more frequently. These cut their whiskers rather singularly sometimes, and leave a narrow strip of their beard on the upper lip, resembling mustachios: the greater part, however, remove the beard altogether, which must often be no easy task. There are no barbers by profession, yet every man is not his own barber, but contrives to shave his neighbour, and is in return shaved by him. Some of the most ludicrous scenes ever exhibited in the islands occur while they are thus employed. Only a few of the chiefs are so far advanced in civilization as to use soap; the farmers cannot understand how it can help to remove the beard, they therefore dispense with it altogether. When the edge of the razor or knife is adjusted, the person to undergo the operation, in order to be quite stationary, lies flat on his back on the ground, sometimes in his house, at other times under the shade of a tree, and his friend kneels down over him, and commences his labour. When he has finished, he lays himself down, and the man who is shaved gets up, and performs the same office for his friend. Sometimes the razor becomes rather dull, and something more than a little additional strength is necessary. A whetstone is then applied to the edge; but if this be not at hand, the man gets up half-shaved, and both go together to the nearest grindstone; and I have beheld that the transition from the grindstone to the chin is sometimes direct, without any intermediate application to the edge of the razor. The hone and the strap, however, have been introduced, and ere long will probably supersede the use of the grindstone, and also the whetstone.

The islanders appear to have paid at all times great attention, not only to cleanliness, but to personal ornaments. On public occasions, their appearance was in a high degree imposing. At their dances, and other places of amusement or festivity, they wore a profusion of ornament, and on ordinary occasions, with the exception of the aged and decrepit, devoted much time to the improvement of their appearance. The hair of the females, which was neatly trimmed, and sometimes appeared in short loose curls, was an object of great attention; the eye-brows were also reduced, or shaped according to their ideas of beauty. The hair was ornamented with elegant native flowers, sometimes exhibited in great profusion and variety, at others with only one or two single jessamine blossoms, or a small wreath interwoven with their black and shining ringlets. They displayed great taste in the use of flowers, and the adorning of their hair. Frequently I have seen them with beautiful wreaths of yellow flowers, worn like fragrant necklaces on their bosoms, and garlands of the same around their brows, or small bunches of the brilliant scarlet *hibiscus rosæ chinensis* fastened in their hair. Though totally unacquainted with what we are accustomed to call artificial flowers, yet the brilliant and varied odoriferous plants, that grew spontaneously among their mountains or their valleys, did not suffice to gratify their wishes; they were therefore accustomed to manufacture a kind of artificial flowers, by extracting the petals and leaflets of the most fragrant plants and flowers, and fastening them with fine native thread, to the wiry stalk of the cocoa-nut leaf, which they saturated with monoi, or scented oil, and wore in each ear, or fixed in the native bonnet, made with the rich yellow cocoa-nut leaf. The men, though unaccustomed to adorn their hair with flowers, were careful of preserving and dressing it. They generally wore it long, and often fastened in a graceful braid on the crown, or on each side of the head, and spent not a small portion of their time in washing and perfuming it with scented oil, combing and adjusting it. When it was short, they sometimes dressed it with the gum of the bread-fruit tree, which gave it a shining appearance, and fixed it as straight as if it had been stiffened with rosin. The open air was the general dressing-place of both sexes; and a group of females might often be seen sitting under the shade of a clump of wide-spreading trees, or in the cool mountain-stream, employing themselves for hours together in arranging the curls of the hair, weaving the wreaths of flowers, and filling the air with their perfumes. Their comb was a rude invention of their own, formed by fixing together thin strips of the bamboo-cane. Their mirror was one supplied by nature, and consisted in the clear water of the stream, contained in a cocoa-nut shell.

The attention of the people to personal decoration rendered looking-glasses valuable articles of trade in their early intercourse with foreigners; and although the habit has very much declined, and their taste with regard to ornament, &c. is materially changed, looking-glasses are still, with many, desirable articles. Those, however, who have furnished them, have often made a mistake in sending, on account of their cheapness, an inferior kind, which, in consequence of a defect in the glass, exhibits the face in a distorted and ludicrous shape. Nothing will more offend a Tahitian than to ask him to look in one of these glasses. They call them *hio maamaa*, foolish glasses, and, instead of purchasing them, would sometimes hardly be induced to accept them as presents.

Since the introduction of Christianity, the use of flowers in the hair, and fragrant oil, has been in a great degree discontinued—partly from the connexion of those ornaments with the evil practices to which they were formerly addicted, and partly from the introduction of European caps and bonnets, the latter being now universally worn.

CHAP. V.

Improved circumstances of the females—Instruction in needlework—Introduction of European clothing—Its influence upon the people—Frequent singularity of their appearance—Development of parental affection—Increased demand for British manufactures—Native hats and bonnets—Reasons for encouraging a desire for European dress, &c.—Sabbath in the South Sea Islands—Occupations of the preceding day—Early morning prayer-meetings—Sabbath schools—Order of divine service—School exercises—Contrast with idolatrous worship.

WHILE the enclosure of plantations and gardens, the erection of neat and commodious dwellings, schools, and the spacious building for the worship of the true God, after the European plan, were rapidly altering the aspect of the settlement, the natives themselves were undergoing a change in appearance, in perfect keeping with this transformation of the surrounding country. The females, no longer exposed to that humiliating neglect to which idolatry had subjected them, enjoyed the comforts of domestic life, the pleasure resulting from the culture of their minds, the ability to read the scriptures, and to write in their own language, in which several excelled the other sex; they also became anxious to engage in employments which are appropriated to their own sex in civilized and Christian communities. The females in Huahine, and the other islands, were therefore taught to work at their needle, and soon made a pleasing proficiency.

The Missionaries' wives had taught some few in Eimeo prior to our arrival; but, until their reception of Christianity, they considered it degrading to attach themselves to the household of the foreigners, or to learn any of their arts and customs; they also thought their own manner of wearing a piece of native or foreign cloth, cast loosely round the body, preferable to the European mode of dress, and consequently had no inducement to learn needlework, or any other kind of female employment. They were, however, now anxious not only to adopt the English style of clothing, but also to be able to make their own dresses. This was a kind of instruction which our wives were competent to impart, even before they had acquired a sufficient knowledge of the language to enable them to teach in the schools. Mrs. Ellis had engaged in it ever since our arrival in Eimeo; and, as soon as we were settled in the Leeward Islands, some were daily occupied in teaching the native females to sew.

In Huahine a large class attended every afternoon from two till five o'clock, alternately at our respective houses, where Mrs. Barff and Mrs. Ellis met, and spent the afternoon pleasantly in each other's society, and unitedly teaching the females by whom they were surrounded. The natives, in general, now considered it a great favour to be taught, though it was sometimes found that they had entertained very incorrect ideas of the motives by which their instructors were influenced. A young woman had attended very regularly for some weeks, and had learned to use her needle as well as could be expected in that time. One Saturday night she presented herself with our native domestics, and begged to be paid her wages for learning to sew! Mrs. E.

said, Why should I pay you? in our country it is customary for those instructed to pay their teachers. The woman answered with some earnestness, You asked me to come and learn—I have been here so long—I have learnt. It must be in some way advantageous to you, or you would not have been so anxious about it; and as I have done what you wished me to do, you ought to pay me for it. She was told that the labour of teaching had been gratuitous, and the advantage resulting was all her own; and appeared satisfied when assured, that now she had learned, she should be regularly paid for the needlework she might do. This, however, at the time to which I now refer, 1819, was a rare occurrence; although, in the earlier periods of the Mission, it had been frequently manifested, not only in regard to needlework, but every department of instruction.

Accustomed only to perform those services that were for the advantage of foreigners, the natives had been usually paid for the same. They could not conceive, notwithstanding the frequent explanations given, why the Missionaries should be so desirous for their learning to read, &c. if they were not, in some way or other, benefited thereby: hence, many of the early scholars expected to be paid for learning, and I believe some for appearing at the chapel. This, however, was only manifested during the time when very few could be induced to attend, and none perhaps came from the influence of that desire for Christian instruction, which attended the general profession of Christianity. After this period, it was only shewn by those who were actuated by a desire to obtain the favour of their superiors.

European cloth, cottons in particular, had long been

favourite articles of barter with the natives, on account of their durability compared with native manufacture, their adaptation to the climate, variegated and showy colours, and the trifling injury they sustained from wet. They no longer traded for ardent spirits, muskets, powder, &c. and were consequently enabled to procure larger quantities of British woven cloth. Hitherto, however, they had generally worn the European cottons, &c. in the native manner, either as a light *tehei*, thrown over the shoulder, a *pareu* wound round the waist; or *ahu buu*, a kind of large scarf or shawl, loosely covering the greater part of the body. They were now desirous to assimilate their dresses in some degree to ours. Mrs. Nott and Mrs. Crook made one or two loose dressing-gowns for Pomare, after a pattern from us. This introduced the fashion, and many of the women made others for their husbands.

The first garment in general use among the females, was a kind of Roman tunic, usually of white or blue calico, these being their favourite colours. It was fastened round the neck with a short collar, which, if possible, was united by a bright gilt or plated button. The sleeves were long and loose, and buttoned at the wrists, while the lower parts reached nearly to the ankles. On the outside of this, they wore the pareu round the waist, and reaching below the knees. The colour of these articles was generally in perfect contrast. When the loose European dress was white, the pareu, worn round the waist on the outside of it, was of dark blue; one end of it was sometimes thrown carelessly over the shoulder, or hung loosely on the arm, heightening the novel and not unpleasing effect produced by their blending, in the apparel of the same individual, the ancient native with the modern European costume. Their dress thus indicated, equally with their half-native and half-foreign dwellings, the peculiar plastic, forming state of the nation, and the advancement of that process which was then constantly imparting to it some fresh impression, and developing new traits of character with rapid and delightful progression.

As the natives experienced the convenience of the new dresses, their desire for them increased, and the long loose dress soon became an every-day garment, while others of a finer texture, made after the European fashion, were procured for holidays and special occasions. From making plain, straight-forward garments, the more expert were anxious to advance still higher; and in process of time, frills appeared round the neck; and, ultimately, caps covered the heads, and shoes and stockings clothed the feet. Our assemblies now assumed quite a civilized appearance, every one, whose means were sufficient to procure it, dressing in a garment of European cloth.

These changes in the exterior of the people were sometimes attended with rather humorous circumstances. I shall not soon forget the first time the queen, and about half a dozen of the chief women of Huahine, appeared in public, wearing the caps which had been sent as a present by some ladies in England. It was some time after the adoption of the English dress. When they first entered with their bonnets on, much surprise was not excited; but when these were removed, and the cap appeared, they viewed each other for some time most significantly, without, however, saying a word, yet each seeming to wonder whether her head, with its new appendages, resembled in appearance that of her neighbour. The attendants, and others who were not so distinguished, after recovering from evident astonishment at seeing the Huahinian ladies for the first time in European caps, were by no means sparing in their remarks. Some observed, they were perhaps designed to keep the head cool; others, to keep it warm; and others supposed they were to preserve it from the flies and musquitoes. All agreed that they looked very strange, and the wearers appeared to think so themselves; but it was supposed to be according to the usage of ladies in England,—and to the despotism of fashion, even here, all minor considerations were rendered subservient.

The desire to obtain foreign clothing was now very great, equal to that with which they sought iron tools; and whenever they procured one article of it, it was worn forthwith, without waiting till the suit was completed. This often rendered their appearance to a European eye exceedingly ludicrous. There was a degree of propriety usually manifested by all classes of the females, in their dress: they either paid more attention to their appearance than the other sex, or were better informed; and the only inconsistency we ever observed was that of a woman's sometimes wearing a coat or jacket belonging to her husband or brother. The men, however, were less scrupulous; and whether it resulted from their fondness of variety, or a supposition that the same clothes, worn in different ways, would appear like distinct articles of dress, I am not able to say; but I have seen a stocking sometimes on the leg, and sometimes on the arm, and a pair of pantaloons worn one part of the day in the proper manner, and during another part thrown over the shoulders, the arms of the wearer stretched through the legs, and the waistband buttoned round the chest.

Their own dress was remarkably simple in its form and appearance, and was generally more or less adapted to their vocation. When employed in agricultural pursuits, or in fishing, in which occupation they were as much in the sea as out of it, the men seldom wore any other dress than their *tihere* or *maro,* a broad girdle passed several times round the body. At other times they wore a *pareu,* which reached from the waist to the calf of the leg. Over the shoulders, when not at work, they wore a loose *ahu buu,* a kind of scarf or mantle, in some degree resembling the Roman toga; or they appeared in the *tiputa,* an article of dress, having an aperture in the centre through which the head is passed, the other parts extending over the shoulders, breast, and back. The tiputa was generally worn by the chiefs and all persons of respectability.

This article is common to all the South Sea Islanders, and resembles in every respect, excepting the material of which it is fabricated, the poncho worn by the aborigines of South America, inhabiting the countries adjacent to the Pacific. The combination of these with some parts of the men's apparel worn in Europe, produced an effect less pleasing than the apparel of the females. Appearance and convenience, however, were not much considered by the Society Islanders, and it was often amusing to see a native *sans culotte,* without waistcoat or shirt, with a maro or pareu round his waist, and a fashionably made black coat on his back. The men are generally above the middle stature, and proportionably stout, and few of the coats, &c. belonging to the captains or officers of vessels touching at the islands were large enough. If, however, they could by any means thrust their large muscular arms through the sleeves, it was thought to fit very well. Notwithstanding the warmth of the climate, they are fond of wearing the coat buttoned; and although when thus fastened it appeared less repulsive to our opinions of propriety, than when, standing open, it exposed the naked breast of the wearer, it was often quite distressing to see the imprisoned and pinioned arms occasionally struggling for liberty, and the perspiration oozing from the pores of the skin, indicating the laborious confinement of the body it enclosed.

These were scenes witnessed immediately after the general adoption of European clothing. Most of those who wear it now, are able to procure at least one complete suit, and consequently appear less singular. In the arrangement, however, of the different articles of a complete dress, they were equally unhappy, and not unfrequently presented an appearance which it was impossible to behold with gravity. A tall man was sometimes seen with a hat and shoes, without stockings, a long surtout black-cloth coat, with the collar turned up and buttoned close to his chin, and over his black coat a white frilled shirt, the collar unbuttoned, and the bosom thrown open, the sleeves drawn up towards the elbows, and the outline of the other parts appearing in strong contrast with the black coat underneath, which reached to his ankles. Such an appearance was more than once presented, and the reason assigned for it was, that the shirt was so much smaller than the coat, that had it, instead of the coat, been put underneath, it would not have been seen. Although exhibited in the person of a chief, the incongruity of such an arrangement furnished matter of ridicule even for the natives themselves, and is now never seen.

European articles of dress are in the greatest demand; this method of clothing being adopted by all whose means enable them to procure either cotton or woollen cloth; and there are few, who, by preparing arrow-root, feeding pigs, manufacturing cocoa-nut oil, or other labour, cannot purchase from the shipping a suit of foreign clothing. I have frequently been delighted to see families of natives going on board the vessels, or repairing to the market-house on shore, with the produce of their labour; and when they have arrived at the place of barter, and the captain or the merchant has spread before them his attractive goods, glossy and bright in all the shining colours of which they are so fond; the parents' eye has often glanced over them, in wonder when and how they were made. They have been seen occasionally looking down to notice what had attracted the attention of a little boy or girl, standing, perhaps, beside them; and if they thought the child could not distinctly see the different pieces, they have lifted it up, that it might look over the table, and then have asked the child which it would like to have. Sometimes the child would smile and hang its head, and fall upon its mother's shoulder, as if it knew not which to choose. At other times it would point to one, upon which the merchant has been directed to cut off so much as would make a frock or gown: it has been folded up and given to the child; and while the parent's eye has marked the pleasure of the child as it held the new frock on its arm, the smile on their own countenances has declared the pleasure they experienced. In many instances I have seen a garment for the mother next selected; and then the father, with the remainder of their native produce, has purchased some articles for himself. Their first effort now is generally to purchase, and to learn to make light clothing for their children; and there are perhaps few parents in the islands who would think of purchasing a garment for themselves, while their little one was destitute.

It is a pleasing fact, which demonstrates unequivocally that the South Sea Islanders are not deficient in capacity, but are capable, when inducement sufficient is offered, of acquiring habits of close industry, that in the islands of Raiatea and Huahine, or any of the stations in the Leeward Islands, there was hardly an adult female, excepting the aged and infirm, who could not use her needle so as to make her own clothes, and those required by other members of the family. I have not had equal opportunity of knowing what progress the females in the Windward Islands have made, but have reason to believe it is highly creditable to their application.

The occupation furnished by the new order of things that has followed the introduction of Christianity, is one of the important sources of their present enjoyment. But this is not the only advantage resulting therefrom. It has opened a new channel for commercial enterprise, and has actually created a market for British manufactures, the consumption of which, among the islands of the Pacific that have received the Gospel, is already considerable: Mr. Stewart estimates that the trade of four American merchants in the Sandwich Islands amounts to one hundred thousand dollars a year; this, however, is a far greater amount than that of all the other islands of Polynesia. The demand will increase in the exact proportion in which industry shall augment the produce of the islands, and the property of their inhabitants. This is a consideration which, though confessedly very inferior to many, ought not to be disregarded by those who take an interest in the transformation of society which is now attending Missionary efforts in various parts of the world, but particularly in such countries as Africa, Madagascar, and the islands of the Pacific.

Shoes and hats are not much less in demand than cottons or woollens; and these also must, for the present, and probably for many years to come, be supplied from England or America. Although the light hats, made with a fine sort of grass, or the bark of a tree, are, in our estimation, remarkably well adapted for the climate, most of the men, making any pretensions to respectability, strive to possess an English hat. We were for a long time surprised at the partiality of the natives for woollen cloth, and hardly knew how to account for it, as it does not altogether arise from its being more durable. At one time, no article of dress was more acceptable to the men than a thick shaggy great coat, which, to us, it was quite oppressive even to behold. Many purchased with avidity a thick blanket, which they would wear as an *ahubuu* over the shoulders, or a pareu round the waist. Frequently, when we have been burdened with the lightest crape or nankeen dress, a native, by no means deficient in corpulency, would walk several miles with an ordinary great coat, without seeming to experience more than usual inconvenience. I never heard them complain of the heat, and the cause of their apparent insensibility to its oppressive influence, is probably to be found in their being early exposed, and constantly habituated to the climate.

Early in the year 1820, another important change took place in the dress of the Society Islanders; affecting not only their appearance, but tending perhaps ultimately to alter their physical structure. This was the introduction of hats and bonnets. If the skulls of those nations that wear no covering on their heads, are thicker than those who do, there is reason to suppose the craniums of the Tahitians will be much thinner in a few generations, than they have been prior to this period; since, from their earliest history, they appear to have gone abroad bareheaded. The inhabitants formerly wore a kind of bonnet, or rather, shade for the eyes, made of the leaves of the cocoa-nut in a variety of forms, many of them tasteful and elegant. They were called *taupoo* or *taumata*, and, as the latter name signifies, were designed to skreen the face or eyes; it being composed of *tau*, to hang upon or over, and *mata*, face or eyes. It was worn on the forehead immediately below the hair, and fastened by a narrow leaflet passing round the back of the head above each of the ears, leaving the whole of the back and upper part of the head entirely exposed.

The first native bonnet we have heard of, as manufactured in the islands, was finished, while we resided in Afareaitu, by Mrs. Ellis. It was made for our infant daughter, with leaflets of the fan-leaved palm, brought from the Marquesas; and the first hat we ever saw that had been made there, was one Mrs. Ellis made for me at Huahine, with the same kind of leaves, which were, platted by a sailor in Eimeo. Hats and bonnets were, however, introduced among the natives by our friends in

Raiatea, with whom many valuable improvements have originated; and the first hats and bonnets ever made in the islands, and worn by the natives, were made by Mrs. Williams and Mrs. Threlkeld in the spring of 1820. Their appearance on the heads of the natives of Raiatea produced no slight sensation there; and the report of their use, as it spread through the islands, occasioned a considerable stir.

Highly approving of whatever had a tendency to civilize the natives, or to furnish them with innocent and useful employment, we rejoiced at their introduction, and endeavoured to persuade the natives of Huahine to follow the example of their Raiatean neighbours; but whether they were influenced by a feeling of pride which made them averse to imitate the Raiateans, or an unwillingness to increase their domestic employments, we do not know; but the females in general, the queen and chief women in particular, seemed at first determined to resist the innovation. The men rejoiced at the idea of making hats; and yet, notwithstanding this, and the repeated offers of Mrs. Barff and Mrs. Ellis to teach the females to plat the leaves of the mau, and to make the plat into bonnets and hats, they were exceedingly averse to learn. Following the example of those in Raiatea, their teachers made bonnets for themselves with the bark of the purau; and though the chief women acknowledged that they looked very well on them, they said they had not yet procured the articles necessary to form a complete European dress—that many were still without shoes and stockings—and that it would be quite ridiculous for the head to be covered with a bonnet after the fashion of the foreigners, while the feet, like those of the islanders in general, were without shoes. A short time afterwards, several of the natives of our island sailed over to Raiatea, and returned with very flattering accounts of the improved appearance of those who wore hats and bonnets. This induced several of the chief women, who had at least one complete English dress, to desire to learn to make them, and ultimately to substitute the European bonnet for the native taumata. A visit which a number of chiefs and their wives, from Raiatea, paid to Huahine, increased their eagerness for this new article of dress—which, when once adopted, was never laid aside.

The desire now became general, and was not confined to those who possessed other articles of foreign dress, it being extended even to such as had none. Thus, wearing a hat and bonnet was the first advance they made towards a more civilized appearance and dress. Our houses were now thronged by individuals anxious to be instructed; and so soon as Mrs. Barff or Mrs. Ellis had taught any of the females, these immediately taught the art to others; and those who excelled in the fineness of their platting, or in putting it together, were fully employed by the chiefs and others, and derived no small emolument from their new avocation. Dress-making and straw-bonnet making, now very profitable employments to a number of females, were certainly the first regular female occupations from civilized society introduced into the islands. The hats and bonnets were at first made with the inner bark of the slender branches of the purau, *hibiscus tiliaceus,* or the leaves of a fine species of rush. The former was beautifully white and glossy, while the latter was of a yellow colour, and much more firm and durable, on which account it was preferred for hats. The only hats I wore in the islands during the subsequent years of my residence there, were made with this material; and in that climate I should never desire any other. The use of hats increased so rapidly, that all the European thread in the islands was soon expended. There were no haberdashers' shops at hand, whence a supply could be procured: recourse was therefore had to native productions. Some employed the long filaments of the dried plantain-stalk; and others split the thin bark of the purau into fine threads or fibres, and, though not equal in strength to twisted thread, both answered remarkably well.

The bonnets were in many instances scarcely finished, when another difficulty met their possessors. They had observed that the wives and daughters of the Missionaries, however plain their dress, wore a riband and strings to their bonnets, and they had often observed a greater profusion of trimmings attached to those worn by the wives of the captains, or the female passengers, in any of the vessels that touched at the islands; they therefore imagined that in point of improvement they might almost as well appear without a bonnet, as with one destitute of these appendages. These, however, it was no easy matter to procure, and they would at that time, certainly, have been the last article a captain or trader would have thought of taking to the South Sea Islands for barter. A few of the chief women were furnished with an English riband, which was considered as valuable as an embroidery of gold would be in some circles of society.

The greater portion of the inhabitants were, however, under the necessity of exercising their ingenuity to provide a substitute. Those they furnished were various, and such perhaps as few English females would have thought of. A part of a black coat, or a soldier's red jacket cut into strips about two inches wide, was greatly esteemed. Next to this, ribands of native cloth, dyed with showy colours, were employed; while others used a string of the bark from a branch of the purau, with the outer rind scraped off, the inner bark washed and bleached, passed round the bonnet, and tied under the chin.

Trimmings are not so scarce now as formerly, but the supply taken is still inadequate to the requirements of the people, among whom bonnets and hats are now so common, that before I left the Leeward Islands, scarce a man, woman, or child was to be seen out of doors without one—many of them possessing two, and sometimes three or four.

They are made entirely by the females, who manufacture not only for themselves, their husbands, and their children, but in some of the stations, several have formed themselves into a kind of society, for the purpose of making bonnets for the poor and the aged, who are unable to make for themselves. They have increased not only in number, but in variety of shape and material. The bonnets are now either sewn together, or woven throughout, after the manner of Leghorns, and are made not only with the leaves of the mau, and the bark of the purau, but of the fine white layers of the inside of the plantain stalk, the leaf of the sugar-cane, and a strong and beautiful species of fine grass.

It may perhaps be supposed, by those who are unacquainted with the circumstances, that the wives of the Missionaries have not acted judiciously in introducing and cherishing a desire for dress and a love of finery. It may be thought that it has a tendency to engender pride, occupy the head and the hands about trifles, to the neglect of more important matters, inducing them to devote to the adorning of the person that time which might with greater advantage be appropriated to the cultivation of morals, and the improvement of their minds. The Missionaries, however, have not, in any degree, introduced the love of finery; they found it there, and cannot be supposed to have produced any change for the worse, in the taste of a people, by whom a black coat fringed round the edge with red feathers was considered a suitable dress even for a high-priest. The most showy English dress they ever saw, would probably, in the estimation of every beholder, appear comparatively plain, when placed by the side of those the natives formerly wore. The splendid appearance of the loose and flowing ahu puu, or the richness of the tiputa, dyed in the bright and favourite scarlet and yellow colours, together with some of their head-dresses of tropic-bird feathers, and garlands of the gayest flowers, gave them certainly an imposing appearance. The former continued to be worn after their renunciation of idolatry; and the Missionaries knew no reason why they should recommend the discontinuance of a dress to which the nation was accustomed, merely on account of its gay appearance.

Convinced it is not in the dress with which the person is invested, but in the feelings of the heart with which that dress is regarded, that the evil exists—and that pride does not consist in the wearing of apparel superior to that to which an individual may have been accustomed, or to that worn by others, provided it be suitable to his circumstances, and the society with which he associates—they did not disapprove of the native dresses. But considering the danger to arise from substituting external adornment for internal worth, and imagining that distinction in dress confers an advantage on its wearer, or entitles him to that which he would not otherwise assume —the Missionaries were led to conclude, that a Tahitian, arrayed in a scarlet and yellow tiputa, or invested in the rich fold of his ahu puu, was perhaps as humble in mind as those who appeared desirous to divest themselves of every exterior ornament.— Their principal aim, however, was to encourage habits of industry; and this, from the heat of the climate, the spontaneous productions of the soil, and other causes, could only be done by the introduction of what might be called artificial wants, which should operate on the native mind with power sufficient to induce labour for their supply. Idleness has been the most fruitful source of many of their vices and sufferings; and when we have seen the females working with their needle, or with the straw for their bonnets, &c. we could not but deem it an occupation far more conducive to their enjoyment, than idleness, or their former unprofitable and often injurious pastimes. It is not to be expected that a people unaccustomed to mental effort should be constantly engaged with their books. They did not relax in their attendance at the school, or any of the meetings for public instruction; and we could not but observe with satisfaction, their altered appearance in all public assemblies, as indicating an improvement in civilization, and an increase of industry.

Their regular and early attendance on the Sabbath, ever has been, and still is, remarkably conspicuous; it is to them a season of holy rest and devotional enjoyment. Excepting in Tahiti, there is now no island on which more than a single Missionary resides, and consequently public preaching only at the station which he occupies. The principal families in most of the islands have removed to the settlement, to enjoy the benefit of regular instruction. Others, however, occupy lands which are at some distance; and even those who have erected their dwellings near the residence of their teacher, having plantations situated in a remote district, are often absent for several days together. Most of them, however, repair to the settlement for the Sabbath; and it is a spectacle that has often gladdened our hearts, when, on the Saturday afternoon, we have seen parties from every direction approaching, by land or by sea, the bay, at the head of which our settlement was formed.

In a walk through the village, on the afternoon of the day preceding the Sabbath, looking along the shore, we have often beheld the light canoe doubling a distant point of land, and, with its native cloth or matting sail, wafted towards the station. Others nearer the shore, with their sails lowered, have been rowed by the men; while the women and children were sitting in the stern, screened from the sun by a temporary awning. Along the coast, many were unlading their canoes, or drawing them upon the beach for security.

The shore presented a scene of activity. The crackling fire or the light column of smoke might be seen rising through the district, and the natives busily engaged in cooking the food for the Sabbath. On account of their food being dressed for the Sabbath on the Saturday, that day is called *mahana maa*, food-day. As the evening approached, multitudes were met returning from the inland streams, whither they had repaired, to bathe after the occupations of the day; the men bringing home their calabashes of water for drinking, or their *aanos* of water for washing the feet; while the females were carrying home bundles of the broad leaves of the hibiscus, which they had gathered, to serve instead of plates for Sabbath meals. On entering the dwellings on the Saturday evening, every thing would appear remarkably neat, orderly, and clean—their food in baskets—their calabashes filled with fresh water—their fruit gathered—and leaves plucked and carefully piled up for use—their clean garments were also laid out ready for the next day. The hours of the evening, instead of being a season of greatest care and hurry, are, I believe, often seasons of preparation—"prelude to hours of holy rest."

The sacred day was not only distinguished by a total cessation from labour, trade or barter, amusements, and worldly pleasure—but no visits were made, no parties of company entertained, no fire lighted, nor food cooked, except in cases of illness. This strict observance of the Sabbath, especially in regard to the latter points, whereby the Tahitian resembled the Jewish more perhaps than the Christian Sabbath, was not directly inculcated by the Missionaries, but resulted from the desire of the natives themselves to suspend, during this day, their ordinary avocations, and also from their imitation of the conduct of the Missionaries in this respect.

We have always been accustomed to have our usual beverage prepared in the morning and afternoon; but this is the only purpose for which, in ordinary seasons, a fire has ever been lighted for any of the Missionary families; and when destitute of these articles, which in the earlier periods of the Mission was often the case, no fire was lighted on the Sabbath; their food was invariably dressed on the preceding day, and the warmth of the climate prevented their requiring fire for any other purpose. In this proceeding they were influenced by a desire that their domestics, and every member of their families, might have an opportunity of attending public worship.

The example, thus furnished by their teachers, has led to the strict and general observance of the Sabbath by the nation at large. Their private devotions are on this, as well as other mornings, usually concluded by sunrise, and shortly afterwards, the greater part of the inhabitants assemble for their Sabbath morning prayer-meeting. Besides a service in English, the Missionaries preach twice in the native language, and visit the Sabbath schools; these services are as many as they are able to undertake: the service at the morning prayer-meeting is therefore performed by the natives. We have, however, sometimes attended, and always with satisfaction.

It is impossible to conceive of the emotions of delight produced by witnessing six or eight hundred natives assembling at this hour in the respective chapels; and, on entering, to see a native, one who was perhaps formerly a warrior or Areoi, or even an idolatrous priest, stand up, and read a psalm or hymn, which the congregation rise, and sing. A portion of the scriptures, in the native language, is then read; and the thanksgivings and petitions of the assembly are offered to Almighty God, with a degree of fervour, propriety, appropriate use of scripture language, and chastened devotional feeling, that is truly astonishing, when it is considered that but a few years before, they were ignorant and barbarous idolaters. A second hymn is sung, another portion of scripture read, and prayer offered by another individual—when the service closes, and the assembly retires.

Soon after eight o'clock the children repair to the Sabbath-schools, those for the boys and girls being distinct. About four hundred usually attend in Fare: they are divided into classes, under native teachers. About a quarter before nine, the congregation begins to assemble, and at nine the morning service commences. I have often heard with pleasure, as I have passed the Sabbath-schools rather earlier perhaps than usual, the praises of the Saviour sung by between three and four hundred juvenile voices, who were thus concluding their morning exercise. The children are then conducted to the chapel, each class led by its respective teacher, the girls walking first, two abreast and hand-in-hand, clothed very generally in European dresses; wearing bonnets made with a fine species of grass, or the bark of a tree; each carrying in her hand a neat little basket, made with similar materials, and containing a catechism, hymn-book, and testament: the little boys following in the same order; more frequently, however, arrayed in the native costume, having a little finely-platted white mat, fringed at the edges, wound round their loins; another of the same kind, or a light scarf, dyed with glowing native colours, passed across their chest, and thrown loosely over their shoulders; their feet naked, and their hair often cut short, but sometimes flowing in ringlets over their open countenances; while their heads were covered with a neat little grass or straw hat, made by their mothers or their sisters.

Before the service began, they were usually led to the

seats appropriated for them in the chapel; and where there have been galleries, these have been occupied by the scholars. Frequently we have been approaching the place of worship at the same time that the schools have entered it, and it has often afforded the sweetest satisfaction to behold a father or a mother, with an infant in the arms, standing under the shade of a tree that grew by the side of the road near the chapel, to see in the line of scholars, a son or daughter pass by. When the object of affection has approached, a smile of pleasure has indicated the satisfaction of the child at the notice taken by the parent, and that smile has been reciprocated by the parent, who, in silent gladness, followed to the house of God.

The morning service commences with singing, during which the congregations stand; a portion of scripture is then read, and prayer offered, the congregation kneeling or standing. This is followed by singing a second time; a sermon is then preached, after which a short hymn is sung, prayer presented, and the benediction given; with which the service closes, between half-past ten and eleven o'clock.

Although the religious exercises are now rather longer than they were when the people first began to attend, they seldom exceed an hour and a half on the Sabbath, and little more than an hour at other times. It has always appeared preferable, even to multiply the services, should that be necessary, than weary the attention of the people by unduly protracting them. When the congregation has dispersed, the children are conducted to the schools in the same order in which they came to the chapel, and are there dismissed by one of their teachers.

In the afternoon they assemble in the schools, and read the scriptures, repeat hymns, or portions of the catechism, and are questioned as to their recollection of the sermon of the forenoon. We have sometimes been surprised at the readiness with which the children have recited the text, divisions, and leading thoughts in a discourse, without having written it down at the time they heard it. Often it has been most cheering to see them thus employed; exhibiting all the native simplicity of childhood, mingled with the indications of no careless exercise of the youthful mind on the important matters of religion. It is always delightful to watch the commencement and progress of mental improvement, and the early efforts of intellect; but it was peculiarly so here. In the Sabbath-schools of the South Sea Islands, the mechanical parts of instruction (namely, learning to read and spell, &c.) are not attended to; the time is wholly occupied in the religious improvement of the pupils, and is generally of a catechetical kind.

Many of the parents attend as spectators at the Sabbath-schools, and it is not easy to conceive the sacred delight they experienced in beholding the improvement of their children, and attending at an exercise often advantageous to their own minds. The greater part of the people, however, spend the middle of the day in their own dwellings. Formerly they were accustomed to sleep, but we believe this practice is by many discontinued.

The public service in the evening commences, in most of the stations, about a quarter before four, and is performed in the same manner as that in the forenoon. Meetings for reading the scriptures and prayer are held at some of the native houses in the evening, and we usually read a sermon in the English language in our own families.

The attendance of the people is regular, and the attention seldom diverted. At first we perceived a great inclination to drowsiness, especially during the afternoon: at this we were not surprised, when we recollected that this was the manner in which they were accustomed to spend several hours every day, and that they were also unaccustomed to fixedness of attention, or exercise of thought on a particular subject, for any length of time. This habit, however, has, we have reason to believe, very greatly diminished in all the islands, and more particularly where congregations are accustomed regularly to assemble.

The scrupulous attention to the outward observance of the Sabbath, may perhaps in some degree be the result of the impression left on the minds of the people by the distinguishing features of their former system, in which all the efficacy of their services consisted in the rigid exactness with which sacred days were kept, and religious ceremonies performed, without the least regard to the motives and dispositions of the devotees. To have kindled a fire, or to have failed in the observance of any rite enjoined, or restriction imposed, during their *tabu*, or sacred seasons, would have been sufficient, not only to have neutralized all the advantages expected from the most costly offerings or tedious services, but would have exposed the offenders to the anger of the god, and perhaps to death as its consequence.

With many, the influence of a system so inflexible has probably operated very powerfully in producing this uniform attention, at least to the outward duties of the Sabbath, the only sacred day now recognized amongst them. With others, there is reason to believe it arises from the influence of example, and the respectability it was at this time supposed to impart to individual character. But with many it originates in far higher motives, and is the result of christian principle in regard to what they consider a duty.

A number of instances, strikingly illustrative of this fact, might be adduced; I will, however, only refer to one. A man came to the Monday evening meeting on one occasion, and said his mind was troubled, as he feared he had done wrong. He was asked in what respect; when he answered, that, on the preceding day, which was the Sabbath, when returning from public worship, he observed that the tide, having risen higher than usual, had washed out to sea a large pair of double canoes, which he had left on the beach. At first he thought of taking a smaller canoe, fetching back the larger ones, and fixing them in a place of security; but while he was deliberating, it occurred to his recollection that it was the Sabbath, and that the scriptures prohibited any work. He therefore allowed the canoes to drift towards the reef, until they were broken on the rocks. But, he added, though he did not work on the Sabbath, his mind was troubled on account of the loss he had sustained, and *that* he thought was wrong. He was immediately told that he would have done right, had he fetched the canoes to the shore on the Sabbath. When, however, it was considered, that perhaps this pair of canoes had cost him nearly twelve months' labour, and that, before they were lost he was comparatively richer than many an English merchant is in the possession of a five or six hundred ton vessel, it appears a remarkable instance of conscientious regard for the Sabbath-day.

Since the abolition of idolatry, no part of the conduct of the South Sea Islanders has impressed the minds of foreign visitants more forcibly than their attention to the observance of the Sabbath. I never saw any, even the most irreligious, or those unfriendly to Missions, who were not constrained to confess that it surpassed all they had heard or imagined could have been exhibited; while others, more favourably disposed, have publicly declared its effect on their own minds.

When Mr. Crook arrived in 1816, the ship reaching Tahiti on the Sabbath, no canoe put off, no native was seen on the beach, no smoke in any part of the district—and they began to apprehend either that the population had been swept off by some contagious disease, or that they had all gone to battle. At length their fears were removed by one of the party, who had been there before, observing, that it was the Sabbath, and that on that day the natives did not launch their canoes, or light their fires, &c. In 1821, Captain Grimes "was surprised at the regularity and good order observed; the children of the Sabbath-school were ushered in by their teachers in their different classes, with as much uniformity as we see in public schools in London." Several masters of South Sea whalers, captains and officers in his majesty's navy, have borne the most decided testimony to these facts. A naval officer, who was at Tahiti in 1822, stated, that he visited the islands under a considerable degree of prejudice against the Missionaries, and suspicion respecting the reported change among the people,—but that his visit had entirely removed both. It was Friday when the vessel arrived; the natives thronged the ship with fowls, fruit, vegetables, &c. for sale, manifesting considerable earnestness and address in the disposal of their goods. The same was continued through the second day; but on the third, to the great astonishment of all on board, no individual came near the ship, assigning, afterwards, as a reason, that it was the Sabbath. On the day following, however, the trade was as brisk as it had been on that of their arrival. Captain Gambier, who visited them in the same year, in the extracts from his journal, which have been published, states, in reference to the manner of attending the duties of the Sabbath among the young, that, "The silence—the order preserved—the devotion and attention paid to the subject, surprised and pleased me beyond measure." "Children," he adds, "are seen bringing their aged parents to the church, that they may partake of the pleasure they derive from the explanation of the Bible." The general attention to the public worship of God, and the exemplary Christian deportment of many of the people, have proved not only delightful, but beneficial to their visitors; and we have the high and grateful satisfaction of knowing, that occasional and transient visits to the christian islands of the Pacific, have been the means of advantage to the visitors; and there are probably many instances of good, which the revelations of the last day alone will disclose.

It is a privilege to visit a country, and a happiness to live in a community, where the Sabbaths are thus spent, and prove to multitudes—

"Foretastes of heaven on earth—pledges of joy
Surpassing fancy's flights and fiction's story,
The preludes of a feast that cannot cloy,
And the bright out-courts of immortal glory!"

This universal observance of the Sabbath-day appears to an Englishman in humiliating contrast with its profanation in many favoured sections of his own country. The contrast is still more striking when compared with the manner in which it is perverted into a season of activity, business, and unwonted gaiety in the pursuit of pleasure, in Catholic countries—but it never appears so surprising as when viewed in comparison with the actual state of the people themselves only a few years ago. No Sabbath had then ever dawned, no happy multitudes met for praise and prayer, no lovely throngs of children gathered in the Sabbath-schools, no inspired page or christian preacher directed their attention to the Lord of the Sabbath; but when the devotees met for public worship, it was under the gloom of dark overshadowing trees, amid the recesses of some rude temple, before some rustic altar, or in the presence of some deity of frightful form and fearful attributes, the offspring of their own imagination.

CHAP. VI.

Public assemblies during the week—Questional and conversational meeting—Topics discussed—The seat of the thoughts and affections—Duty of prayer—Scripture biography and history—The first parents of mankind—Paradise—Origin of moral evil—Satanic influence—A future state—Condition of those who had died idolaters—The Sabbath—Inquiries respecting England—The doctrine of the resurrection—Visits to Maeva—Description of the aoa—Legend connected with its origin—Considered sacred—Cloth made with its bark—Manufacture of native cloth—Variety of kinds--Methods of dyeing—Native matting—Different articles of household furniture.

THE religious services of a general kind, among the natives, during the week, are not numerous. There is one lecture, which is on Wednesday evening.—Numbers assemble at this time, and the exercise we have reason to believe is useful, in keeping alive that interest in matters of religion, which might be diminished by the secular engagements of the week. The following account of one of these meetings is given by Captain Gambier, in the extracts of his journal.

"On Wednesday afternoon we attended a native divine service. It was begun with a hymn; then Mr. Nott, who did duty, prayed extempore for some length, and then read a passage from the Scripture, upon which he preached with great fluency in the Otaheitan language. The church was well attended, though not so full as on Sundays, when it is crowded. Almost all the women, young and old, were habited in the European manner. The most perfect order reigned the whole time of the service. The devout attention these poor people paid to what was going forward, and the earnestness with which they listened to their teacher, would shame an English congregation. I declare, I never saw any thing to equal it! Objects of the greatest curiosity at all other times, they paid no sort of attention to, during the solemnity of their worship. After it was over, crowds, as usual, gathered round, to look at our uniforms, to them so new and uncommon. I looked round very often during the sermon, and saw not one of the congregation flag in their attention to it. Every face was directed to the preacher, and each countenance strongly marked with sincerity and pleasure. I had heard of the success of the Missionaries before I came to Otaheite, and, after making great allowance for exaggeration in the accounts they had sent home, there remained sufficient to lead me to anticipate that they had done a great deal. But I now declare, their accounts were beyond measure modest, and, far from colouring their success, they had not described it equal to what I found it. It is impossible to describe the sensations experienced on seeing the poor natives of Otaheite walking to a Protestant church in the most orderly and decent manner, with their books in their hands, and most of them dressed in European clothes.—Having just quitted the Marquesas, where we saw the very state the Otaheitans were in at the time of their first visitors, we of course saw the change to great advantage; and the magnitude of it is so astonishing, that all has the appearance of a dream. When, however, fully convinced of the reality, the hand of an Almighty Providence is distinctly acknowledged."

There are special meetings, held once a week, for the instruction of those who desire to make a public profession of the Christian faith by baptism, and another for the candidates for communion. In addition to these, there is a public meeting for general conversation, or rather for answering the questions of the people, held every Monday afternoon or evening.

This meeting originated in that held on the 26th of July 1813, for the purpose of writing the names of those who were desirous of publicly professing Christianity; and was designed for the more particular instruction of such individuals, though it has since assumed a more general character. This has been one of the most important and efficient means of promoting general and religious improvement in the islands. The greater part of the inhabitants of the settlement in which it is held, and many from remote districts, having assembled in the place of worship; we usually took our seats near a table at one end of the building. Soon after the Missionaries have entered, a native, perhaps in some distant part of the house, stands up, and, addressing them by name, asks a question, states a difficulty that may have perplexed his mind, begs an explanation of a passage of Scripture, or makes an inquiry relative to some subject or portion of the sacred volume, &c. Our answers generally lead to farther questions, either from the first inquirer, or other individuals in the assembly. The conversation is sometimes continued until a late hour; and both the queries and the replies are usually listened to with great attention. We always endeavoured to divest these meetings of all formality and reserve; and studied to render them engaging, by accompanying our illustrations with suitable anecdotes, and encouraging in the people the most unembarrassed confidence; requesting them to present all their difficulties, and solicit explanations or directions.

This meeting has always been highly interesting, and has generally indicated the progressive improvement of the people. The subjects discussed are perhaps less miscellaneous than they were some years ago, when the people were totally uninformed in all the first principles of Christianity; and the nature of these meetings in some of the stations has, perhaps, undergone a slight change. They are, however, productive of important benefit.

Subjects of every kind were formerly discussed, and questions brought forward relative to the discipline of children, the forming of connexions, and the whole of their domestic economy, agriculture, trade, or barter, legislature, war and politics, history and science, as connected with the natural phenomena by which they were surrounded; and occasionally what might be termed the first efforts of philosophical research, in their partially enlightened minds.

When the political questions referred to their foreign relations, or their intercourse with other islands, we sometimes allowed them to be entertained; but whenever they were connected with any civil proceedings, or the internal government of the island, although the person who introduced it was not interrupted during his speech, the matter was always referred to the king and chiefs, for whose consideration he was directed to present it at a convenient season, unless the chiefs, who were generally present, wished it to be then discussed.

One of the most curious and interesting topics of

conversation, frequently introduced by the more thinking or inquisitive among them, was, the seat of the affections, and the locality of intellect. Their ideas and ours were totally at variance on this point; and, from the very nature of the subject, it was impossible to demonstrate the accuracy of one or the other. No part in the system of Drs. Gall and Spurzheim ever obtained among them; and so far from being phrenologists, they did not imagine the brain to be even the seat of thought. The frequent eulogy pronounced by us on an oration or action, in which intellect and right feeling are developed, viz. "that it is creditable alike to the head and the heart" of the speaker or actor, would have been altogether unintelligible to them. The only exception to the prevailing opinion, which deprives the head or brain of all connexion with the exercise of the mind, is the term for headach, which is *tahoa*, and is also used to signify confusion of noise, and perplexity from attention to a multitude of objects at the same time.

The phraseology employed in speaking of the seat of the intellect and the affections, presents another analogy between the idiom of their language, and that of the ancient Hebrews. When speaking of mental or moral exercises, they invariably employ terms for which the English word "bowels" is perhaps the best translation: hence they say, *te manao o te obu*, or *i roto i te obu*; the thought of the *bowels*, or within the *bowels*; *te hinaaro o te aau*, the desire of the bowels; *te riri o te aau*, the anger of the *bowels*. Although bowels is, perhaps, the best single word for *obu* or *aau*, in the signification of which we have not been able to discover any difference, it does not convey the full meaning of the word *aau*. In some places it might be rendered *heart* according to our idiom, as in the thoughts of the heart or mind—the desire of the mind, or soul—or, the anger. of the soul. For soul and spirit, however, they have distinct terms, *varua*, and the ancient word *vaiti;* but it does not appear that they were accustomed to consider the soul or spirit as experiencing, in conjunction with the body, either mental or moral sensations. All the varied passions and the mental exercises of which they were sensible, they spoke of as connected with the *aau* or *obu*, a term literally signifying the whole of the abdominal viscera—for each separate organ in which, they have a distinct name.

To the head they attributed nothing in connexion with intellect, nor to the heart with regard to moral feeling. To the organ which in the language of anatomy would be called the heart, they attributed no other susceptibilities than those which are common to other parts of the body. This led them generally to contend that the thoughts were in the body, and not in the brain; stating, in proof of the accuracy of their opinion, that the bowels or stomach were affected or agitated by desire, fear, joy, sorrow, surprise, and all strong affections or exercises of the mind. They were, probably, confirmed in this definition by the fact of such being the belief of their ancestors.

In reply, we usually informed them, that we were accustomed to speak of the heart as the seat of the affections and moral principles, though by the heart we often meant nearly the same as they intended by the word *aau* or *obu*, but that we considered our sensations and mental perceptions to be connected with the brain. It was in vain that we endeavoured to shew the reasonableness of this opinion, by pointing out and explaining the connexion between the nerves pervading the several organs of sense, and the brain—the cessation and interruption of mental sensation and exercise, when the nerves of the brain were permanently injured—or when the line of nerves extending from an organ to the brain was broken. They usually answered, they would believe it because we said so, but that they did not understand it: indeed it was not to be expected that they should, as their knowledge of the anatomy of the human frame was exceedingly limited. They had no idea even of the existence of nerves, and it was necessary to introduce into their language a word by which they might be designated. Discussions of this nature, though adapted to interest the people, and encourage the exercise of intellect, were probably more amusing than profitable; and notwithstanding the diversified subjects presented, their inquiries generally referred to the new order of things which Christianity had introduced.

In reference to this, while they were sometimes trivial, and perhaps ludicrous, they were often deeply interesting and highly important, and not unfrequently difficult and perplexing. I have written many of them down at the time; others have been recorded by my companions: a selection will convey a more correct idea of their mode of thinking and expression, than any general description.

Many of their questions referred to the exercise of prayer, for punctual attendance to which they have been uniformly distinguished. Prayer for Divine direction accompanied their earliest inquiries on the subject of religion; and when in any district even two or three were desirous of becoming the disciples of Jesus Christ, they were accustomed to associate together for this purpose. Private prayer has long been almost universal, as well as the practice of imploring a blessing on their food; and although they at first asked whether they must not learn to pray in the English language? whether God would not be angry if they should use incorrect expressions in prayer? or whether, when they had retired to their gardens, or the bushes adjacent to their dwellings, and were there engaged in prayer, their attention should be diverted by an intruder, they should leave off or continue? Sometimes they would ask, whether engaging in conversation, and praying, with very wicked persons, such as had been murderers, &c. would not appear in some degree sanctioning or extenuating their crimes? With more frequency, however, and greater eagerness, they often inquired how they could prevent evil thoughts arising in their minds during seasons of devotion—how they could avoid repeating words of prayer unattended by devotional desires—and how they could at all times engage the heart in this exercise? I recollect a father and a mother asking with ardent solicitude, whether it would be right to take their little boy, or girl, with them to the bushes or the garden, talk with it in this retirement, and teach it there to pray to God. Prayer in their families was regularly observed; and among the many inquiries in reference to this subject, it was once asked, whether Jesus Christ had family prayer with his disciples; whether among their own domestic establishments, in the event of the sickness or absence of the husband, the wife should not convene the family, and perform this important duty?

Portions of scripture history and biography were among the most engaging subjects of inquiry, especially those contained in the Old Testament. Those in the New Testament also interested them. On one occasion, they asked what the heavy burdens were that our Lord accused the Scribes and Pharisees of binding on men's shoulders; and what was meant by "Let the dead bury their dead." At another time they inquired who were the Scribes, so often mentioned by the Saviour; and asked if they were the secretaries of the Auxiliary Missionary Societies in Jerusalem? &c. This arose from the circumstance of the word, which in English is translated *scribe*, being in Tahitian rendered *writer*, and the secretaries of the native Missionary societies being the only individuals among them thus designated.

The usages and customs prevailing among the ancient Jews were often topics of conversation, and more than once they have, with evident sincerity, inquired if their repentance would not be more acceptable to God, were they to rend their garments, and cover their heads with ashes, or gird themselves with sackcloth, than simply expressing their penitence when employed in prayers. This question, with those frequently asked relative to the consequences of mistakes or interruption in prayer, probably arose from the impression left by the system of idolatry they had so recently abandoned, whose only excellency consisted in the correctness of mere external form and ceremony.

In all their idol worship, however large or costly the sacrifices that had been offered, and however near its close the most protracted ceremony might be, if the priest omitted or misplaced any word in the prayers with which it was always accompanied, or if his attention was diverted by any means, so that the prayer was *hai*, or broken, the whole was rendered unavailable, he must prepare other victims, and repeat his prayers over from the commencement.

The history of our first parents was frequently brought forward. Sometimes they wanted to know what was the colour of Adam and Eve's skin, or what language they spoke: with regard to the former, their opinions were in accordance with those of the late Bishop Heber; they said it was very likely they were brown or olive-coloured, and, as their descendants, or the descendants of Noah, travelled to hotter climates, they became darker; while those, whose information had removed the belief that our colour was the effect of disease, acknowledged the plausibility of our ancestors having become white from the influence of cold, and a clouded atmosphere, whereby they were shaded from the sun.

More important matters concerning them were however often the subjects of inquiry. They felt interested in their destiny, and asked whether, after the fall, and expulsion from Paradise, they had repented and obtained pardon; and at one time, when, in answer to this question, it had been stated that there was reason to believe that they had obtained forgiveness, and were now in heaven, the native immediately inquired further, how Adam's crime could affect his posterity, after the guilt contracted by it had been removed even from the perpetrators of that crime? The origin of moral evil was sometimes introduced. It has been asked at meetings where I have been present, Would Satan have tempted Eve, or would man have fallen, if God had not forbidden our first parents to eat of the fruit of the tree of knowledge? To which it was answered, That if God had not made that the peculiar test of their obedience, Satan would have found some other medium through which to tempt them to sin.

A man once asked, What caused the angels in heaven to sin, or Satan to become a wicked spirit? He was told that pride was the cause of his fall, but that how pride entered heaven was not revealed. Another once proposed the following query: You say God is a holy and a powerful being, that Satan is the cause of a vast increase of moral evil or wickedness in the world, by exciting or disposing men to sin. If Satan be only a dependent creature, and the cause of so much evil, which is displeasing to God, why does not God kill Satan at once, and thereby prevent all the evil of which he is the author? In answer, he was told that the facts of Satan's dependence on, or subjection to the Almighty, and his yet being permitted to tempt even to evil, were undeniable from the declarations of scripture, and the experience of every one accustomed to observe the operations of his own mind. Such an observer would often find himself exposed to an influence that could be attributed only to satanic agency; but that why he was permitted to exert this influence on man, was not made known in the Bible. We always stated plainly, that it was the contents of that volume which we came to teach them; that the existence of this baneful and often fatal influence was too extensively felt to allow of its being questioned; that the antidote to the evil it might have already inflicted, and the preservative against its future effects, were pointed out; and that it was wiser, and far more important, to apply to those remedies, than to indulge in unprofitable speculations relative to its origin.

The duration of sufferings inflicted on the wicked in the future state, was occasionally introduced; and more than once I have heard them ask, if none of their ancestors, nor any of the former inhabitants of the islands, had gone to heaven? This, to us and to them, was one of the most distressing discussions upon which we ever entered. To them it was peculiarly so; for we may naturally suppose, the recollection of the individuals whom many of them had perhaps poisoned, murdered without provocation, slain in battle, or killed for sacrifice, would on these occasions forcibly recur to their minds; and at these times, many a parent's heart must have been rent with anguish, to us inconceivable, at the remembrance of those children in whose blood their hands had been imbrued. Besides these sources of intensely painful reflection, there is something overwhelming in the thought of relatives and friends removed from the world of hope and probation, having their doom irrevocably fixed! Hence we could perceive a degree of painful emotion among the people, whenever the subject was introduced; and although less intimately affected by this inquiry than those around us, it was to us a most appalling subject—one on which we could not dwell with composure. This feeling, on their parts, also, has been at times almost overpowering, and has either suspended our conversation, or induced an abrupt transition to some other topic.

This is a most distressing consideration, and is a subject often brought before a Missionary's mind, from the circumstances into which his engagements lead him, and the intimate connexion of his every effort with the future and eternal destinies of those around him; while it furnishes, next to the love of Christ, one of the most powerful incentives to devotedness and unabated effort. Well might one now engaged in this work exclaim, "Five hundred millions of souls,* who are represented as being unenlightened! I cannot, if I would, give up the idea of being a Missionary, while I reflect upon this vast number of my fellow-sinners, who are perishing for lack of knowledge. 'Five hundred millions!' intrudes itself upon my mind wherever I go, and however I am employed. When I go to bed, it is the last thing that recurs to my memory; if I awake in the night, it is to meditate upon it alone; and in the morning, it is generally the first thing that occupies my thoughts."

What mind, under the influence of the unequivocal declarations of the sacred volume, and an acquaintance with the true condition of the heathen, can calmly entertain the thought of the perishing millions of the pagan world?

We always told those who inquired, that it was not for us to say, what was the actual state of the departed; that of those who died in infancy, we were permitted to cherish the soothing and consolatory hope of their felicity; that those who survived infancy, had not been without the admonitions of conscience, which had borne a faithful testimony to the character of all their actions; and that on the evidence of that witness they would be acquitted or convicted at the awful bar of God. At the same time assuring them, that whatever crimes they might have to answer for, rejection of the gospel would not be one; though this would, perhaps, involve the heaviest condemnation on their descendants, if by them that gospel was neglected or despised.

* It is estimated that there are more than six hundred millions destitute of the knowledge of the gospel.

Many of their inquiries related to the proper observance of the Sabbath and under what circumstances it would be proper to launch a canoe or undertake a voyage? This resulted from the king's sister being taken ill at Afareaitu, while we were residing there; and the natives wishing to send word to her relations, but hesitating because it was the Sabbath. A man once came and said, that while he was attending public worship, a pig broke into his garden; that on his return, he saw him devouring the sweet potatoes, sugar-cane, taro, and other productions, in which pine-apples were probably included, but that he did not drive it out, because he was convinced it would immediately return, unless he repaired the broken fence, and *that* he supposed was a kind of labour prohibited on the Sabbath. He therefore allowed the pig to remain till he was satisfied, and did not mend the fence till the following morning. He, however, wished to know, and the people in general were evidently interested in the inquiry—whether, in the event of a similar occurrence at any future period, he should do wrong in driving out the animal, and repairing the fence. He was told that the most secure way would be to keep the fence in good repair, but that if pigs should break in on the Sabbath, they ought by all means to be driven out, and the breaches they had made, so far repaired as to secure the enclosure till the following day. A chief of Huahine once asked me whether it would be right, supposing he were walking in his garden on that day, and saw ripe plantains hanging from the trees that grew by the side of the path, to gather and eat them: I answered, that I thought it would not be wrong. I felt inclined to do so, said he, last Sabbath, when walking in my garden; but on reflecting that I had other fruit ready plucked and prepared, I hesitated,—not because I believed it would be in itself sinful, but lest my attendants should notice it, and do so too, and it should become a general practice with the people to go to their gardens, and gather fruit to eat on the Sabbath, which would be very unfavourable to the proper observance of that sacred day.

Their inquiries referred not only to historical, biographical, and other facts connected with the sacred volume, but to those relating to other nations of the earth. The extent of territory, number of inhabitants, colour, language, religion, of the different countries of whom they had heard from occasional visiters, were topics of conversation at these meetings, together with the efforts of Christians to propagate the gospel among them. But the most interesting of these, referred to England; and although their recollections of Captain Cook were generally more indistinct, and very different from those entertained by the Sandwich Islanders, he was often alluded to; and we were asked, if any members of his family still survived, and whether they would ever come to the islands. The cities, towns, houses, carriages, dress, and manners of the English, the royal state of king George, the numbers in his army, the evolutions of his troops, the laws of the kingdom, the punishment of crimes, the principles of commerce, and the extent and variety of manufactures, were at different times brought forward.

Numbers of the natives had indeed visited England, but their observation had been so limited, or their accounts so contradictory and exaggerated, that their countrymen knew not what to believe, and not unfrequently, when any of these had returned, the substance of their reports was brought to the questioning meeting, to receive our confirmation or explanation. The religious character and observances of the English were usually matters of great interest. The dimensions and number of our cathedrals, churches, and chapels, the size of the congregations, the proportion of the population that attended public worship, and the order of the services, were often topics of inquiry or conversation. The experience of those who were true Christians in England, was also introduced; and their remarks on this point, especially when they first became interested in the subject of religion themselves, were often rather amusing. "How happy the Christians in England must be," they would sometimes say.— "So many teachers, so many books, the whole of the Bible in their language, and no idolatry, they must have little else to do but to praise God. Their crimes have never been like ours; they never offered human sacrifices, murdered their infants, &c. Do they ever repent? have they any thing to repent of?" It was, however, only those who were recently awakened to a sense of the enormity of these crimes, and were but very partially informed as to the true state of England, that ever asked such questions as these.

The doctrine of the resurrection of the body has ever appeared to them, as it did when announced by the apostle to the civilized philosophers of Athens, or the august rulers in the Roman hall of judgment, as a fact astounding or incredible. Of another world, and the existence of the soul in that world after the dissolution of the body, they appear at all times to have entertained some indistinct ideas; but the reanimation of the mouldering bodies of the dead, never seems, even in their wildest flights of imagination, to have occurred to them. When first declared by the Missionaries, it merely awakened astonishment, and was considered as one among the many novel and striking facts connected with the doctrines which the new religion unfolded. But as the subject was more frequently brought under their notice in public discourse, or in reading the Scriptures, and their minds were more attentively exercised upon it in connexion with their ancestry, themselves, and their descendants, it appeared invested with more than ordinary difficulty; bordering, to their apprehension, on impossibility. On this, as well as other equally important points, their queries, from native simplicity and entire ignorance, were sometimes both puerile and amusing.

A number of the attendants on the queen's sister, soon after their reception of Christianity, came to the meeting, and stated that one of their friends had died a few days before, and that they had buried the corpse according to their ancient manner, not laying it straight in a coffin, as Christians were accustomed to do, but placing it in a sitting posture, with the face between the knees, the hands under the thighs, and the whole body bound round with cords. Since the interment, (they added,) they had been thinking about the resurrection, and wished to know how the body would then appear, whether, if left in that manner, it would not rise deformed, and whether they had not better disinter the corpse, and deposit it in a straight or horizontal position. A suitable reply was of course returned. They were directed to let it remain undisturbed—that probably long before the resurrection it would be so completely dissolved, and mingled with the surrounding earth, that no trace would be left of the form in which it had been deposited.

Questions of this kind were only presented during the first stages of their christian progress, and they were not frequent. In general their inquiries were exceedingly interesting. The time when, the means by which, the attending circumstances, and the manner of the resurrection, the recognition of friends, the identity of the bodies of adults, and whether the souls of infants would be united to infant bodies, and whether they would be as inferior in the future state, as their powers and faculties appeared in this, often furnished matter for hours of interesting conversation.

There were, however, other points of inquiry peculiarly affecting to themselves. Many of their relatives or countrymen had been devoured by sharks; a limb or large portion of the fleshy part of the body of others, had been destroyed by these savage fish. A constant attendant on these meetings at Afareaitu had, while we resided there, one side of his face torn off, and eaten by one. The sharks, that had eaten men, were perhaps afterwards caught, and became food for the natives, who might themselves be devoured by other sharks. Cannibalism, though some deny its having been practised among themselves, is supposed to have existed in one of the islands at least, and is known, and universally acknowledged to prevail among those by which they are surrounded; and it is not considered by them improbable that some of their own countrymen have been eaten by the islanders among whom they have, from stress of weather, been cast. The men who had eaten their fellow-men, might, and perhaps often were, (as many of the cannibals inhabit the low coralline islands, and live by fishing,) eaten by sharks, which would sometimes be caught and eaten by the inhabitants of distant islands.

After urging these, and corresponding inquiries which had exercised their minds, they would ask, After all these processes of new combination, will the original parts of every human body be reunited at the resurrection? &c. On such occasions, the truth of the doctrine of the resurrection was exhibited, as demonstrated by the resurrection of Lazarus and of Christ; the identity of our Lord's body, by his subsequent intercourse with the disciples, especially with Thomas; and the certainty of the general resurrection presented, as deduced from the numerous and explicit declarations of Scripture, and the reasoning of the inspired writers. The identity of the body was stated as being consistent with the character and moral government of God, which appeared to require that the same body which had suffered for or in his cause on earth, should be glorified in heaven; and that the same body which in union with the soul had been employed in rebellion and vice, should suffer the just consequences in a future state. The declarations of Scripture on this momentous point, always appeared to be satisfactory; and although the circumstances of the resurrection, and the manner by which parts of the same body would be united, &c. were inquiries pursued with deepest interest, we generally found them terminate in expressions of desire that they might be prepared, rise with glorified bodies, and come forth from their graves "to the resurrection of life."

In connexion with this subject, and others of a similar kind, the most important referred to what might be called their Christian experience—the effect of texts of Scripture committed to memory, in stimulating to duty, and restraining from sin. Often they would ask, "How can we attain true repentance, and a change of heart? How may we know that we are not deceiving ourselves? How can we be preserved from forsaking God and committing sin? We desire genuine faith; where can we obtain it? Once they observed—Adam fell in Paradise, and angels fell even in heaven itself; how then can we be preserved from sinning against God? Tell us how we may be safe from Satan?—how we may be safe for heaven, and secure of admission there?" I refrain from comments on the numerous inquiries brought forward at these meetings, which have been proved essentially serviceable to the nation—stimulating inquiry, giving a proper direction to their researches after truth, expanding the mind and strengthening the intellect, yet restraining them within the limitations of revelation. Their inquiries shew, if evidence were wanting, that their mental capabilities are not contemptible, and demonstrate the influence of the highest order of Christian principles upon the exercise of the mind and the affections of the heart.

Mr. Orsmond having removed to Raiatea in the close of 1816, and Mr. Nott to Tahiti in the summer of 1819, those of us who remained at Huahine endeavoured to engage as extensively as possible in the instruction of the people. We therefore went to several of the remote villages to preach to the people on the Sabbath, and frequently visited the district of Maeva, situated on the margin of an extensive lake, surrounded by most rich and varied scenery.

Among the singularly beautiful and diversified vegetable productions that adorn the banks of the lake, the sacred *aoa* deserves particular attention; it stands near the large temple of Tane, at Tamapua, and is one of the most ancient and extensive that I have met with in the islands. In its growth, the aoa resembles the banian tree of the East, and is probably a variety of the species. The bark has a light tinge and shining appearance, the leaf lance-shaped and small, of a beautiful pea-green colour. It is an evergreen, and is propagated by slips or branches, which readily take root. When the stem of the young tree is about two or three inches in diameter, the bark, immediately below the branches, which generally spread from the trunk, about six feet above the ground, begins to open near the lower part of the limbs. A number of fine yellow pointed roots protrude, and increase in size and length every year. The branches grow horizontally, and rather bending than otherwise: from different parts of these, fibres shoot forth through the bursting bark, and hang like fine dark-brown threads. The habits of growth in these pendulous roots are singular. Sometimes they appear like a single line, or rope, reaching from the highest branches nearly to the ground, where they terminate in a bunch of spreading fibres, not unlike a tassel. At other times, while there is one principal fibre, a number of others branch off from this at unequal distances, from its insertion in the bough above, and terminate in a cluster of small fibres. The different threads are sometimes separate from each other for a considerable distance, and, near the bottom, unite in one single root.

As soon as these depending fibres reach the ground, they take root, and, in the course of a number of years, become solid stems, covered with a bark resembling that of the original tree, and forming so many natural pillars to the progressively extending branches above.

By this singular process, the aoa, at Tamapua, appears more like a clump or grove than a single tree. The original stem was joined by one or two, of such dimensions, that it was not easy to distinguish the parent from the offspring, and the fibres that had united with the ground, and thus became so many trunks or stems of the tree, covered a space many yards in circumference. The lateral branches continue to extend, and tendrils of every length and size are seen in all directions depending from them, appearing as if in time it would cover the face of the country with a forest, which yet should be but one tree.

The most remarkable appearance, however, which the aoa presents, is when it grows near some of the high mountain precipices that often occur in the islands. A short distance from Buaoa, where the rocks are exceedingly steep, and almost perpendicular for a hundred feet, or more, an aoa appears to have been planted near the foot of the rocky pile, and the tender fibres protruded from the branches, being nearer the rocks at the side than the ground below, have been attracted towards the precipice. From this, fresh nourishment has been derived, the tree has continued to ascend and throw out new fibres still higher, till it has reached the top. Here a branching tree has flourished, exhibiting all the peculiarities of the aoa; while the root, and that part growing along the face of the rock, resembles a strong interwoven hedge, extending from the base to the summit of the precipice.

The account of the origin of this tree is one of the

most fabulous of native legends: it states that the moon is diversified with hill and valley like our earth, that it is adorned with trees, and among these the aoa, the shadow of whose spreading branches, the Polynesians suppose, occasions the dark parts in her surface. In ancient times, they state that a bird flew to the moon, and plucked the berries of the aoa; these are smaller than grapes; the bird readily carried them, and flying over the islands, dropped some of the seeds, which, germinating in the soil, produced the aoa tree. It was considered sacred, and frequently planted in the neighbourhood or precincts of the marae. The large one at Tamapua was supposed to be a frequent resort of the god; and the human sacrifices offered in the temple beneath, were usually suspended among its branches. A common imprecation was, *E tau oe i te amaa toro i momona;* "You will hang in the branch stretching towards Momona," a spot beneath one of the principal branches.

The cottage erected for my lodging, when visiting the district of Maeva, was within a few yards of this tree, and the chapel stood within the precincts of the ancient idol temple. I have often conversed with the people respecting it, and they have said that the most appalling horrors filled their minds, if they approached the tree or the pile after dark; that when they have pushed their light canoe along the adjacent lake, they have, as an act of reverence, uncovered the upper part of their persons, and almost trembled with fear till they had safely passed. Considering the dark and gloomy appearance of the temple, the wide extent, dense shade, fantastic shape, and grotesque appearance of the tortuous fibrous roots or trunks of the aoa, it is not surprising that the superstitious natives should be under the influence of those feelings, especially if it be recollected that there were few seasons when a human body was not suspended from some of the branches, to propitiate the deity of the place.

The aoa was not entirely devoted to the nurture of that debasing superstition by which the people were oppressed. With the thin slender twigs or young branches of this tree, a strong kind of cloth was made, which they called *ora*, or *aoa*, and which, on account of its durability, was highly esteemed.

Garments made with the bark of a tree constituted the principal article of native dress, prior to the introduction of foreign cloth. It is manufactured chiefly by females, and was one of their most frequent employments. The name for cloth, among the Tahitians, is *ahu*. The Sandwich Island word *tapa*, is, we believe, never used in this sense, but signifies a part of the human body. In the manufacture of their cloth, the natives of the South Sea Islands use a greater variety of materials than their neighbours in the northern group: the bark of the different varieties of *wauti*, or paper mulberry, being the only article used by the latter; while the former employ not only the bark of the paper mulberry, which they call *auti*, but also that of the aoa and of the bread-fruit.

The process of manufacture is much the same in all, though some kinds are sooner finished than others. When the bark from the branches of the bread-fruit or auti is used, the outer green or brown rind is scraped off with a shell; it is then slightly beaten, and allowed to ferment, or is macerated in water. A stout piece of wood, resembling a beam, twenty or thirty feet long, and from six to nine inches square, with a groove cut in the under side, is fixed on the ground; across this, the bark is laid and beaten with a heavy mallet of casuarina or iron-wood. The mallet is usually fifteen or eighteen inches long, about two inches square, and round at one end, for the purpose of being held firmly. The sides of the mallet are grooved; one side very coarse or large, the opposite side exceedingly fine. One of the remaining sides is generally cut in chequers or small squares, and the other is plain or ribbed. The

Tihitian Cloth Mallet.

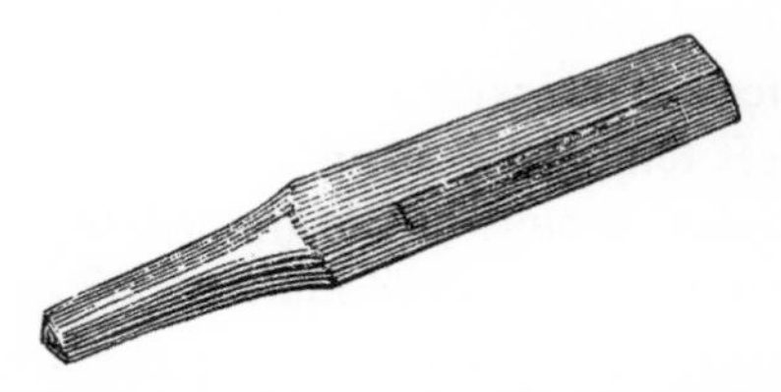

bark is placed lengthwise across the long piece of wood, and beaten first with the rough side of the mallet, and then with those parts that are finer.

Vegetable gum is rarely employed; in general, the resinous matter in the bark is sufficiently adhesive. The fibres of the bark are most completely interwoven by the frequent beating with the grooved or chequered side of the mallet; and when the piece is finished, the texture of the cloth is often remarkably fine and even; and the inequalities occasioned by the fine grooves, or small squares, give it the appearance of woven cloth. During the process of its manufacture, the cloth is kept saturated with moisture, and carefully wrapped in thick green leaves every time the workwomen leave off; but as soon as it is finished, they spread it to dry in the sun, and bleach it according to the purpose for which it is designed. The *ore*, or cloth, made with the bark of the aoa, is usually very thin, and of a dark brown colour;

that made with the bark of the bread-fruit and a mixture of the auti, is of a light brown and sometimes fawn colour; but the finest and most valuable kind is called hobu. It is made principally, and sometimes entirely, with the bark of the paper mulberry, and is bleached till it is beautifully white. This is chiefly worn by the females.

It is astonishing that they should be able, by a process so simple, to make bales, containing sometimes two hundred yards of cloth, four yards wide; the whole in one single piece, made with strips of bark seldom above four or five feet long, and, when spread open, not more than an inch and a half broad—joined together simply by beating it with the grooved mallet. When sufficiently bleached and dried, it is folded along the whole length, rolled up into a bale, and covered with a piece of matting, —this is called *ruru vehe.* The wealth of a chief is sometimes estimated by the number of these covered bales which he possesses. The more valuable kinds of cloth are rolled up in the same way, covered with matting or cloth of an inferior kind, and generally suspended from some part of the roof of the chief's house. The estimation in which it was held has been greatly diminished since they have become acquainted with European cloth, and large quantities are now seldom made. It is, however, still an article in general use among the lower classes of society, and the mother yet continues to beat her parure, or native pareu, for herself and children.

A number of smaller pieces are still made, among which the tiputa is one of the most valuable. It is prepared by beating a number of layers of cloth to gether, to render it thicker than the common cloth: for the outside layer, they select a stout branch of the auti, or bread-fruit, about an inch and a half in diameter: this they prepare with great attention, and, having beaten it to the usual width and length, which is about ten feet long and three feet wide, they fix it on the outside, and attach it to the others by rubbing a small portion of arrow-root on the inner side, before beating it together. The tiputa of the Tahitians corresponds exactly with the poncho of the South Americans. It is rather longer, but is worn in the same manner, having a hole cut in the centre, through which, when worn, the head is passed; while the garment hangs down over the shoulders, breast, and back, usually reaching, both before and behind, as low as the knees. Next to the tiputa, the ahufara is a general article of dress. These are either square like a shawl, or resemble a scarf. They are sometimes larger, and correspond with a counterpane more than a shawl, and are always exceedingly splendid and rich in their colours.

The natives of the Society Islands have a variety of vegetable dyes, and display more taste in the variations and patterns of the cloth, than in any other use of colours. Much of the common cloth is dyed either with the bark of the aito, *casuarina*, or tiairi, *aleurites*. This gives it a kind of dark red or chocolate colour, and is supposed to add to its durability. The leaves of the arum are sometimes used, but brilliant red and yellow are their favourite hues. The former, which they call mati, is prepared by mixing the milky juice of the small berry of the mati, *ficus prolixa*, with the leaves of the tou, a species of *cordia*. When the dye is prepared by this combination, it is absorbed on the fibres of a kind of rush, and dried for use. It produces a most brilliant scarlet dye, which, when preserved with a varnish of gum, retains its brightness till the garment is worn out. The yellow is prepared from the inner bark of the root of the nono, *morinda citrifolia,* and though far more fugitive than the scarlet of the mati, is an exceedingly bright colour. The yellow dye is prepared by infusing the bark of the root in water, in which the cloth is allowed to remain till completely saturated, when it is dried in the sun. The mati, or scarlet dye, is moistened with water, and laid on the dry cloth. Their patterns are fixed with the scarlet dye on a yellow ground, and were formerly altogether devoid of uniformity or regularity, yet still exhibiting considerable taste. They now fix a border round the ahufara, and arrange the patterns in different parts. Nature, and not art, supplies the pattern. They select some of the most delicate and beautiful ferns, or the hibiscus flowers: when the dye is prepared, they lay the leaf, or the flower, carefully on the dye; as soon as the surface is covered with the colouring matter, they fix the stained leaf or flower, with its leaflets or petals correctly adjusted on the cloth, and press it gradually and regularly down. When it is removed, the impression is often beautiful and clear.

The scarf or shawl, and the tiputa, are the only dresses prepared in this way, and it is difficult to conceive of the dazzling and imposing appearance of such a dress, loosely folded round the person of a handsome chieftain of the South Sea Islands, who perfectly understands how to exhibit it to the best advantage. This kind of cloth is made better by the Tahitians than any other inhabitants of the Pacific. It is not, however, equal to the wairiirii of the Sandwich Islanders. Much of this cloth, beautifully painted, is now employed in their houses for bed and window curtains, &c. Several kinds of strong cloth are finished with a kind of gum or varnish, for the purpose of rendering them impervious.

But in the fabrication of glazed cloth, the natives of the Austral Islands, especially those of Rurutu, excel all with whom I am acquainted. Some of the pieces of cloth are thirty or forty yards square, exceedingly thick, and glazed on both sides, resembling on the upper side the English oil-cloth table-covers. It must have required immense labour to prepare it, yet it was abundant when they were first discovered. It is usually red on one side, and black on the other, the latter being highly varnished with a vegetable gum.

In the manufacture of cloth, the females of all ranks were employed; and the queen, and wives of the chiefs of the highest rank, strove to excel in some department—in the elegance of the pattern, or the brilliancy of the colour. They are fond of society, and worked in large parties, in open and temporary houses erected for the purpose. Visiting one of these houses at Eimeo, I saw sixteen or twenty females all employed. The queen sat in the midst, surrounded by several chief women, each with a mallet in her hand, beating the bark that was spread before her. The queen worked as diligently and cheerfully as any present.

The spar or square piece of wood on which the bark is beaten, being hollow on the under-side, every stroke produces a loud sound, and the noise occasioned by sixteen or twenty mallets going at one time, was to me almost deafening; while the queen and her friends seemed not only insensible to any inconvenience from it, but quite amused at its apparent effect on us. The sound of the cloth-beating mallet is not disagreeable, where heard at a distance, in some of the retired valleys, indicating the abode of industry and peace; but in the cloth-houses it is hardly possible to endure it.

As the wives or daughters of the chiefs take a pride in manufacturing superior cloth, the queen would often have felt it derogatory to her rank, if any other females in the island could have finished a piece of cloth better than herself. I remember in the island of Huahine, when a native once passed by, wearing a beautiful ahufara, hearing one native woman remark to another—What a finely printed shawl that is! The figures on it are like the work, or the marking, of the queen! This desire, among persons in high stations, to excel in departments of labour, is what we have always admired. This feeling probably led Pomare to bestow so much attention on his hand-writing, and induced the king of the Sandwich Islands to request that we would not teach any of the people till we had fully instructed him in reading and writing.

The ahu, or cloth made with the bark of a tree, although exceedingly perishable when compared with European woven cloth, yet furnished, while it lasted, a light and loose dress, adapted to the climate, and the habits of the people. The duration of a Tahitian dress depended upon the materials with which it was made, the aoa being considered the strongest. Only the highly varnished kinds were proof against wet. The beauty of the various kinds of painted cloth was soon marred, and the texture destroyed, by the rain, as they were kept together simply by the adhesion of the interwoven fibres of the bark. Notwithstanding this, a tiputa, or a good strong pareu, when preserved from wet, would last several months. Though the native cloth worn by the inhabitants was made by the women, there were some kinds used in the temples, in the service of the idols, which were made by men, and which it was necessary, according to the declarations of the priest, should be beaten during the night.

Although the manufacture of cloth was formerly the principal, it was not the only occupation of the females. Many of the people, especially the raateiras, or secondary chiefs, wore a kind of mat made with the bark of the hibiscus, which they call purau; and the preparation of this, as well as the beds or sleeping mats, occupied much of the time of the females. Great attention was paid to the manufacture of these fine mats. They chose for this purpose, the young shoots of the hibiscus, and having peeled off the bark, and immersed it in water, placed it on a board, the outer rind being scraped off with a smooth shell. The strips of bark were an inch or an inch and a half wide, and about four feet long, and when spread out and dry, looked like so many white ribands. The bark was slit into narrow strips frequently less than the eighth of an inch wide. They were woven by the hand, and without any loom or machinery. They commenced the weaving at one corner, and having extended it to the proper width, which was usually three or four feet, continued the work till the mat was about nine or ten feet long, when the projecting ends of the bark were carefully removed, and a fine fringe worked round the edges.—Only half the pieces of bark used in weaving were split into narrow strips throughout their whole length. The others were slit five or six inches at the ends where they commenced, while the remaining part was rolled up like a riband. These they unrolled, and extended the slits as the weaving advanced, until the whole was complete. When first finished, they are of a beautifully white colour, and are worn only by the men, either bound round the loins as a pareu, or with an aperture in the centre as a tiputa or poncho, and sometimes as a mantle thrown loosely over the shoulder. Their appearance is light and elegant, and they are remarkably durable, though they become yellow from exposure to the weather.

The inhabitants of the Palliser Islands, to the eastward of Tahiti, exceed the Society Islanders in the quality of their mats, which are made of a tough white rush or grass, exceedingly fine and beautiful. They frequently manufacture a sort of girdle, called Tiheri, six inches in width, and sometimes twenty yards in length, but remarkably fine and even, being woven by the hand, but with a degree of regularity rivalling the productions of the loom. They are highly valued by the Tahitians, and are a principal article of commerce between the inhabitants of the different islands.

The sails for their canoes, and beds on which they sleep, are a coarser kind of matting made with the leaves of different varieties of palm, or pandanus, found in the islands. Some kinds grow spontaneously, others are cultivated for their leaves. The matting sails are much lighter than canvass, but far less durable. The size and quality of the sleeping mats is regulated by the skill of the manufacturer, or the rank of the proprietor. Those who excel in making them, use very fine ones themselves. They are all woven by the hand, yet finished with remarkable regularity and neatness.

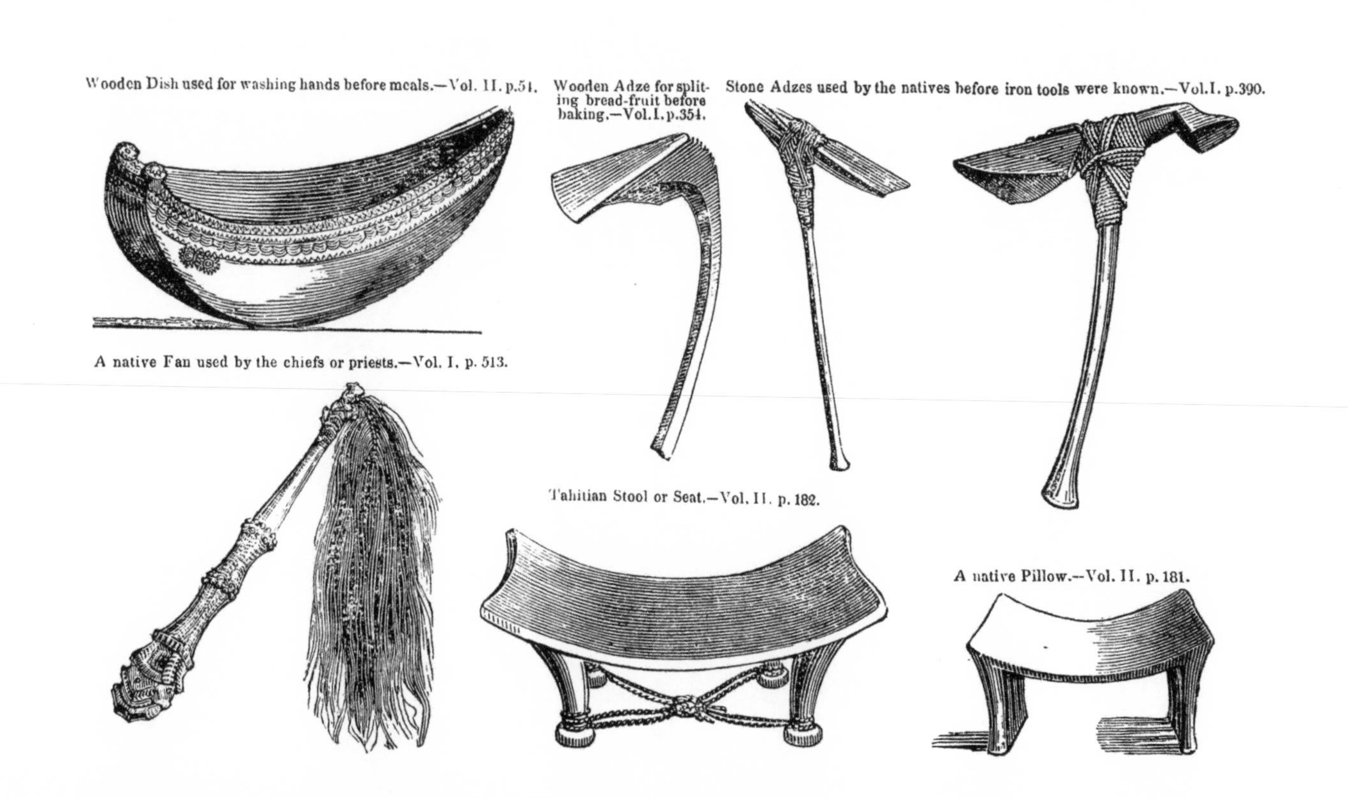

Wooden Dish used for washing hands before meals.—Vol. II. p.51.

Wooden Adze for splitting bread-fruit before baking.—Vol. I. p.354.

Stone Adzes used by the natives before iron tools were known.—Vol. I. p.390.

A native Fan used by the chiefs or priests.—Vol. I. p. 513.

Tahitian Stool or Seat.—Vol. II. p. 182.

A native Pillow.—Vol. II. p. 181.

The ordinary mats are not more than six feet wide, and nine or twelve feet long, but some are twelve feet wide, and sixty or eighty, or even a hundred yards long. Mats of this size, however, are only made for high chiefs, and in the preparation, perhaps, the females of several districts have been employed. They are kept rolled up, and suspended in some part of the chief's dwelling, more for the purpose of displaying his wealth, and the number of his dependents, than for actual use.

The kinds of leaf least liable to crack, are selected, and, for the purpose of sleeping upon, or even spreading on a floor, the use to which we generally applied them, the mats look neat, and last a considerable time. Several kinds of fine matting, ornamented with bright stained rushes interwoven with the others, were formerly made as articles of dress for the kings, or presents to the gods; but in this department of labour they were always inferior to the Sandwich Islanders, whose variegated mats are superior to any I have seen in the Pacific. Weaving of mats, with beating and staining of cloth, was the chief occupation of the females. A large portion of the property of the people consisted in mats and cloth, which also constituted part of their household furniture.

A variety of other articles were, however, necessary to the furnishing of their houses, but these were manufactured by the men. Next to a sleeping mat, a pillow was considered essential. This was of hard wood, and often exceedingly rude, though sometimes ingeniously wrought, resembling a short low stool, nine inches or a foot in length, and four or five inches high. The upper side was curved, to admit the head; the whole pillow, which they call tuaurua, is cut out of a single piece. Upon the bare wood they reclined their heads at night, and slept as soundly as the inhabitants of more civilized parts would do on the softest down.

In general, they sat cross-legged on mats spread on the floor; but occasionally used a stool, which they called *iri* or *nohoraa.* This resembled the pillow in shape, and, though much larger, was made out of a single piece of wood. The tamanu, or *callophyllum*, was usually selected, and immense trees must have been cut down for this purpose. I have seen iris four or five feet long, three feet wide, and at each end three feet six inches high; yet the whole cut out of one solid piece of timber. The upper part was curved, and the extremes being highest, the seat resembled the concave side of a crescent, so that, however large it might be, only one sat on it at a time. The iri was finely polished, and the wood, in its grain and colour resembling the best kinds of mahogony, rendered it, although destitute of carving or other ornament, a handsome piece of furniture in a chieftain's dwelling. The rank of the host was often indicated by the size of this seat, which was used on public occasions, or for the accommodation of a distinguished guest. Those in more ordinary use were low, and less curved, but always made out of a single piece of wood.

Next to these, their weapons, drums, and other musical instruments, were their most important furniture; a great portion, however, of what might be called their household furniture, was appropriated to the preparation or preservation of their food.

The *umete*, or dish, was the principal. Sometimes it was exceedingly large, resembling a canoe or boat more than a dish for food. It was frequently made with the wood of the tamanu, exceedingly well polished; some were six or eight feet long, a foot and a half wide, and twelve inches deep, these belonged only to the chiefs, and were used for the preparation of arrow-root, cocoa-nut milk, &c. on occasions of public festivity. The umetes in ordinary use were oval, about two or three feet long, eighteen inches wide, and of varied depth. They were supported by four feet, cut out of the same piece of wood, and serve not only for the preparation of their food, but as dishes, upon which it is placed when taken from the oven.

The *papahia* is extensively used. It is a low solid block or stool, supported by four short legs, and smoothly polished on the top. It is cut out of one piece of wood, and is used instead of a mortar for pounding bread-fruit, plantains, or bruising taro; which is done by placing these upon the papahia, and beating them with a short stone pestle called a *penu*. This is usually made with a black sort of basalt, found chiefly in the island of Maurua, the most western of the group. The penu is sometimes constructed from a species of porous coral.*

The water used for washing their feet is kept in bottles called *aano,* made from the shells of large and full-grown cocoa-nuts. That which they drink is contained in calabashes, which are much larger than any I ever saw used for the same purpose in the Sandwich Islands, but destitute of ornament. They are kept in nets of cinet, and suspended from some part of the dwelling.

The drinking cups are made with the cocoa-nut shell after it is full grown, but before it is perfectly ripe. The shell is then soft, and is scraped until much thinner than a saucer, and frequently transparent. They are of

* A fine specimen of that kind of *penu* which I procured at Rurutu, is deposited in the Missionary museum at Austin Friars.

a yellow colour, and plain, though the cups formerly used for drinking ava were carved. These are the principal utensils in the preparation of their food; they are kept remarkably clean, and, when not in use, suspended from some part of the dwelling, or hung upon a stand.

The *fata*, or stand, is a single light post planted in the floor, with one or two projections, and a notch on the top, from which the calabashes of water, baskets of food, umetes, &c. are suspended. Great labour was formerly bestowed on this piece of furniture, and the fata pua was considered an ornament to the house in which it was erected. About a foot from the ground, a projection extended six or eight inches wide, completely round, flat on the top, but concave on the under side, in order to prevent rats or mice from ascending and gaining access to the food. Their only knife was a piece of bamboo-cane, with which they would cut up a pig, dog, or fish, with great facility.

CHAP. VII.

Station at Maeva—Appearance of the lake and surrounding scenery—Ruins of temples, and other vestiges of idolatry—General view of Polynesian mythology—Ideas relative to the origin of the world—Polytheism—Traditionary theogony—Taaroa supreme deity—Different orders of gods—Oro, &c. gods of the wind, the ocean, &c.—Gods of artificers and fishermen—Oramatuas, or demons—Emblems—Images—Uru, or feathers—Temples—Worship—Prayers—Offerings—Sacrifices—Occasional and stated festivals and worship—Rau-mata-vehi-raa Maui-fata—Rites for recovery from sickness—Offering of first-fruits—The Pae Atua—The ripening of the year, a religious ceremony—Singular rites attending its close.

As soon as we had acquired a sufficient knowledge of the native language to engage in public teaching, while we alternately performed the regular services at the settlement in Fare, we formed branch stations in different parts of the island.

Two were commenced on the west and southern coast, viz. one in the beautiful, fertile, and formerly populous valley of *Mahapu*, and the other in the extensive district of *Parea*. Schools were opened under approved native teachers at each of these places. In the former three hundred scholars were instructed by *Narii*, a well qualified teacher. The inhabitants erected neat places of worship. Mr. Barff performed divine service at each station alternately every other Sabbath; and between three and four hundred attended.

A similar branch-station was commenced at *Tamabua*, a populous and central village in the district of Maeva, on the borders of a beautiful and extensive *roto*, or lake, of the same name, in the northern part of the island. Here a school was opened by Tiori, an intelligent native, and three hundred and eighty adults and children were taught. A commodious native chapel was built, and a cottage for the accommodation of the Missionary who visited them.

It was a considerable distance from Fare; I went on the Saturday afternoon, and spent the Sabbath at Maeva, where upwards of four hundred usually attended public worship. We continued our labours at these stations until the summer of 1820, when the greater part of the residents were induced to remove to the settlement at Fare harbour. Some of the happiest seasons I have enjoyed in Missionary occupations, were in connexion with my occasional services at this place. The scenery of the adjacent country is remarkably fine. The lake of Maeva is five miles in length, and of unequal breadth, though often two miles wide. Unagitated by the long rolling billows of the Pacific, and seldom ruffled by the northern and eastern breezes, from which it is sheltered by mountains, its surface was often smooth as a polished mirror, reflecting the groves around, and the heavens above. It abounds with fish. These not only supply the inhabitants of the border of the lake with the means of subsistence, but, when viewed from the light canoe, as they sported in the depths beneath, or leaped above its surface, enlivened its solitude. On the eastern side, a number of streams rose among the mountains, and, winding their way through the valleys, at length united with its waters. On this side, though the ascent from its margin to the distant mountains was generally gradual, it was sometimes abrupt and bold: the rocky precipices, adorned with pendulous and creeping plants, rich in verdant foliage or clustering flowers, rose almost perpendicularly from the water; the hills were ornamented with clumps of the graceful cypress-shaped casuarina; and in the narrow border of lowland, that in many parts extended from the shores to the foot of the mountains, the *hibiscus tiliaceus*, the *betonica splendida*, the *inocarpus*, and other trees of larger growth, reared their majestic forms, and spread their stately branches, clothed with dark and glossy foliage, while round their gigantic stems, and spread from bough to bough, the beautiful and large bell-flowering convolvolus, was often hung in wild luxuriant wreaths.

The walk from Fare to the head of the lake was delightful; for more than a mile, it was actually under what the natives call the *maru uru*, bread-fruit shade, large groves of this useful tree growing on each side of the path. A number of small plantations give variety to the wild scenery, and many of the *raatiras*, or inferior chiefs, have erected their dwellings near the path. Hautia had, when we first arrived, a noble house standing at the southern end of the lake. Along the eastern shore, small villages were seen amidst a grove of cocoanut and bread-fruit trees. A succession of agreeable sensations has arisen in my mind on a Saturday afternoon, when passing along the lake in my canoe, which was paddled by two native attendants. I have seen the columns of smoke curling up among the bread-fruit trees, where the inhabitants were dressing their food for the following day. Sometimes I have received their salutations from the shore; and, in contrast with their peaceful dwellings, and their present occupation, I have often been struck with the appearance of the villages, the dilapidated family maraes, or idol-temples, mouldering in ruins on almost every projecting point.

The western side of this extensive lake is bordered by a low flat tract of land, in many places a mile wide, extending from south to north. At the northern extremity of this beautiful piece of water, there is a narrow channel, by which it communicates with the sea. The western side, though very different from the opposite shore, adds to the variety of the scenery; it is thickly wooded, and among the trees that reach the highest perfection, the cocoa-nut, waving its crown of elegant leaves, and the no less elegant casuarina, whose boughs hang in arches over the water, are most conspicuous. The eastern side was doubtless originally the shore of the sea, and the lake filled by its waters, while the low border of the land on the opposite side constituted the reef. After the reef reached the level of the sea at high-water, it ceased to ascend, but spread horizontally; fragments of coral, and pieces of wood, were thrown upon its widened surface, till at length it resisted the shock of the ocean, and the waves rolled back without overflowing it. Every year increased the substances accumulated on its surface; vegetation at length commenced, and the process of organization and decomposition, accelerated by the humidity of the atmosphere and the warmth of the climate, formed the mould, in which the trees, at present covering it, spread their roots and find their nourishment.

But the most conspicuous and picturesque object, in connexion with the lake scenery, is *moua tabu*, sacred or devoted mountain, which rises on the eastern shore near the northern end. It is a beautiful and almost regular cone, partially covered with trees and bushes, even to its summit, while the shining basaltic or volcanic rocks, occasionally projecting through the cypress or pine-growing casuarina, add to the novel and agreeable diversity which its figure produces. The northern shore of the *roto*, or lake, of Maeva, was the favourite residence of many of the native kings. Here, also, the chief who governed the island after the last visit of Captain Cook, resided, and erected a house for Mai, or Omai, that he might be near him. The shores, and even the smooth surface of the lake itself, have been the scene of some of the most sanguinary battles that have been fought between rival parties on the island, or the people of Huahine and those of Raiatea and Borabora. Near its margin, on a rising ground, the ruins of one of the largest artificial fortifications in the group still remain.

But it is not so distinguished by any of these as by the vestiges of the ancient superstition of the island which every where abound. Temples to the gods of the water were erected on every point of land, and family maraes in almost every grove, while the extensive national temples of Oro and Tane stood near the northern extremity of the lake, where the greater number of human sacrifices were offered, where the idols were usually kept, and the national religious assemblies convened.

Every object around the lake, and every monument of art or labour, in the district of Maeva, bore marks of its connexion with their ancient religion. I have often visited the ruins of the large national temple of Tane, and the site of the house of Oro, and in my intercourse with the people of Maeva, at the meeting for inquiries, these were frequent topics of conversation.

It may, perhaps, be well, in this place, to introduce some account of the mythology of the islanders, and the principal features of their system of idolatry, although so many incidental notices of the same have already occurred, that it has been in some degree anticipated.

Like that of all the ancient idolatrous nations, the mythology of the South Sea Islanders is but an assemblage of obscure fables brought by the first settlers, or originated in remarkable facts of their own history, and handed down by tradition through successive generations. If so much that is mysterious and fabulous has been mingled with the history of those nations among whom hieroglyphics or the use of letters has prevailed, it might be expected to exist in a greater degree, where oral communication, and that often under the fantastic garb of rude poetry, is the only mode of preserving the traditional knowledge of former times.

Distinguished, however, as the Polynesian mythology is by confusion and absurdity, it is not more so than the ancient systems of some of the most enlightened and cultivated pagan nations, of the past or present time. It was not more characterized by mystery and fable, than by its abominations and its cruelty. Its objects of worship were sometimes monsters of iniquity. They had "lords many and gods many," but seldom attributed to them any moral attributes. Among the multitude of their gods, there was no one whom they regarded as a supreme intelligence or presiding spiritual being, possessing any moral perfections, resembling those which are inseparable from every sentiment we entertain of the true God.

Like the most ancient nations, they ascribe the origin of all things to a state of chaos, or darkness, and even the first existence of their principal deities refer to this source. Taaroa, Oro, and Tane, with other deities of the highest order, are on this account said to be *fanau po*, born of Night. But the origin of the gods, and their priority of existence in comparison with the formation of the earth, being a matter of uncertainty even among the native priests, involves the whole in the greatest obscurity. Taaroa, the Tanaroa of the Hawaians, and the Tangaroa of the Western Isles, is generally spoken of by the Tahitians as the first principal god, uncreated, and existing from the beginning, or from the time he emerged from the *po*, or world of darkness.

Several of their taata-paari, or wise men, pretend that, according to other traditions, Taaroa was only a man who was deified after death. By some he is spoken of as the progenitor of the other gods, the creator of the heavens, the earth, the sea, man, beasts, fowls, fishes, &c.; while by others it is stated, that the existence of the land, or the universe, was anterior to that of the gods.

There does not appear to be any thing in the Tahitian mythology corresponding with the doctrine of the Trinity, or the Hindoo tradition of Brahma, Vishnou, and Siva. Taaroa was the former and the father of the gods; Oro was his first son: but there were three classes or orders between Taaroa and Oro. As in the theogony of the ancients, a bird was a frequent emblem of deity; and in the body of a bird they supposed the god often approached the marae, where it left the bird, and entered the *too*, or image, through which it was supposed to communicate with the priest.

The gods and men, the animals, the air, earth, and sea, were by some supposed to originate in the procreative power of the gods. One of the legends of their origin and descent, furnished to some of the Missionaries, by whom it has been recorded, states, that Taaroa was born of Night, or proceeded from Chaos, and was not made by any other god. His consort, Ofeufeumaiterai, also uncreated, proceeded from the *po*, or night. Oro, the great national idol of Tahiti, Eimeo, and some of the Leeward Islands, was the son of Taaroa and Ofeufeumaiterai. Oro took a goddess to wife, who became the mother of two sons. These four male and two female deities constituted the whole of their highest rank of divinities, according to the traditions of the priests of Tahiti—though the late king informed Mr. Nott that there was another god, superior to them all, whose name was Rumia; he did not, however, meet with any of their priests or bards who knew any thing about him.

Raa was also ranked among the principal deities; although inferior to Taaroa and Oro, he was supposed to be an independent being; but nothing of consequence is ascribed to him in the native fables. His wife, Otupapa, who was also a divinity, bore him three sons and two daughters. Tane, the tutelar idol of Huahine, was also numbered among the uncreated gods, considered as having proceeded from the state of Night, or Chaos. His goddess was called Taufairei. They were the parents of eight sons, who were all classed with the most powerful gods, and received the highest honours. Among the sons of Tane was Temeharo, the tutelar deity of Pomare's family.

The most popular traditions in the Leeward Islands differed in several minor points from the above, which prevailed in the windward group. According to one, for which I am indebted to my friend Mr. Barff, Taaroa, who was supreme here as well as in Tahiti, was said to be *Toivi*, or without parents, and to have existed from eternity. He was supposed to have a body, but it was invisible to mortals. After innumerable seasons had passed away, he cast his *paa*, shell or body, as birds do their feathers, or serpents their skins; and by this means, after intervals of innumerable seasons, his body was renewed. In the *reva*, or highest heavens, he dwelt alone. His first act was the creation of Hina, who is also called his daughter. Countless ages passed away, when Taaroa and his daughter made the heavens, the earth, and the sea. The foundation of the world was a solid rock; which, with every part of the creation, Taaroa was supposed to sustain by his invisible power.—It is stated, that the Friendly Islanders suppose that the earth is supported on the shoulders of one of their gods, and that when an earthquake takes place, he is transferring it from one shoulder to the other.

Having, with the assistance of Hina, made the heavens, earth, and sea, Taaroa *oriori*, or created, the gods. The first was Rootane, the god of peace. The second was Toahitu, in shape like a dog; he saved such as were in danger of falling from rocks and trees. *Te fatu* (the lord) was the third. *Teiria*, (the indignant,) a god of war, was the fourth. The fifth, who was said to have had a bald head, was called Ruanuu. The sixth was a god of war. The seventh, Tuaraatai, Mr. Barff thinks was the Polynesian Neptune. The eighth was Rimaroa, (long arms,) a god of war. The ninth in order were the gods of idiots, who were always considered as inspired. The tenth was *Tearii tabu tura*, another Mars. These were created by Taaroa, and constituted the first order of divinities.

A second class were also created, inferior to these, and employed as heralds between the gods and men. The third order seems to have been the descendants of Raa, were numerous and varied in their character, some being gods of war, others among the Esculapiuses of the nation.

Oro was the first of the fourth class, and seems to have been the medium of connexion between celestial and terrestrial beings. Taaroa was his father. The shadow of a bread-fruit leaf, shaken by the power of the arm of Taaroa, passed over Hina, and she afterwards became the mother of Oro. Hina, it is said, abode in Opoa at the time of his birth; hence that was honoured as the place of his nativity, and became celebrated for his worship. Taaroa afterwards created the wife of Oro, and their children were also gods.

After the birth of Oro, Taaroa had other sons, who were called brothers of Oro, among whom were the gods of the Areois. These were the four orders of celestial beings worshipped in the Leeward Islands. The different classes only have been mentioned; an enumeration of the individual deities, and their offices or attributes, would be tedious and useless.

These objects of fear and worship were exceedingly numerous, and may be termed the chief deities of the Polynesians. There was an intermediate class between the principal divinities and the gods of particular localities or professions, but they are not supposed to have existed from the beginning, or to have been born of Night. Their origin is veiled in obscurity, but they are often described as having been renowned men, who after death were deified by their descendants. Roo, Tane, Teiri, probably Tairi principal idol of the Sandwich Islanders, Tefatu, Ruanuu, Moe, Teepa, Puaua, Tefatuture, Opaevai, Haana, and Taumure. These all received the homage of the people, and were on all public occasions acknowledged among Tahiti's gods.

Their gods of the ocean were not less numerous; this was to be expected amongst a people dwelling in islands deriving a great part of their sustenance from the sea, and almost amphibious in their habits. The names of fourteen principal marine divinities were communicated by the first Missionaries; others have been subsequently added, but it is unnecessary to enumerate them here. They are not supposed by the people to be of equal antiquity with the *akua fauau po,* or night-born gods.

They were probably men who had excelled their contemporaries in nautical adventure or exploit, and were deified by their descendants. Hiro is conspicuous amongst them, although not exclusively a god of the sea. The most romantic accounts are given in their *aai,* or tales, of his adventures, his voyages, his combat with the gods of the tempests, his descent to the depth of the ocean, and residence at the bottom of the abyss, his intercourse with the monsters there, by whom he was lulled to sleep in a cavern of the ocean, while the god of the winds raised a violent storm, to destroy a ship in which his friends were voyaging. Destruction seemed to them inevitable—they invoked his aid—a friendly spirit entered the cavern in which he was reposing, roused him from his slumbers, and informed him of their danger. He rose to the surface of the waters, rebuked the spirit of the storm, and his followers reached their destined port in safety.

The period of his adventures is probably the most recent of any thus preserved, as there are more places connected with his name in the Leeward Islands than with any other. A pile of rock in Tahaa is called the Dogs of Hiro; a mountain ridge has received the appellation of the Pahi, or Ship of Hiro; and a large basaltic rock near the summit of a mountain in Huahine, is called the Hoe or Paddle of Hiro.

Tuaraatai and Ruahatu, however, appear to have been the principal marine deities. Whether this distinction resulted from any superiority they were supposed to possess, or from the conspicuous part the latter sustains in their tradition of the deluge, is not known; but their names are frequently mentioned. They were generally called akua mao, or shark gods; not that the shark was itself the god, but the natives supposed the marine gods employed the sharks as the agents of their vengeance, in punishing transgressors.

The large blue shark was the only kind supposed to be employed by the gods; and a variety of the most strange and fabulous accounts of the deeds they have performed are related by their priests. These voracious animals were said always to recognize a priest on board any canoe, to come at his call, retire at his bidding, and to spare him in the event of a wreck, though they might devour his companions, especially if they were not his maru, or worshippers. I have been repeatedly told by an intelligent man, formerly a priest of an akua mao, that the shark through which his god was manifested, swimming in the sea, carried either him or his father on its back from Raiatea to Huahine, a distance of twenty miles. The shark was not the only fish the Tahitians considered sacred. In addition to these, they had gods who were supposed to preside over the fisheries, and to direct to their coasts the various shoals by which they were periodically visited. Tahauru was the principal among these; but there were five or six others, whose aid the fishermen were accustomed to invoke, either before launching their canoes, or while engaged at sea. Matatini was the god of fishing-net makers.

Next in number and importance to the gods of the sea, and the aerial regions, frequently worshipped under the figure of a bird, were those of the *peho te moua te pari e te faa*, the valleys, the mountains, the precipices, and the dells or ravines. The names of twelve of the principal of these are preserved by the Missionaries—I have them by me—but as few of them are indicative of the character or attributes of these gods, their insertion is unnecessary.

I have often thought, when listening to their fabulous accounts of the adventures of their gods, which, when prosecuting our researches in their language, manners, customs, &c. we have sometimes with difficulty induced them to repeat, that, had they been acquainted with letters, these would have furnished ample materials for legends rivalling in splendour of machinery, and magnificence of achievement, the dazzling mythology of the eastern nations. Rude as their traditions were, in the gigantic exploits they detail, and the bold and varied imagery they employ, they are often invested with an air of romance, which shews that the people possessed no inferior powers of imagination.

By their rude mythology, their lovely islands were made a sort of fairy-land, and all the spells of enchantment were thrown over its varied scenes. The sentiment of the poet that—

> "Millions of spiritual creatures walk the earth
> Unseen, both when we wake, and when we sleep,"

was one familiar to their minds; and it is impossible not to feel interested in a people who were accustomed to consider themselves surrounded by invisible intelligences, and who recognized in the rising sun—the mild and silver moon—the shooting star—the meteor's transient flame—the ocean's roar—the tempest's blast, or the evening breeze—the movements of mighty spirits. The mountain's summit, and the fleecy mists that hang upon its brows—the rocky defile—the foaming cataract—and the lonely dell—were all regarded as the abode or resort of these invisible beings.

An eclipse of the moon filled them with dismay; they supposed the planet was *natua*, or under the influence of the spell of some evil spirit that was destroying it. Hence they repaired to the temple, and offered prayers for the moon's release. The shape and stability of their islands they regarded as depending on the influence of spirits. The high and rocky obelisks, and detached pieces of mountain, were viewed as monuments of their power.

The large mountain on the left-hand side of the entrance to Opunohu, or Taloo harbour, which separates this bay from Cook's harbour, and is only united to the island by a narrow isthmus, was ascribed by tradition to the operations of those spirits, who, like the spirits in most other parts of the world, prefer the hours of darkness for their achievements. This mountain, it is stated, was formerly united with the mountains of the interior, and yielded in magnitude to none; but one night, the spirits of the place determined to remove it to the Leeward Islands, nearly one hundred miles distant, and accordingly began their operations, but had scarcely detached it from the main land, when the dawn of day discovered their proceedings, and obliged them to leave it where it now stands, forming the two bays already named. An aperture in the upper part of a mountain near Afareaitu, which appears from the lowland like a hole made by a cannon-ball, but which is eight or nine feet in diameter, is said to have been made by the passage of a spear, hurled by one of these supernatural beings.

Amusement was in part the business of a Tahitian's life; and with their games, as well as with every other institution, idolatry was connected Many were called sacred games, and over almost every one, the gods were supposed to exercise a control, though the people do not appear to have been such ancient gamblers as the Hawaiians were. Five or six gods were imagined to preside over the upaupa, or games, of which Urataetae was one of the principal.

The most benevolent of their gods were Roo or Tane, Temaru, Feimata, and Teruharuhatai. These were invoked by the tahua faatere, or expelling priests; and were supposed to exert their influence in restraining the effects of sorcery, or expelling the evil spirits, which, from the incantations of the sorcerer, had entered the sufferer. They had also patron deities of the healing art. Tama and Tetuahuruhuru were the gods of surgery; and their assistance was implored in reducing dislocations, healing fractures, bruises, &c.; while Oititi, or Rearea, was their Esculapius or god of physic.

In addition to these, were gods who presided over the mechanic arts. The first was Oihanu or Ofanu, the god of husbandry; the chief of the others was Taneetehia, the god of carpenters, builders, canoe-wrights, and all who wrought in wood. Nenia and Topca the gods of those who thatched houses, and especially of those who finished the angles where the thatch of each side joined. With these, others of a more repulsive character might be associated, but I shall only mention Heva the god of ghosts and apparitions, and Hiro the god of thieves. To the list from which the above are taken, including nearly one hundred of the objects formerly worshipped by the nation, a number of the principal family idols of the king and chiefs might be added, as every family of any antiquity or rank had its tutelar idol.

The general name by which they were distinguished was *atua*, which is perhaps most appropriately translated god. This word is totally different in its meaning, as well as sound, from the word *varua*, spirit, although that is sometimes applied to the gods: when the people were accustomed to speak disrespectfully of them, they called them *varaua ino*, bad or evil spirits. It is also different in its signification from the word which is used to designate an image, and the spirits of departed children or relations, and frequently those evil genii to whom the sorcerers addressed their incantations.

Atua, or akua, is the name for god, without any exception, throughout the whole of the eastern part of Polynesia. The first *a* appears to be a component part of the word, though in many sentences it is omitted, in consequence of the preceding word terminating in a vowel. It is then pronounced tua; and though but little light is thus thrown on the origin of the people, it is interesting to trace the correspondence between the *taata* or *tangata*, first man, in Polynesia, and *tangatanga*, a principal deity among the South Americans; the *atua*, or *tua*, of the South Sea islanders, and the *tev*, which is said to be the word for god in the Aztec or Mexican language, the *deviyo* of the Singhalese, and the *deva* of the Sanscrit.

The objects of worship among the Tahitians, next to the *atua* or gods, were the *oramatuas tiis* or spirits. These were supposed to reside in the *po*, or world of night, and were never invoked but by wizards or sorcerers, who implored their aid for the destruction of an enemy, or the injury of some person whom they were hired to destroy. They were considered a different order of beings from the gods, a kind of intermediate class between them and the human race, though in their prayers all the attributes of the gods were ascribed to them. The *oramatuas* were the spirits of departed fathers, mothers, brothers, sisters, children, &c. The natives were greatly afraid of them, and presented offerings, to avoid being cursed or destroyed, when they were employed by the sorcerers.

They seem to have been regarded as a sort of demons. In the Leeward Islands, the chief oramatuas were spirits of departed warriors, who had distinguished themselves by ferocity and murder, attributes of character usually supposed to belong to these evil genii. Each celebrated tii was honoured with an image, through which it was supposed his influence was exerted. The spirits of the reigning chiefs were united to this class, and the skulls of deceased rulers, kept with the images, were honoured with the same worship. Some idea of what was regarded as their ruling passion, may be inferred from the fearful apprehensions constantly entertained by all classes. They were supposed to be exceedingly irritable and cruel, avenging with death the slightest insult or neglect, and were kept within the precincts of the temple. In the marae of *Tane* at Maeva, the ruins of their abode were still standing, when I last visited the place. It was a house built upon a number of large strong poles, which raised the floor ten or twelve feet from the ground. They were thus elevated, to keep them out of the way of men, as it was imagined they were constantly strangling, or otherwise destroying, the chiefs and people. To prevent this, they were also treated with great respect; men were appointed constantly to attend them, and to keep them wrapped in the choicest kinds of cloth, to take them out whenever there was a pae atua, or general exhibition of the gods; to anoint them frequently with fragrant oil; and to sleep in the house with them at night. All this was done, to keep them pacified. And though the office of calming the angry spirits was honourable, it was regarded as dangerous, for if, during the night or at any other time, these keepers were guilty of the least impropriety, it was supposed the spirits of the images, or the skulls, would hurl them headlong from their high abodes, and break their necks in the fall. The figures marked No. 7, in the engraving of the Idols, represent the images of two tiis or oramatuas; whose form and appearance convey no inappropriate exhibition of their imagined malignity of disposition.

Among the animate objects of their worship, they included a number of birds as well as fishes, especially a species of heron, a kingfisher, and one or two kinds of woodpecker, accustomed to frequent the sacred trees growing in the precincts of the temple. These birds were considered sacred, and usually fed upon the sacrifices. The natives imagined the god was embodied in the bird, when it approached the temple to feast upon the offering; and hence they supposed their presents were grateful to their deities. The cries of those birds were also regarded as the responses of the gods to the prayers of the priests.

They supposed their gods were powerful spiritual beings, in some degree acquainted with the events of this world, and generally governing its affairs; never exercising any thing like benevolence towards even their most devoted followers, but requiring homage and obedience, with constant offerings; denouncing their anger, and dispensing destruction on all who either refused or hesitated to comply. But while the people supposed they were spiritual beings, they manufactured images either as representations of their form, and emblems of their character, or as the vehicle or instrument through which their communications might be made unto the god, and his will revealed to them.

The idols were either rough unpolished logs of the aito, or casuarina tree, wrapped in numerous folds of sacred cloth; rudely carved wooden images; or shapeless pieces covered with curiously netted cinet, of finely braided cocoa-nut husk, and ornamented with red feathers. They varied in size, some being six or eight feet long, others not more than as many inches. Those representing the spirits, they called *tii;* and those representing the national or family gods, *toos.* Into these they supposed the god entered at certain seasons, or in answer to the prayers of the priests. During this indwelling of the gods, they imagined even the images were very powerful: but when the spirit had departed, though they were among the most sacred things, their extraordinary powers were gone.

I had repeated conversations with a *tahua-tarai-too*, a maker of gods, whom I met with on a visit to Raiatea. As he appeared a serious inquirer after truth, and I could place some confidence in what he related, I was anxious to know his own opinion as to the idols it had been his business to make,—whether he really believed they were the powerful beings which the natives supposed; and if so, what constituted their great power over the other parts of the tree from which they were hewn? He assured me, that although at times he thought it was all deception, and only practised his trade to obtain the property he received for his work; yet at other times he really thought the gods he himself had made, were powerful beings. It was not, he said, from the alteration his tools had effected in the appearance of the wood, or the carving with which they were ornamented, but because they had been taken to the temple, and were filled with the atua, that they became so powerful. The images of aito-wood were only exceeded in durability by those of stone. Some of the latter were calcareous or siliceous, but the greater part were rude, uncarved, angular columns of basalt, various in size, and destitute of carving or polish; they were clothed or ornamented with native cloth.

The sacred flag was also used in processions, and regarded as an emblem of their deities.

Throughout Polynesia, the ordinary medium of communicating or extending supernatural powers, was the red feather of a small bird found in many of the islands, and the beautiful long tail-feathers of the *tropic*, or man-of-war bird. For these feathers the gods were supposed to have a strong predilection; they were the most valuable offerings that could be presented; to them the power or influence of the god was imparted, and through them transferred to the objects to which they might be attached. Among the numerous ceremonies observed, the *paeatua* was one of the most conspicuous. On these occasions, the gods were all brought out of the temple, the sacred coverings removed, scented oils were applied to the images, and they were exposed to the sun. At these seasons, the parties who wished their emblems of deity to be impregnated with the essence of the gods, repaired to the ceremony with a number of red feathers, which they delivered to the officiating priest.

The wooden idols being generally hollow, the feathers were deposited in the inside of the image, which was filled with them. Many idols, however, were solid pieces of wood, bound or covered with finely-braided cinet, of the fibres of the cocoa-nut husk; to these the feathers were attached on the outside by small fibrous bands. In return for the feathers thus united to the god, the parties received two or three of the same kind, which had been deposited on a former festival in the inside of the wooden or inner fold of the cinet idol. These feathers were thought to possess all the properties of the images to which they had been attached, and a supernatural influence was supposed to be infused into them. They were carefully wound round with very fine cinet, the extremities alone remaining visible. When this was done, the new-made gods were placed before the larger images from which they had been taken; and, lest their detachment should induce the god to withhold his power, the priest addressed a prayer to the principal deities, requesting them to abide in the red feathers before them. At the close of his *ubu*, or invocation, he declared that they were dwelt in or inhabited, (by the gods,) and delivered them to the parties who had brought the red feathers. The feathers taken home, were deposited in small bamboo-canes, excepting when addressed in prayer. If prosperity attended their owner, it was attributed to their influence, and they were usually honoured with a *too*, or image, into which they were inwrought; and subsequently, perhaps, an altar and a rude temple were erected for them. In the event, however, of their being attached to an image, this must be taken to the large temple, that the supreme idols might sanction the transfer of their influence.

Their temples were either national, local, or domestic. The former were the depositories of their principal idols, and the scenes of all great festivals; the second were those belonging to the several districts; and the third, such as were appropriated to the worship of family gods. *Marae* was the name for temple, in the South Sea Islands. All were uncovered, and resembled oratories rather than temples. The national places of worship were designated by distinct appellations. *Tabu-tabu-a-tea* was the name of several in the South Sea Islands, especially of those belonging to the king; the word may mean wide-spread sacredness. The national temples consisted of a number of distinct maraes, altars, and sacred dormitories, appropriated to the chief pagan divinities, and included in one large stone enclosure of considerable extent. Several of the distinct-temples contained smaller inner-courts, within which the gods were kept. The form of the interior or area of their temples was frequently that of a square or a parallelogram, the sides of which extended forty or fifty feet. Two sides of this space were enclosed by a high stone wall; the front was protected by a low fence; and opposite, a solid pyramidal structure was raised, in front of which the images were kept, and the altars fixed. These piles were often immense. That which formed one side of the square of the large temple in Atehuru, according to Mr. Wilson, by whom it was visited when in a state of preservation, was two hundred and seventy feet long, ninety-four wide at the base, and fifty feet high, being at the summit one hundred and eighty feet long, and six wide. A flight of steps led to its summit; the bottom step was six feet high. The outer stones of the pyramid, composed of coral and basalt, were laid with great care, and hewn or squared with immense labour, especially the *tiavâ*, or corner stones.

National Temple.

Within the enclosure, the houses of the priests, and keepers of the idols, were erected. Ruins of temples are found in every situation: on the summit of a hill, as at Maeva, on the extremity of a point of land projecting into the sea; or in the recesses of an extensive and overshadowing grove. The trees growing within the walls, and around the temple, were sacred; these were the tall cypress-like casuarina, the *tamanu*, or callophyllum, *mero*, or thespesia, and the *tou*, or cordia. These were, excepting the casuarina-trees, of large foliage and exuberant growth, their interwoven and dark umbrageous branches frequently excluded the rays of the sun; and the contrast between the bright glare of a tropical day, and the sombre gloom in the depths of these groves, was peculiarly striking. The fantastic contortions in the trunks and tortuous branches of the aged trees, the plaintive and moaning sound of the wind passing through the leaves of the casuarina, often resembling the wild notes of the Eolian harp—and the dark walls of the temple, with the grotesque and horrific appearance of the idols—combined to inspire extraordinary emotions of superstitious terror, and to nurture that deep feeling of dread which characterized the worshippers of Tahiti's sanguinary deities.

The priests of the national temples were a distinct class; the office of the priesthood was hereditary in all its departments. In the family, according to the patriarchal usage, the father was the priest in the village or district; the family of the priest was sacred, and his office was held by one who was also a chief. The king was sometimes the priest of the nation, and the highest sacerdotal dignity was often possessed by some member of the reigning family. The intimate connexion between their false religion and political despotism, is, however, most distinctly shewn in the fact of the king's personifying the god, and receiving the offerings brought to the temple, and the prayers of the supplicants, which have been frequently presented to Tamatoa, the present king of Raiatea. The only motives by which they were influenced in their religious homage, or service, were, with very few exceptions, superstitious fear, revenge towards their enemies, a desire to avert the dreadful consequences of the anger of the gods, and to secure their sanction and aid in the commission of the grossest crimes.

Their worship consisted in preferring prayers, presenting offerings, and sacrificing victims. Their *ubus*, or prayers, though occasionally brief, were often exceedingly protracted, containing many repetitions, and appearing as if the suppliants thought they should be heard for their much speaking. The petitioner did not address the god standing or prostrate, but knelt on one knee, sat cross-legged, or in a crouching position, on a broad flat stone, leaning his back against an upright basaltic column, at the extremity of a smooth pavement, usually six or ten yards from the front of the idol. He threw down a branch of sacred mero on the pavement before the image or altar, and began his *tarotaro*, or invocation, preparatory to the offering of his prayer. *Pure* is the designation of prayer, and *haamore* that for praise, or worship.

Small pieces of *niau*, or cocoa-nut leaf, were suspended in different parts of the temple, to remind the priest of the order to be observed. They usually addressed the god in a shrill, unpleasant, or chanting tone of voice, though at times the worship was extremely boisterous. That which I have often heard in the northern islands was peculiarly so; and on these occasions, when we have induced the priest to repeat any of the prayers, they have always recited them in these tones.

Nothing can exceed the horror they have of their former worship. An instance of this kind occurred at Parea with an old blind priest of the fisherman's temple there. When the majority of the inhabitants embraced Christianity, he declared he would not abandon the idols, nor unite in the worship of the God of the Christians; and in order to shew his determination, on the Sabbath-day, when the people went to the chapel, he went out to work in, I think, a part of the ground belonging to his temple: while thus engaged in mending a fence, a bough struck his eyes, and not only inflicted great pain, but deprived him of his sight, and, like Elymas, he was obliged to be led home. This circumstance deeply affected his mind; he became a firm believer in the true God, maintained an upright and resigned frame of mind, and, when baptized, adopted the name of Paul, from the similarity in the means employed in humbling and converting him, and those used to bring the apostle to a sense of the power and mercy of the Saviour. He died in 1824. Two or three years before this, I visited his residence; and, in company with others, attended him to the temple of which he had been priest. After examining its ruins, we requested him to recite, simply for our information relative to the nature of the former worship, one of the prayers he had been accustomed to offer there. After great persuasion, he consented, and, assuming the crouching position, or sitting as it were on his heels, he commenced, in a shrill, tremulous tone, to repeat the names of the gods, &c., but he was soon seized with a violent trembling and evident alarm, and declared he durst not, he could not proceed. Corresponding apprehensions have often been manifested by others, and we have seldom induced any to recite their idolatrous prayers.

Their offerings included every kind of valuable property: the fowls of the air, the fishes of the sea, the beasts of the field, and the fruits of the earth, together with their choicest manufactures, were presented. The sacrifice was frequently called *Taraehara,* a compound term, signifying disentangling from guilt; from *tara,* to untie or loosen, and *hara,* guilt. The animals were taken either in part or entire. The fruits and other eatables were generally, but not always, dressed. Portions of the fowls, pigs, or fish, considered sacred, dressed with sacred fire within the temple, were offered; the remainder furnished a banquet for the priests and other sacred persons, who were privileged to eat of the sacrifices. Those portions appropriated to the gods were deposited on the *fata* or altar, which was of wood. Domestic altars, or those erected near the corpse of a departed friend, were small square wicker structures; while those in the public temple were large, and usually eight or ten feet high. The surface of the altar was supported by a number of wooden posts or pillars, often curiously carved, and highly polished. The altars were covered with sacred boughs, and ornamented with a border or fringe of rich yellow plantain leaves. The pigs, &c. when presented alive, received the sacred mark, and ranged the district at liberty; when slain, they were exceedingly anxious to avoid breaking a bone, or disfigur ing the animal. One method of killing them was by holding the pig upright on its legs, placing a strong stick horizontally under its throat, and another across upon its neck, and then pressing them together until the animal was strangled. Another plan was, by bleeding the pig to death, washing the carcase with the blood, and then placing it in a crouching position on the altar. Offerings and sacrifices of every kind, whether dressed or not, were placed upon the altar, and remained there, until by decomposition they were consumed. The heat of the climate, and frequent rain, accelerated this process, yet the atmosphere in the vicinity of the maraes was frequently most offensive.

Altar, and Offerings.

Animals, fruits, &c. were not the only articles presented to their idols; the most affecting part of their sacrificing was the frequent immolation of human victims. These sacrifices, in the technical language of the priests, were called *fish*. They were offered in seasons of war, at great national festivals, during the illness of their rulers, and on the erection of their temples. I have been informed by several of the inhabitants of Maeva, that the foundation of some of the buildings, for the abode of their gods, was actually laid in human sacrifices; that every pillar, supporting the roof of one of the sacred houses at Maeva, was planted upon the body of a man, who had been offered as a victim to the sanguinary deity about to be deposited there. The unhappy wretches selected were either captives taken in war, or individuals who had rendered themselves obnoxious to the chiefs or the priests. When they were wanted, a stone was, at the request of the priest, sent by the king to the chief of the district from which the victims were required. If the stone was received, it was an indication of an intention to comply with the requisition. It is a singular fact, that the cruelty of the practice extended not only to individuals, but to families and districts. When an individual had been taken as a sacrifice, the family to which he belonged was regarded as *tabu* or devoted; and when another was required, it was more frequently taken from that family than any other: and a district from which sacrifices had been taken, was, in the same way, considered as devoted; and hence, when it was known that any ceremonies were near, on which human sacrifices were usually offered, the members of tabu families, or others who had reason to fear they were selected, fled to the mountains, and hid themselves in the dens and caverns till the ceremony was over. At a public meeting in Raiatea, Paumoana, a native chieftain, alluded to this practice in terms resembling these:—How great our dread of our former gods! Are there not some here who have fled from their houses, to avoid being taken for sacrifices? Yes! I know the cave in which they were concealed.

In general, the victim was unconscious of his doom, until suddenly stunned by a blow from a club or a stone, sometimes from the hand of the very chief on whom he was depending as a guest for the rights of hospitality. He was usually murdered on the spot—his body placed in a long basket of cocoa-nut leaves, and carried to the temple. Here it was offered, not by consuming it with fire, but by placing it before the idol. The priest, in dedicating it, took out one of the eyes, placed it on a plantain leaf, and handed it to the king, who raised it to his mouth as if desirous to eat it, but passed it to one of the priests or attendants, stationed near him for the purpose of receiving it. At intervals during the prayers some of the hair was plucked off, and placed before the god; and when the ceremony was over, the body was wrapped in the basket of cocoa-nut leaves, and frequently deposited on the branches of an adjacent tree, After remaining a considerable time, it was taken down, and the bones buried beneath the rude pavement of the marae. These horrid rites were not unfrequent, and the number offered at their great festivals was truly appalling.

The seasons of worship were both stated and occasional. The latter were those in which the gods were sought under national calamities, as the desolation of war, or the alarming illness of the king or chiefs. In addition to the rites connected with actual war, there were two that followed its termination. The principal of these, *Rau ma ta vehi raa*, was designed to purify the land from the defilement occasioned by the incursions or devastations of an enemy, who had perhaps ravaged the country, demolished the temples, destroyed or mutilated the idols, broken down the altars, and used as fuel the *unus*, or curiously carved pieces of wood, marking the sacred places of interment, and emblematical of their tii's or spirits. Preparatory to this ceremony, the temples were rebuilt, new altars reared, new images, inspired or inhabited by the gods, placed in the maraes, and fresh unus erected.

At the close of the rites in the new temples, the parties repaired to the sea-beach, where the chief priest offered a short prayer, and the people dragged a small net of cocoa-nut leaves through a shallow part of the sea, and usually detached small fragments of coral from the bottom, which were brought to the shore. These were denominated fish, and were delivered to the priest, who conveyed them to the temple, and deposited them on the altar, offering at the same time an *ubu* or prayer, to induce the gods to cleanse the land from pollution, that it might be pure as the coral fresh from the sea. It was now supposed safe to abide on the soil, and appropriate its produce to the purposes of support; but had not this ceremony been performed, death would have been anticipated.

The *maui fata*, altar-raising, was connected with the preceding rites. No human victim was slain, but numbers of pigs, with abundance of plantains, &c. were placed upon the altars, which were newly ornamented with branches of the sacred mero, and yellow leaves of the cocoa-nut tree. These rites extended to every marae in the island, and were designed to secure rain and fertility for the country gained by conquest, or recovered from invasion.

Besides these, the chief occasional services were those connected with the illness of their rulers. This was supposed to be inflicted by the gods for some offence of the chiefs or people. Long and frequent prayers were offered, to avert their anger, and prevent death. But, supposing the gods were always influenced by the same motives as themselves, they imagined that the efficacy of their prayers would be in exact proportion to the value of the offerings with which they were accompanied. Hence, when the symptoms of disease were violent and alarming, if the sufferer was a chief of rank, the fruits of whole fields of plantains, and a hundred or more pigs, have been taken to the marae, and frequently, besides these, a number of men, *with ropes round their necks*, have been also led to the temple, and presented before the idol. The prayers of the priests have often been interrupted by the ejaculatory addresses of the men, calling by name, and exclaiming "Be not angry, or let thy wrath be appeased; here we are: look on us, and be satisfied," &c. It does not appear that these men were actually sacrificed, but probably they appeared in this humiliating manner, with ropes about their necks, to propitiate the deity, and to shew their readiness to die, if it should be required.

While these ceremonies were observed, the progress of the disease was marked, by the friends of the afflicted, with intense anxiety. If recovery followed, it was attributed to the pacification of the deities; but if the disease increased, or terminated fatally, the god was regarded as inexorable, and was usually banished from the temple, and his image destroyed.

Religious rites were connected with almost every act of their lives. An *ubu* or prayer was offered before they ate their food, when they tilled their ground, planted their gardens, built their houses, launched their canoes, cast their nets, and commenced or concluded a journey. Numerous ceremonies were performed at the birth of any child of rank, at marriages, and interments. The first fish taken periodically on their shores, together with a number of kinds regarded as sacred, were conveyed to the altar. The first-fruits of their orchards and gardens were also *taumaha*, or offered, with a portion of their live-stock, which consisted of pigs, dogs, and fowls, as it was supposed death would be inflicted on the owner or the occupant of the land, from which the god should not receive such acknowledgment.

Altar and Unus.

The *bure arii,* a ceremony in which the king acknowledged the supremacy of the gods, was one attended with considerable pomp and show; but one of the principal stated festivals was the *pae atua,* which was held every three moons. On these occasions all the idols were brought out from their sacred depository, and *meheu,* or exposed to the sun; the cloth in which they had been kept was removed, and the feathers in the inside of the hollow idols were taken out. The images were then anointed with fragrant oil; new feathers, brought by their worshippers, were deposited in the inside of the hollow idols, and folded in new sacred cloth: after a number of ceremonies, they were carried back to their dormitories in the temple. Large quantities of food were provided for the entertainment, which followed the religious rites of the pae atua.

The most singular of their stated ceremonies was the *maoa raa matahiti,* ripening or completing of the year. This festival was regularly observed in Huahine: although I do not know that it was universal, vast multitudes assembled. In general, the men only engaged in pagan festivals; but men, women, and children, attended at this; the females, however, were not allowed to enter the sacred enclosure. A sumptuous banquet was held annually at the time of its observance, which was regulated by the blossoming of reeds.

Their rites and worship were in many respects singular, but in none more so than in the ripening of the year, which was regarded as a kind of annual acknowledgment to the gods. When the prayers were finished at the marae, and the banquet ended, a usage prevailed much resembling the popish custom of mass for souls in purgatory. Each individual returned to his home, or to his family marae, there to offer special prayers for the spirits of departed relatives, that they might be liberated from the *po*, or state of night, and ascend to *rohutunoanoa*, the mount Meru of Polynesia, or return to this world, by entering into the body of some inhabitant of earth.

They did not suppose, according to the generally received doctrine of transmigration, that the spirits who entered the body of some dweller upon earth, would permanently remain there, but only come and inspire the person to declare future events, or execute any other commission from the supernatural beings on whom they imagined they were constantly dependent.

CHAP. VIII.

Description of Polynesian idols—Human sacrifices—Anthropophagism—Islands in which it prevails—Motives and circumstances under which it is practised—Traditiou of its existence in Sir Charles Sanders' Island—Extensive prevalence of Sorcery and Divination—Views of the natives on the subject of satanic influence—Demons—Imprecations—Modes of incantation—Horrid and fatal effects supposed to result from sorcery—Impotency of enchantment on Europeans—Native remedies for sorcery—Native oracles—Means of inspiration—Effects on the priest inspired—Manner of delivering the responses—Circumstances at Rurutu and Huahine—Intercourse between the priest and the god—Augury by the death of victims—Divination for the detection of theft.

The system of idolatry, which prevailed among a people separated from the majority of their species by trackless oceans, breathing a salubrious air, inhabiting a beautiful and fertile country, and possessing the means, not only of subsistence but of comfort, in an unusual degree, presents a most affecting exhibition of imbecility, absurdity, and degradation. Whether we consider its influence over the individual, the family, or the nation, through the whole period of life—its oppressive exactions, its frequent and foolish rites, its murderous sacrifices, the engines of its power, and the objects of its homage and its dread—it is impossible to contemplate it without augmented thankfulness for the blessings of revelation, and increased compassion for those inhabiting the dark places of the earth.

The idols of the heathen are in general appropriate emblems of the beings they worship and fear; and if we contemplate those of the South Sea Islanders, they present to our notice all that is adapted to awaken our pity. The idols of Tahiti were generally shapeless pieces of wood, from one to four feet long, covered with finely braided cinet of cocoa-nut fibres, ornamented with scarlet feathers. Oro was a straight log of hard casuarina wood, six feet in length, uncarved, but decorated with feathers. The gods of some of the adjacent islands exhibit a greater variety of form and structure. The accompanying plate contains several of these.

The two figures in the centre, No. 1. exhibit a front and profile view of Taaroa, the supreme deity of Polynesia; who is generally regarded as the creator of the world, and the parent of gods and men. The image from which these views were taken, is nearly four feet high, and twelve or fifteen inches broad, carved out of a solid piece of close, white, durable wood. In addition to the number of images or demigods forming the features of his face, and studding the outside of his body, and which were designed to shew the multitudes of gods that had proceeded from him; his body is hollow, and when taken from the temple, in which for many generations he had been worshipped, a number of small idols were found in the cavity. They had perhaps been deposited there, to imbibe his supernatural powers, prior to their being removed to a distance, to receive, as his representatives, divine honours. The opening to the cavity was at the back; the whole of which, as shewn in the profile view, might be removed. The image to the right, No. 3. is another representation of Taaroa. No. 5. is Terongo, one of the principal gods, and his three sons. No. 2. is an image of Tebuakina, three sons of Rongo, a principal deity in the Harvey Islands. The name is probably analogous to Orono in Hawaii, though distinct from Oro in Tahiti. No. 6. exhibits a sacred ornament of a canoe from the island of Huahine. The two figures at the top, are images worshipped by fishermen, or those frequenting the sea. The two small idols at the lower corners of the plate, No. 7. are images of oramatuas, or demons. The others are gods from the Harvey Islands. The gods of Rarotogna were some of them much larger; Mr. Bourne, in 1825, saw fourteen about twenty feet long, and six feet wide.

Such were the objects the inhabitants of these islands were accustomed to supplicate; and to appease or avert the anger of which, they devoted not only every valuable article they possessed, but murdered their fellow creatures, and offered their blood. Human victims were sacrificed to Taaroa, Oro, and several others. It has been supposed, that the circumstance of the priests' offering the eye, the most precious part of the victim, to the king, who appeared to eat it, indicated their having formerly devoured the men they had sacrificed. I do not regard this fact as affording any very strong evidence, although I have not the least doubt that the inhabitants of several of the South Sea Islands have eaten human beings.

From the many favourable traits in their character, we have been unwilling to believe they had ever been cannibals; the conviction of our mistake has, however, been impressed by evidence so various and multiplied as to preclude uncertainty. Their mythology leads them to suppose, that the spirits of the dead are eaten by the gods or demons; and that the spiritual part of their sacrifices is eaten by the spirit of the idol before whom it is presented. Birds resorting to the temple, were said to feed upon the bodies of the human sacrifices and it was imagined the god approached the temple in the bird, and thus devoured the victims placed upon the altar. In some of the islands, "man-eater" was an epithet of the principal deities; and it was probably in connexion with this, that the king, who often personated the god, appeared to eat the human eye. Part of some human victims were eaten by the priests.

The Marquesans are known to be cannibals; the inhabitants of the Palliser or Pearl Islands, in the immediate neighbourhood of Tahiti, to the eastward, are the same. A most affecting instance of their anthropophagism is related by recent visitors; who state that a captive female child, pining with hunger, on begging a morsel of food from the cruel and conquering invaders of her native island, was supplied by a piece of her own father's body!

The bodies of prisoners in war, or enemies slain in battle, appear to have been eaten by most of the Harvey Islanders, who reside a short distance to the west of the Society group. There were several inducements to this horrid practice. The New Zealanders ate the bodies of their enemies, that they might imbibe their courage, &c. Hence, they exulted in their banquet on a celebrated warrior; supposing that, when they had devoured his flesh, they should be imbued with his valiant and daring spirit. I am not certain that this was the motive by which the eastern Polynesians were influenced, but one principal design of their wars was to obtain men to eat. Hence, when dwelling in their encampment, and clearing the brushwood, &c. from the place in which they expected to engage the enemy, they animated each other to the work in the following terms, "Clear away well, that we may kill and eat, and have a good feast to-day." To "kill and eat," was the haughty warrior's threat; and to be "killed and eaten," the dread of the vanquished and the exile. In the island of Rarotogna, they cut off the heads of the slain, piled them in heaps within the temple, and furnished the banquet of victory with their bodies.

The desire of revenge, or the satisfaction resulting from actually devouring an enemy, was not their only motive. The craving of nature, and the pangs of famine, often led to this unnatural crime. It was the frequent inducement in the Marquesas, and also in the Harvey Islands. In Maute, Metiaro, and Atiu, seasons of scarcity are severely felt; and, to satisfy their hunger, a number of persons, at the hour of midnight, have stolen a man from a neighbouring residence, killed, and eaten him at once. Mr. Bourne, who visited the islands in 1825, states, that members of the same family are not safe; and so awful is their wretchedness, that this horrid cruelty is practised towards those who, in civilized communities, are the objects of most endearing attachment: the husband has preyed upon the body of his wife, and the parent upon his child, in a most revolting manner, without subjecting it to any previous preparation. These facts are too painful and barbarous to admit detail. Another, and perhaps more criminal motive than either revenge or want, led some to the perpetration of these appalling deeds: this was, the indulgence of their depraved and vitiated appetite.

In the little island of Tapuaemanu, between Eimeo and Huahine, tradition states that there were formerly cannibals. In the reign of Tamatafetu, an ancestor of the present ruler, it is related, that when a man of stout or corpulent habit went to the island, or low-land on the reef, he was seldom heard of afterwards. The people of the island imagined those thus missing were destroyed by the sharks: but for many years, the servants of the king followed them to the island on the reef, and having murdered, baked them there. When the bodies were baked, they wrapped them in leaves of the hibiscus and plantain, as they were accustomed to wrap their eels, or other fish, taken and cooked on the island; they then carried them to the interior, where the king and his servants feasted on them. Their deeds were at length discovered by Feito, the wife of the king. She was in the house on one occasion, and, as they supposed, asleep, when she overheard the king and his servants planning the death of Tebuoroo, her brother. Anxious to save her brother's life, she revealed to him the purpose of the chief. He communicated it to the raatiras, or farmers, who immediately repaired to the marae of Taaroa, to inquire what they should do; and left with a unanimous determination to destroy their chief. Two men, Mehoura and Raiteanui, were appointed to hide themselves near his place of bathing; and when the chief came to bathe, they killed him with stones. A native of this island related the above statement within the last two years, at a public meeting held near the place where it is reported to have occurred, and afterwards in private stated that it was according to their traditions. Mr. Barff, to whom I am indebted for the tradition, adds, "The people at large affirmed it to be true." This unnatural crime does not appear to have been general; and the above is the only direct account that we have of its existence in what are properly the Society Islands. It is not probable that it will ever be revived, and, at a recent public meeting, in alluding to it, as illustrative of the former, and contrasting it with the present state of the people, the native speaker concluded by saying, "Behold, under the gospel of Jesus Christ, this land, where man-eaters have dwelt, has become a land of neighbours and of brethren."

No people in the world, in ancient or modern times, appear to have been more superstitious than the South Sea Islanders, or to have been more entirely under the influence of dread from imaginary demons, or supernatural beings. They had not only their major but their minor demons, or spirits, and all the minute ramifications of idolatry. Sorcery and witchcraft were extensively practised. By this art, the sorcerers pretended to be able to inflict the most painful maladies, and to deprive of life the victims of their mysterious rites.

It is unnecessary now to inquire whether satanic agency affects the bodies of men. We know this was the fact at the time our Saviour appeared on earth. Many of the natives of these islands are firmly persuaded, that while they were idolaters, their bodies were subject to most excruciating sufferings, from the direct operation of satanic power. In this opinion they might be mistaken, and that which they regarded as the effect of super-human agency, might be only the influence of imagination, or the result of poison. But considering the undisputed exercise of such an influence, recognized in the declarations and miracles of our Lord and of his apostles, existing not only in heathen, but Jewish society, and considering, in connexion with this, the undisputed dominion, moral and intellectual, which the powers of darkness held over those that were entirely devoted to the god of this world, it does not appear impossible, or inconsistent with the supreme government of God, that these subordinate powers should be permitted to exert an influence over their persons, and that communities, so wholly given to idolatry of the most murderous and diabolical kinds, should be considered corporeally, as well as spiritually, to be lying "in the wicked one." In addition to the firm belief which many who were sorcerers, or agents of the infernal powers, and others who were the victims of incantation, still maintain, some of the early Missionaries are disposed to think this was the fact. Since the natives have embraced Christianity, they believe they are now exempt from an influence, to which they were subject during the reign of the evil spirit.

Individuals, among the most intelligent of the people, sometimes express their deliberate conviction, that it is because they live under the dispensation or government of Jesus Christ, that they are now exempt from those bodily sufferings to which they were exposed while they were willing and zealous devotees of the devil. It is, I believe, also an indisputable fact, that those kinds of violent, terrific, and fatal corporeal agony, which they attributed to this agency, have altogether ceased, since the subversion of that system, of which it was so dreadful a part. I am not prepared to pronounce the opinions many of the natives still hold, as altogether imaginative: at the same time, the facts that have come to my knowledge, during my residence among them, have led me to desire the most satisfactory evidence for rejecting their sentiments.

Witchcraft and sorcery they considered the peculiar

province of an inferior order of supernatural beings. The names of the principal oramatuas were, Mau-ri, Bua-rai, and Tea-fao. They were considered the most malignant of beings, exceedingly irritable and implacable; they were not confined to the skulls of departed warriors, or the images made for them, but were occasionally supposed to resort to the shells from the sea-shore, especially a beautiful kind of murex, the *murex ramoces*. These shells were kept by the sorcerers, and the peculiar singing noise perceived on applying the valve to the ear, was imagined to proceed from the demon it contained.

These were the kinds of beings invoked by the wizards or sorcerers. Different names were applied to their arts, according to the rites employed, or the effects produced. Tahu, or tahutahu, natinatiaha or pifao, were the general terms employed, both for sorcery and the performance of it. *Tahu*, in general, signifies to kindle, and is much the same in import as *ahikuni*, the word for sorcery in the Sandwich Islands. The application of fire was common to both. *Natinati* signifies involved, entangled, and knotted: *aha*, is cinet; and the persons afflicted with this, were supposed to be possessed by a demon, who was twisting and knotting their inside, and thus occasioning most excruciating pain and death. *Pifao* signifies a hook or barb; and is also indicative of the condition of those, under the visitation of evil spirits, who were holding them in agony, as severe as if transfixed by a barbed spear or hook.

Incantations sometimes commenced with an imprecation or curse, either by the priest or the offended party, and it was usually denounced in the name of the gods of the party, or of the king, or some oramatua. This was generally employed in revenge for an injury or insult, which the party using the imprecation imagined they had received; and the poor people entertained the greatest horror of this mode of vengeance, as it was generally considered fatal, unless, by engaging a more powerful demon, its effects could be counteracted.

This dreadful system of iniquity, and demon tyranny, was complex and intricate. The party using sorcery against another, whose destruction they designed, employed a tahutahu, or a taata-obu-tara, whose influence with the demons procured their co-operation in the murderous design, and was supposed to induce the *tii*, or spirit, to enter into the victim of their malice.

Prayers, offerings, and the accustomed mysteries, however numerous, were not sufficient for this purpose. It was necessary to secure something connected with the body of the object of vengeance. The parings of the nails, a lock of the hair, the saliva from the mouth, or other secretions from the body, or else a portion of the food which the person was to eat. This was considered as the vehicle by which the demon entered the person, who afterwards became possessed. It was called the *tubu*, growing, or causing to grow. When procured, the *tara* was performed; the sorcerer took the hair, saliva, or other substance that had belonged to his victim, to his house or marae, performed his incantations over it, and offered his prayers; the demon was then supposed to enter the tubu, and through it the individual, who suffered from the enchantment. If it was a portion of food, similar ceremonies were observed, and the piece of bread-fruit, fish, &c., supposed by this process to be impregnated by the demon, was placed in the basket of the person for whom it was designed; and, if eaten, inevitable destruction followed.

The use of the portable spittoon by the Sandwich Island chiefs, in which the saliva was carefully deposited, carried by a confidential servant, and buried every morning, and the custom of the Tahitians in scrupulously burning or burying the hair when cut off, and also furnishing to each individual his distinct basket for food, originated in their dread of sorcery by any of these means. When the tara had been performed, and the tubu secured, the effects were violent, and death speedy. The most acute agonies and terrific distortions of the body were often experienced; the wretched sufferer appeared in a state of frantic madness, or, as they expressed it, torn by the evil spirit, while he foamed and writhed under his dreadful power.

On one occasion, Mr. Nott sent two native boys, who were his servants, from Eimeo to Tahiti, for *taro*, or arum-roots. The man, under whose care it was growing, was a sorcerer: he was from home, I believe—but the boys, according to the directions they had received, went to the field, and procured the roots for which they had been sent. Before they had departed, the person who had charge of the field returned, and was so enraged, that he pronounced the most dreadful imprecations upon one, if not both of them, threatening them with the *pifao*. The boys returned to Eimeo, but apparently took no notice of the threatening. One of them was shortly afterwards taken ill; and the imprecation of the sorcerer being made known to his friends, it was immediately concluded that he was possessed by the evil spirit. Alarming symptoms rapidly increased, and some of the Missionaries went to see him in this state. On entering the place where he lay, a most appalling spectacle was presented. The youth was lying on the ground, writhing in anguish, foaming at the mouth, his eyes apparently ready to start from his head, his countenance exhibiting every form of terrific distortion and pain, his limbs agitated with the most violent and involuntary convulsions. The friends of the boy were standing round, filled with horror at what they considered the effects of the malignant demon; and the sufferer shortly afterwards expired in dreadful agonies. In general, the effects of incantations were more gradual in their progress, and less sudden, though equally fatal in their termination.

The belief of the people in the power of the sorcerers remained unshaken, until the renunciation of idolatry, and the whole population were consequently kept in most humiliating and slavish fear of the demons. No rank or class was supposed to be exempt from their fatal influence. The young prince of Taiarabu, Te-arii-na-vaho-roa, brother of the late king, was by many of the people considered as destroyed, by Metia, a prophet of Oro, and a celebrated sorcerer, who had sometimes been known to threaten even the king himself with the effects of his indignation. "Give up, give up," was the language he on one occasion employed, when addressing the king, "lest I bend my strong bow;" in allusion, it is supposed, to his pretended influence with the demon. Whole families were sometimes destroyed. In Huahine, out of eight, one individual alone survives; seven, it is imagined, having been cut off by one sorcerer.

The imprecation was seldom openly denounced, unless the agent of the powers of darkness imagined his victim had little prospect of escape, and that his family were not likely to avenge his death. In general, these mysteries were conducted with that secrecy, which best comported with such works of darkness. Occasionally the tahutahu employed his influence with the evil spirit, to revenge some insult or injury he or his relatives had received; but he more frequently exercised it for hire. From his employers he received his fee and his directions, and having procured the tubu, or instrument of acting on his victim, repaired to his own rude marae, performed his diabolical rites, delivered over the individual to the demon whom he invoked, imploring the spirit to enter into the wretch, and inflict the most dreadful bodily sufferings, terminate at length the mortal existence, and then hurry the spirit to the po, or state of night, and there pursue the dreadful work of torture. These were the infernal labours of the .tahutahu or the pifao, the wizard or the sorcerer; and these, according to the superstitions of the people, their terrific results.

It is possible that in some instances these sufferings may have been the effects of imagination, and a deep impression on the mind of the afflicted individual, that he was selected as the victim of some insatiable demon's rage. Imagining he was already delivered to his grasp, hope was abandoned, death deemed inevitable, and the infatuated sufferer became the victim of despair. It is also possible that poison, of which the natives had several kinds, vegetable and animal, (some few of which they have stated capable of destroying human life,) might have produced the violent convulsions that sometimes preceded dissolution. It is probable that into the piece of food, over which the sorcerer performed his incantations, he introduced a portion of poison, which would prove fatal to the individual by whom it was eaten. Indeed, some of the sorcerers, since their conversion to Christianity, and one of them on his death-bed, confessed that this had been practised, and that they supposed the poison had occasioned the death which had been attributed to their incantations. Others, however, still express their belief, that they were so completely under the dominion of the evil spirit, that his power extended to the body as well as to the mind. I offer no opinion on this matter, but confine myself to stating the sentiments of the people, and some of the facts connected with the same. It has been a subject of very frequent conversation with several of the most reflecting among the natives, who, since they have become Christians, have expressed their deliberate belief that their bodies were subject to satanic agency.

It is a singular fact, that while the practice continued, with all its supernatural influence, among the natives, the sorcerers invariably confessed that incantations were harmless when employed upon Europeans: several have more than once been threatened with sorcery, and there is reason to believe it has been put to the test upon them. They have always declared, that they could not prevail with the white men, because such were under the keeping of a more powerful Being than the spirits *they* could engage against them, and therefore were secure. The native Missionaries, in different islands, have also been threatened with sorcery from the idolaters among whom they have endeavoured to introduce Christianity. They have always defied the sorcerers and their demons, telling them that Jehovah would protect them from their machinations; and though frequently exposed to incantations, have never sustained the slightest injury.

The sentiments entertained by the natives relative to the character of these supernatural beings, led them to imagine them such as they were themselves, only endowed with greater powers. They supposed that in all their actions they were influenced by motives exactly corresponding with those that operated upon their own minds; hence they believed, that even spirits could be diverted from their purposes by the offer of a larger bribe than they had received to carry it into effect, or that the efforts of one tii could be neutralized or counteracted by another more powerful.

Under the influence of these opinions, when any one was suffering from incantations, if he or his friends possessed property, they immediately employed another sorcerer. This person was frequently called a *faatre,* causing to move or slide, who, on receiving his fee, was generally desired, first to discover who had practised the incantations which it was supposed had induced the sufferings : as soon as he had accomplished this, he was employed, with more costly presents, to engage the aid of his demons, that the agony and death they had endeavoured to inflict upon the subject of their malignant efforts, might revert to themselves—and if the demon employed by the second party was equally powerful with that employed by the first, and their presents more valu able, it was generally supposed that they were successful.

How affecting is the view these usages afford, of the mythology of these rude untutored children of nature! How debasing their ideas of those beings on whom they considered themselves dependent, and whose services they regarded as the principal business of their lives!—how degrading and brutalizing such sentiments, and how powerful their effect must have been, in cherishing that deadly hatred which often found but too congenial a home in their bosoms! They were led to imagine that these super-human beings were engaged in perpetual conflict with each other, employing their dreadful powers, at the instigation of their priests, in afflicting with deepest misery, and ultimately destroying, the devotees of some rival demon.

A mythology so complicated, and a system of idolatry so extensive and powerful, as that which prevailed in the South Sea Islands, led the people not only to consider themselves as attended and governed by the gods, but also induced them to seek their direction, and submit to their decision, in every event of interest or importance. Every island had its oracle; and divination, in various forms, was almost universally practised by the priests.

In many respects, the oracles of the Polynesians resembled those of the ancients; in some they differed. Oro, the great national idol, was generally supposed to give the responses to the priests, who sought to know the will of the gods, or the issue of events; and Opoa, being considered as the birth-place of this god, was among the most celebrated oracles of the people. It does not appear that there were any persons specially appointed to consult the gods. The priest, who officiated in other services, presented the offerings, and proposed the inquiries of those who thus sought supernatural direction.

No event of importance was determined, nor any enterprise of hazard or consequence undertaken, without, in the first instance, inquiring of the gods its result. The priest was directed, as they expressed it, to spread the matter before the idol, and to wait the intimation of his will, or the prediction of its consequences. The priest, who was called *taura*, or *tairoiro*, repaired to the temple, presented the offerings, and proposed the inquiry, while the parties by whom he was employed anxiously waited his report.

In the Sandwich Islands, the king, personating the god, uttered the responses of the oracle, from his concealment in a frame of wicker-work. In the southern islands, the priest usually addressed the image, into which it was imagined the god entered when any one came to inquire his will. Sometimes the priest slept all night near the idol, expecting his communication in a dream; at other times it was given in the cry of a bird, whose resort was in the precincts of his temple; in the sighing of the breeze among the entwining branches of the tall and slender trees around the temple; or in the shrill, squeaking articulations of some of the priests. When the priest returned to those by whom he had been employed, if an unfavourable answer had been given, the project was at once abandoned, however favourable other circumstances might appear. If the answer was propitious, arrangements were forthwith made for its prosecution; but if no answer had been given, no further steps were then taken, it was considered to be restrained by the idol, and was left in abeyance with him.

Appearing to the priest in a dream of the night, though a frequent, was neither the only nor the principal mode by which the god intimated his will. He frequently entered the priest, who, inflated as it were with the divinity, ceased to act or speak as a voluntary agent, but moved and spoke as entirely under supernatural influence. In this respect there was a striking resemblance between the rude oracles of the Polynesians, and those of the celebrated nations of ancient Greece.

As soon as the god was supposed to have entered the priest, the latter became violently agitated, and worked himself up to the highest pitch of apparent frenzy, the muscles of the limbs seemed convulsed, the body swelled, the countenance became terrific, the features distorted, and the eyes wild and strained. In this state he often rolled on the earth, foaming at the mouth, as if labouring under the influence of the divinity by whom he was possessed, and, in shrill cries, and violent and often indistinct sounds, revealed the will of the god. The priests, who were attending, and versed in the mysteries, received, and reported to the people, the declarations which had been thus received.

When the priest had uttered the response of the oracle, the violent paroxysm gradually subsided, and comparative composure ensued. The god did not, however, always leave him as soon as the communication had been made. Sometimes the same *taura*, or priest, continued for two or three days possessed by the spirit or deity; a piece of peculiar native cloth, worn round one arm, was an indication of inspiration, or of the indwelling of the god in the individual who wore it. The acts of the man during this period were considered as those of the god, and hence the greatest attention was paid to his expressions, and the whole of his deportment.

In the year 1808, during the civil war between the king and rebel chiefs, of whom Taute was the leader, the priest of Oro, who was known to be not only attached to the king's interests, but a personal friend of Pomare, left the royal camp, and went over to that of the enemy. Many of Pomare's friends endeavoured to persuade him to remain with them, but no one dared to use force, as it was supposed that he acted under the inspiration of the Oro. This circumstance greatly discouraged the king and his friends, and probably prepared the way for their discomfiture, and retreat from the island, as they supposed the god had forsaken them, and fought with their enemies.

On another occasion, of more recent date, the god and the prophet were not treated with quite so much respect, but were rather rudely handled. The natives of Rurutu having determined to renounce idolatry, it was proposed by the native teachers that the people should meet together at the sacred enclosure, near the idol temple, where both sexes would unitedly partake of those kinds of food which had heretofore been regarded as sacred, and the eating of which by any female, especially in such a place, the gods would have punished with death.

At a previous meeting, Auura, one of the chiefs, had told a priest, who pretended to be inspired, that he was the very foundation of the deceit, and that he should never deceive them again. The priests, however, appeared at the appointed meeting; and one of them, pretending to be inspired, began denouncing, in the name of his god, the most awful punishment upon those that had violated the sacred place. One or two of the natives of Raiatea went up to him, and told him to desist, and not attempt to deceive them any longer, that the people would not tolerate their imposition. The priest answered, that it was the god that was within him, and that he was the god. When *uruhia*, (under the inspiration of the spirit,) the priest was always considered as sacred as the god, and was called, during this period, *atua*, god, though at other times only denominated *taura*, or priest. Finding him determined to persist in his imprecations, one of the christian boatmen from Raiatea said, "If the god is in, we will try and pinch, or twist, him out." Immediately seizing the priest, who already began to shew symptoms of violent convulsive muscular action, they prevented his throwing himself on the ground. For a long time, the priest and one of the Raiateans struggled together; when the god, insulted at the rude liberty taken with his servant, left him, and the priest in silence retired from the assembly.

When one of the priests was exhibiting all the violent gestures of inspiration in Huahine, a bystander observed, that it was all deceit, and that if they were to open the body of the priest, they should not find any god within. The multitude, however, appeared struck with horror at the startling proposal, and seemed to think the individual who had dared to utter it would not escape the signal vengeance of the powerful spirit.

Although so much ceremony, and such extraordinary effects, attended the public or formal intercourse between the god and the people, through the medium of the priest, the communications between the priest and the god were sometimes of an opposite character, and ludicrously colloquial. Mr. Davies, when itinerating round the island of Eimeo, in the early part of his missionary labour in that island, arrived at a village near Tiatae-pua, where he endeavoured to purchase provisions from the inhabitants. Vegetables were procured with facility, but the only animals were a number of fowls, and these belonged to the priest of the adjacent temple. Application was made to this individual, who looked at the articles (scissors, looking-glasses, &c.) offered in exchange, and seemed desirous to barter his fowls for them, but he said they belonged to the god, having been presented as offerings, and that without his leave he dare not part with any.

Again he examined the articles, and then said he would go and ask if the god was willing to part with any of the fowls. He proceeded to the temple, whither he was followed by Mr. Davies, who heard his address to the object of hope and fear, in words to the following effect:—"O my atua, or god, here is some good property, knives, scissors, looking-glasses, &c. *e hoo paha vau, na moa na taua;* perhaps I may sell some of the fowls belonging to us two, for it. It will be good property for you and me." After waiting a few moments, he pretended to receive an answer in the affirmative, and returned, stating that the god had consented to the appropriation. The sacred fowls were accordingly hunted by a number of boys and dogs, and several secured, and sold for the above-mentioned articles.

The oracle was not the only method by which the people were accustomed to consult the gods; nor was the inspiration of the priests the exclusive manner by which supernatural direction was revealed to the people. Divination, or augury, was practised in a variety of modes, and by these means it was thought that future events were made known, and information was communicated. Much of their augury was connected with the sacrifices they offered. The diviners noticed the manner in which the victims died, their appearance after death, &c. and by these means determined what was the will of the god.—They had also a singular method of cutting a cocoa-nut, and, by minutely examining its parts, of ascertaining their portentous indications. These ceremonies were generally practised in the temple.

There were others, however, performed elsewhere, as the *patu,* which consisted in dividing a ripe cocoa-nut into two equal parts, taking the half opposite to that to which the stalk was attached, and proceeding with it in a canoe to some distance from the shore; here the priest offered his prayers; and then placing the cocoa-nut in the sea, continuing his prayers, and narrowly watching its descent, he thereby pretended to ascertain the result of any measures in which those by whom he was employed were interested. The patu was frequently resorted to while negociations for peace were carried on between parties who had been engaged in war. Divination was employed to discover the cause or author of sickness, or to ascertain the fate of a fleet or a canoe that might have commenced a distant or hazardous voyage. This latter was often used in the islands to the westward of the Society group.

The natives had also recourse to several kinds of divination, for discovering the perpetrators of acts of injury, especially theft. Among these was a kind of water ordeal. It resembled in a great degree the *wai harru* of the Hawaiians. When the parties who had been robbed wished to use this method of discovering the thief, they sent for a priest, who, on being informed of the circumstances connected with the theft, offered prayers to his demon. He now directed a hole to be dug in the floor of the house, and filled with water; then taking a young plantain in his hand, he stood over the hole, and offered his prayers to the god, whom he invoked, and who, if propitious, was supposed to conduct the spirit of the thief to the house, and place it over the water The image of the spirit, which they imagined resembled the person of the man, was, according to their account, reflected in the water, and being perceived by the priest, he named the individual, or the parties, who had committed the theft, stating that the god had shewn him the image in the water. The priests were rather careful how they fixed upon an individual, as the accused had but slight prospect of escaping, if unable to falsify the charge; but when he could do this, the credit of the god and the influence of the priest were materially diminished.

Sometimes the priest, after the first attempt, declared that no answer had been returned, and deferred till the following day the repetition of his enchantments. The report, however, that this measure had been resorted to, generally spread among the people, and the thief, alarmed at the consequences of having the gods engaged against him, usually returned the stolen property under cover of the night, and by this superseded the necessity for any further inquiries.—Like the oracles among the nations of antiquity, which gradually declined after the propagation of Christianity, the divinations and spells of the South Sea Islanders have been laid aside since their reception of the gospel. The only oracle they now consult is the Sacred Volume; and multitudes, there is reason to believe, give to its divine communications unreserved credence, and yield to its requirements the most conscientious obedience.

CHAP. IX.

Increased desire for books—Application from the blind—Account of Hiro, an idolatrous priest—Methods of distributing the Scriptures—Dangerous voyages—Motives influencing to desires for the Scriptures—Character of the translation—Cause of delay in baptizing the converts—General view of the ordinance—Baptism of the king—Preparatory instructions—First baptism in Huahine—Mode of applying the water—Introduction of Christian names—Baptism of infants—Impressions on the minds of the parents—Interesting state of the people—Extensive prevalence of a severe epidemic.

A NUMBER of elementary books, and several hundred copies of St. Luke's Gospel, printed at Eimeo, and reserved for the Leward Islands, had been distributed among the people. But these were soon found inadequate to meet their daily increasing wants; and the great desire of all classes for books, furnished a powerful stimulus to hasten the printing, and we were soon enabled to furnish a supply of spelling-books.

I have often been amused with the perseverance and ingenuity manifested by the people to procure books, or at least a substitute for them. The bark of the auti, or paper mulberry, was frequently beaten to a pulp, spread out on a board, and wrought and dried with great care, till it resembled a coarse sort of card. This was sometimes cut into pieces about the size of the leaves of a book; and upon these, with a reed cut in the shape of a pen, and immersed in red or purple vegetable dye, the alphabet, syllabic, and reading lessons of the spelling-book, and the Scripture extracts usually read in the school, have been neatly and correctly copied. Sometimes the whole was accurately written on one broad sheet of paper—like native cloth, and, after the manner of the ancients, carefully rolled up, except when used. This was often the only kind of book that the natives in remote districts possessed; and many families have, without any other lessons, acquired a proficiency, that has enabled them to read at once a printed copy of the Scriptures. It has also gratified us, as indicative of the estimation in which the people held every portion of the word of God, and their desire to possess it, to behold them anxiously preserving even the smallest piece of paper, and writing on it texts of the Scripture which they had heard in the place of worship.

These detached scraps of paper containing the sacred texts, were not, like the phylacteries of the Jews, bound on the forehead, or attached to the border of the garment, but carefully kept in a neat little basket. The possessor of such an envied treasure might often be seen sitting on the grass, with his little basket beside him, reading, to his companions around, these portions of the scripture. I have a number in the hand-writing of the natives, some of which they have brought, to have them more fully explained, or to inquire what connexion they bore to parts with which they might be better acquainted.—Their use was, however, superseded by the printing of the Gospel of St. Matthew, an edition of upwards of two thousand copies of which was finished in less than eighteen months after our arrival in Huahine.

The people were anxious to receive them, and multitudes thronged the place where they were preparing, for some time before they were ready. The district of Fare presented a scene strongly resembling that which Afareaitu had exhibited when the first portion of the sacred volume was printed there; and many said they could not sleep, from the apprehension of not obtaining a copy. It was not easy to distribute them to the greatest advantage, and we determined to give a copy to none but such as could read; but so importunate were many, that we could not abide by our resolution. Sometimes those who were scholars induced their chiefs to apply for a number of copies, guaranteeing their payment, and their suitable appropriation. From this representation, many were given to the different chiefs; but we found it desirable afterwards, in order to insure the most advantageous distribution, to give only to those who we ourselves were satisfied could read.

Several blind persons applied at the different stations, earnestly soliciting books, stating, that though they could not read, they could hear and remember as well as those who could see. To have denied to those suffering natural darkness the means of obtaining spiritual light, when we had every reason to believe they were sincere in their expression of desire for it, would have been cruel; and we rejoice in having been honoured of God to communicate the gospel, as the servants of Him who—

> ——" from thick films shall purge the visual ray,
> And on the sightless eye-ball pour the day.

It is a most pleasing fact, that, in the South Sea Islands, a number of blind persons have not only had their understandings enlightened by the preceptive parts of Scripture truth, but that to many it has proved "the light of life," more valuable than natural light, as the soul is more precious than the body, and eternity more important than time. Some have died, and we have reason to believe have entered those realms of day, where night and darkness are unknown.

One remarkable instance occurred during the year in which I left the islands. The native name of the individual to whom I allude was *Hiro.* He was the priest of one of the principal temples of Parea, in the lesser peninsula of the island, or *Huahine iti.* He was a priest of Hiro, the god of plunderers and thieves, and, in perfect accordance with the spirit of his office, was the captain or leader of a band of robbers, who spread terror through the surrounding country. He was one of the first and most determined opposers of Christianity in Huahine; reproaching its adherents, defying the power, and disclaiming the authority, of its Author. But, like Saul of Tarsus, he found it hard to resist.

He was in the prime and vigour of manhood, being at the time between thirty and forty years of age. When the number of Christians increased in his neighbourhood, and the Sabbath-day was first publicly observed, in order to shew his utter contempt of christian institutions, he determined to profane that day "in defiance of Jehovah." He repaired for this purpose to the grounds in the neighbourhood of the temple, and engaged in erecting a fence; but while thus employed, his career of impiety was suddenly arrested. The twig of a tree came in contact with his eyes; almost instant blindness followed; and he was led home by his affrighted companions, who considered it a visitation from the Almighty.

I had frequent interviews with him afterwards, one in the precincts of his own temple, which I visited in company with Messrs. Bennet, Tyerman, and Barff. I have already mentioned it. His spirit was subdued; he subsequently became a humble, and, we trust, sincere disciple of that blessed Redeemer whom he had persecuted. He died trusting in the merits of Christ for acceptance with God the Father. The history of the conversion of the great apostle to the Gentiles interested and affected him much; and though the scales on his bodily eyes were not removed, but his blindness continued until his death, such was the impression which analogy of circumstances produced, that when he presented himself for baptism, he desired to be called *Paul.*

Other instances of spiritual illumination, equally pleasing, now exist both in the Society and Sandwich Islands, in reference to individuals suffering one of the most distressing and hopeless privations to which humanity is exposed. Some of our most interesting conversations with the natives have been with such. "My eyes," said a blind man one day to Mr. Williams, "behold no attractive objects when I am engaged in prayer, or hearing the word of God; and yet my heart wanders, and my thoughts are often engaged on other subjects. My eyes see not another man's property, &c.; and yet, when I hear it spoken of, my heart covets it. The objects that tempt others to sin, are unseen by me; but my imagination creates objects of sin, which often occupy my thoughts."

The experience of Bartimeus Lalana, a native of the Sandwich Islands, is also remarkably interesting and satisfactory. Blindness is not more common among the Polynesians than with the inhabitants of other countries; yet there are numbers of aged persons who have lost their sight; and the influence of that sympathy which this affliction always awakens in a Christian bosom, is now excited in the natives themselves, though formerly the blind were objects of neglect and ridicule. There is now connected with the Missionary station at Bunaauïa, or Burder's Point, a blind man, who can repeat correctly half the Gospel by John, though it has not long been printed.

When we have been distributing the Scriptures, two or three fine boys or girls have come, begging for copies, though they could not read; assuring us, they were learning; and, when they have failed, they have entreated that we would write their names on the books, and reserve them till they were able to read. To our satisfaction, in this request they have often been joined by their parents, who have offered payment for the copies. We have usually complied with their wishes, and have witnessed the most entire confidence on their part, as it regarded the ultimate accomplishment of their wishes, when once their names have been written.

It was necessary to select some public place for the distribution of the books; the school-room was fixed upon, and, on the day appointed, the place was actually thronged until the copies were expended. In their application at our own houses, we found it impossible to restrain the people; they filled our yards and gardens, and thronged every window, sometimes to such a degree, that one of the Missionaries, Mr. Bicknell, found it necessary to fasten the lower doors and windows of his house, and retire to the chamber. The natives then procured long bamboo-canes, and, fastening their measure of oil, the price of the book, to one end, lifted it up to the window. Mr. Bicknell was so influenced by the ingenuity and determination of the contrivance, that he distributed a number of copies, by fixing them in a slit or notch in the end of the cane presented at his window.

When the edition issued from the press in Huahine, the relative proportion for Raiatea, Tahaa, and Borabora, was sent to the Missionaries residing with the people; but the supply was too small, and numbers of the disappointed individuals, supposing they should find a greater abundance at Huahine, came, when the wind was fair, twenty or thirty miles in their canoes, several of which were such small and fragile barks as quite astonished us. I was really surprised at the temerity of the individuals who had committed themselves to the mercy of the waves of the largest ocean in the world, in the hollowed trunk of a tree, twelve or twenty feet long; the sides of which, when the men were in it, were not more than four or five inches above the surface of the water.

It would be too much to suppose that they were all influenced by the highest motives, in the desire they thus manifested for the sacred volume; but while some probably sought it only as an article of property in high and general esteem, others were undoubtedly actuated by a conviction that it was able to make them wise unto salvation, through faith which is in Christ Jesus.

The intensity of ardour manifested by many at first, has, as might be expected, subsided: yet still the Scriptures are earnestly sought, and highly prized, by a great portion of the adult population.

The whole of the New Testament has been translated and printed, not indeed in a uniform volume, but in detached portions, which many of the natives have bound up together. Separate portions of the Old Testament have also been translated, and some of the books are printed; it is to be hoped that a uniform edition of the

Bible will, at no very distant period, be circulated among the people. Whether or not any of the Apocryphal books will ever assume a Polynesian dress, it is impossible to say, but at present it is improbable.

The dialects spoken by the tribes inhabiting the different groups in the South Sea, being strictly analogous to each other, it was hoped that the Tahitian translation of the Scriptures would have answered for the whole; there is, however, reason to fear that distinct translations will be necessary, not only for the Sandwich Islands, the Marquesas, and Tongatabu, but also for the Harvey Islands, which are not more than 600 or 700 miles distant from the Society Isles. So strong a resemblance, however, exists between the dialects, that the Tahitian translation will require only slight variations, the idioms and structure of the language being, in all their distinguishing features, the same.

When the uncultivated nature of the language, into which the Scriptures have been translated, is considered, connected with the remembrance that it is only by the labours of the Missionaries that it has been within the last few years reduced to a system, and employed in a written form, it cannot be expected that these books, more than any other first translations, should be altogether faultless. The knowledge of the Missionaries themselves in the language, notwithstanding thirty years' attention to it, is constantly increasing; and, compared with future translations which their successors or well-educated natives may make the present will perhaps appear imperfect. Nevertheless, from the qualifications of the translators, their unquestionable integrity, and united patient attention to the preparation of every work, I believe the only imperfections that may be found, will refer to minor points of style in idiom or language Some of the Missionaries excel in acquaintance with the original languages, others with the native dialect, and every copy is inspected by all, before going to the press.

The year 1819 is also distinguished in the annals of the South Sea Islanders, by the administration of the rite of baptism to the first Christian converts in the islands Pomare and others made a profession of Christianity in 1813; names were written down; the change became general during the same year; persecution raged with violence in 1814; the inhabitants of Tahiti and Eimeo embraced the gospel in 1815, and those of the remaining group in 1816; and it certainly appears singular that none should have been baptized until 1819. This delay, however, did not arise from any doubts in the minds of the Missionaries as to the nature of the ordinance itself, the proper subjects of it, or the manner in which it was to be administered; on all these points they were agreed. It arose from a variety of circumstances, peculiar in their kind, local in their influence, and such as they could neither foresee nor control.

At first, their continuance and their existence were very uncertain, in consequence of the efforts of the idolaters, and the war that followed; afterwards the conduct of the king, who, on his first profession, they would not have hesitated to baptize, was such, as to induce them to fear that his baptism would injure the Christian cause among the people; and subsequently, as they were on the point of separating and forming distinct stations, it was thought best to defer it till they should have entered upon the fields of their permanent labour, where they hoped to gather around them congregations of converts, administer the rite of baptism, and form Christian churches.

The Missionaries considered the proper subjects for the ordinance to be those who professed their faith in Christ, and the children of such individuals: but considerable difficulty was experienced in determining what the moral or religious qualifications of the adults ought to be, and the connexion that should exist between their baptism and admission to the sacred communion. Although we read different authors on the subject, their views were seldom altogether adapted to our circumstances, and I believe we derived but little real assistance from any.

We desired to bow only to the authority of scripture, and to follow implicitly its directions. We considered our circumstances by no means dissimilar to those of the individuals for whose guidance those directions were primarily given. Having the commission of our Lord to his disciples for our warrant, and the conduct of his apostles in the execution of it for our model; we hope we have been enabled to proceed according to the divine will, and in such a manner as to secure the approbation of the Christian churches by which we had been sent to preach Christ among the gentiles. Our situation at this time was regarded as most critical, and our procedure in this respect such, as it was presumed would have an important bearing on future generations.

Happily, however, for us, and for all placed in similar circumstances, the terms of the commission are unequivocal and explicit; and we could not but perceive, that by the same warrant, in virtue of which we preach the gospel, and, as the word is rendered in the Tahitian, *proselyted* those among whom we laboured, we were also bound to baptize in the name of the Father, and the Son, and the Holy Ghost. The intimate connexion between the administration of this rite by the apostles, and the reception of the gospel on the part of those to whom they preached, also convinced us of the design of our Lord, that it should follow the belief in the testimony concerning him, which we were commissioned to deliver. Hence, it was regarded as our duty to baptize those who desired to become the disciples of Christ, as well as to instruct them concerning his will.

We did not apprehend that there was any spiritual virtue or efficacy connected with, or contained in, baptism, nor did we consider any spiritual blessings communicated by it, much less that most important of all—the one thing needful, a regeneration of the heart. It appeared designed, by the great Head of the church, to occupy that place in the dispensation of the New Testament, which circumcision did in that of the Old. The acts of desiring and receiving baptism, on the part of the subject of it, were viewed as a public and solemn renunciation of paganism, and a declaration of discipleship with Christ; and the circumstance of baptism was regarded as constituting the grand, public, and open line of demarkation, between the idolatrous, and the thus separated or christian portions of the community. While we thus felt ourselves bound to baptize those who, like the Ethiopian eunuch, and those to whom Philip preached in Samaria, professed their belief in the Saviour, and the grand truths of the Christian system, we also felt that it was desirable to receive suitable evidence of the sincerity of such profession.

As to the degree of evidence that should be required, there was a very considerable difference of opinion. A few of our number supposed that no adults should receive this initiatory rite, but such as, there was every reason to believe, were regenerated persons; and that a general belief in the testimony that Christ was the Saviour of men, and a desire to receive farther instruction, however sincere it might be, should be accompanied with an experience of that change of heart, which these truths, under the special influences of the Holy Spirit, are adapted to produce; and, in short, that such only should be baptized as would be at once unhesitatingly admitted to the Lord's supper.

The majority, however, of the Missionaries were of opinion that the ordinances were totally distinct, and that though it was proper that every church member should have been baptized, yet it did not follow that every one who had received such rite was thereby admitted to church fellowship. Satisfactory evidence of sincerity in belief that Jehovah was the true God, and Jesus Christ the only Saviour, was considered a sufficient warrant for its administration to those who required it.

No one, however, at any time desired to exercise undue influence over the opinions of his coadjutors; and, although uniformity was desirable, we did not think it important to sacrifice much for sameness of sentiment or practice in this respect. After repeated and prayerful deliberation, recognizing, and aiming to act, upon the broad and liberal principles upon which the institution, under whose patronage we laboured, was founded, it was mutually agreed that each Missionary should, in his own station, pursue that course which appeared to him most in accordance with the declarations of scripture.

In two of the stations, or perhaps three, the Missionaries have baptized those only whom they had reason to believe had been baptized by the Holy Ghost, and were Christians in the strictest sense of the term; the children of such persons they also baptized. In the other stations, the Missionaries have administered this rite to all whom they had reason to believe sincere in profession of discipleship, without requiring evidence of their having experienced a decisive spiritual change. In this respect some slight difference prevailed, but on every other point there has been perfect uniformity in their proceedings.

The first public baptism that occurred in the islands took place in the Royal Mission Chapel at Papaoa, in Tahiti, on the 16th of July, 1819. Pomare, the king of the island, was the individual to whom, in the midst of what, but a few years before, had been a scoffing, ignorant, obstinate, cruel, and idolatrous nation, that rite was administered. It was the Sabbath-day. The congregation in the chapel, though less numerous than during the services of the previous week, amounted to between four and five thousand. The subject of discourse was appropriate, Matt. xxviii. 18—20. At the close of the sermons, the Missionaries gathered round the central pulpit; the ceremony commenced with singing. Mr. Bicknell, one of the Missionaries who had arrived in the Duff, implored the Divine blessing, and then, assisted by Mr. Henry, the only other senior Missionary at Tahiti, poured the water on his head, baptizing him "in the name of the Father, and the Son, and the Holy Ghost." The venerable Missionary then addressed the king, not without agitation, yet with firmness, "entreating him to walk worthy of his high profession, in the conspicuous station he held before angels, men, and God himself." Mr. Henry addressed the people, and Mr. Wilson implored the Divine benediction, that what had been done on earth might be ratified in heaven.

Although the subsequent conduct of Pomare was a matter of the deepest regret to his best friends, yet there was something in the ceremony unusually imposing; and the emotions associated with it, must have been intense and interesting, especially to the two elder Missionaries, who had performed the rite. He had been identified with the chief events of their lives; upwards of two and twenty years had rolled by since the providence of God first brought them acquainted with him on the shores of Matavai; and in connexion with that interview which memory would, probably, present in strong and vivid colours on this occasion, they, perhaps, recollected the opinion formed of him, by the humane commander of the Duff, that he appeared the last person likely to receive the gospel. Yet amid the thickest darkness that had ever veiled their prospects, through him the first cheering ray of dawning light had broken upon them: he was their first convert; in every difficulty, he had been their steady friend; in every labour, a ready coadjutor; and had now publicly professed that his faith was grounded on that rock whereon their own was fixed, and his hopes, with theirs, derived from one common source. What intense and mingled hopes and fears must have pervaded their hearts! what hallowed joy must they have felt in anticipation of his being with them an heir of immortality, chastened with appalling, and not ungrounded fears, that after all he might become a cast-away.

Numbers, both adults and children, were subsequently baptized in the Windward Islands, but it was not until some months after, that the ordinance was dispensed to any in the Leeward or Society group.

It was in Huahine that the first, from among those who had renounced paganism in the Leeward Islands, were thus initiated into the outward church of Christ. Huahine was a new station, and few of the inhabitants, when we landed, knew much more of Christianity than its name. Fifteen months had elapsed since our arrival, and during that period, we had made the doctrines and general precepts of the gospel the topics of our discourses, among a people who had every thing to learn. Many of them now came forward, declaring their desire to become altogether the disciples of the Saviour, to make a public profession of faith in him by baptism, and to seek instruction in all his will. We found that, had we been so disposed, we could no longer defer the rite, with regard at least to some of those who applied.

Anxious that it should be on their part a reasonable act, and that, before being received, it should be understood, we proposed to meet one afternoon every week, with those who desired to be baptized. At this meeting we endeavoured to instruct them in the nature, origin, design, and subjects of the ordinance, together with the duties of those who should receive it. There was no wish on our parts to baptize by stratagem, as some of the popish Missionaries have done, but we sought to make the people well acquainted with the matter in all its bearings.—At the first weekly assemblies, between twenty and thirty of the most promising of the converts attended, afterwards the numbers exceeded four or five hundred.

In the instructions given, the Scriptures, and the Scriptures only, were our guide; and we endeavoured to inculcate the doctrine as we found it there, and as if it had never been controverted. Our warrant for its adminis tration we derived from our Lord's commission to the first Missionaries, which was also our own. In its nature, we instructed them not to consider baptism as possessing any saving efficacy, or conferring any spiritual benefit, but being on our parts a duty connected with our office, and on theirs a public declaration of discipleship or proselytism to the Christian faith; designed to teach all, their moral defilement in the sight of God, and their need of that washing of regeneration, and spiritual purification, which it figuratively signified.

The duties of those who desired it were also inculcated, and the necessity that existed not only for their renunciation of every open idolatrous practice, and attention to instruction in the principles, but a deportment accordant with the precepts of Christianity, and the conspicuous situation in which this very act would place them, before those by whom they were surrounded. We also informed them, that it appeared to us from the Scripture, that the ordinance was designed for believers and their children, and therefore directed that, as they desired them to be brought up in the Christian faith, they should dedicate them to Jehovah by baptism. It was found necessary, at the same time, plainly to caution them against supposing there was in baptism any thing meritorious, or on account of which they would receive any special blessing from God, other than that which would follow general obedience to his word. This was the more requisite, as there was reason to apprehend, that from the influence of a system in which strict observance of rites and ceremonies, without regard to motive or moral character, was all that was necessary, they might rest satisfied with having received the mere external declaratory rite. We also endeavoured carefully to avoid holding out any prospect of distinction, or temporal advantage, as an inducement to the people to apply for baptism, but constantly and plainly represented its observance as only an act of obedience to Him whom they professed to desire for their Master and their Lord, and who had promised that his people should be baptized with the Holy Ghost.

This weekly meeting was designed to answer another purpose, that of affording us the means of judging of the sincerity of the candidates, as well as of imparting to them necessary instruction. After several months had been occupied in devoting one afternoon in the week to their instruction, it was deemed proper to baptize a number of candidates, and two of their children.

It was now necessary to determine upon the mode: this had never appeared to us the most important part of the matter. We should not have objected to immerse any individuals who had themselves desired it. But as the Scriptures are not decisive on this point, and though it is stated that Philip and the Eunuch went down to the water, or into the water, yet it was not in this act, but in the application of water in the name of the Trinity, that we considered baptism to consist: in such application, it is not stated that the Eunuch was immersed. Hence, we did not explain this, or other passages of similar import, as signifying immersion—and consequently the converts did not desire it. But had one of our own number thought it right to have administered this rite by immersion, I do not think we should have said he acted wrong in so doing. In this respect, however, there was no difference of opinion, and consequently a perfect uniformity of practice prevailed. With regard to the other modes, we did not think it was very material whether we poured or sprinkled the element upon the individual.

The 12th of September, 1819, was fixed for the baptism of the first converts in Huahine. It was also the Sabbath. A suitable discourse was delivered in the morning to a numerous congregation, who thronged the chapel. Mr. Davies, being the senior Missionary at the station, officiated, assisted by Mr. Barff and myself.

The climate in the South Sea Islands is remarkably fine, the weather warm, the streams abundant, and the waters clear as crystal; and, had we been disposed to perform the service in the open air, under the shade of a spreading grove, we had every facility for so doing. The converts might have been led into the river, and, standing on the bank or in the stream ourselves, we might have applied its waters to their persons, using the words prescribed. On such occasions, the most delightful scenes of which it is possible for imagination to conceive, would have been presented; scenes similar perhaps to those often witnessed by the disciples in the days of the apostles; and for the sake of effect, and the associations they would have awakened, I have sometimes for a moment wished we had. But the wish has only been momentary, for whatever might have been the impression of such a scene, or the emotions enkindled, they would not have been attended with any valuable practical result. On the present, therefore, and every subsequent occasion, the rite was administered before the whole congregation, in the place of worship.

During the ordinary morning worship, the approved candidates sat in front of the pulpit. At its close, they kept their places, and, after imploring the divine blessing upon the service, we proceeded to its performance. Their profession of faith in Christ, and desire to be instructed in his word, had been received at a preceding meeting; and it was only necessary now, after a short address to the whole, to ask the name of each adult, and the parents the names of their children.

This, Mr. Davies did,—beginning with Mahine, the principal hereditary chieftain of the island. Having received his reply, Mr. Davies immersed his hand in a vessel of water, which Mr. Barff or myself held by his side, and then holding his hand over the crown or forehead of the chief, while the water from his hand flowed or fell upon Mahine's head, Mr. Davies pronounced aloud, with distinctness and solemnity, *Mahine e tapape du vau ia oe i te ioa o te Medua, eo te Tamaidi, eo te Varua maitai:* "Mahine, I apply water to you in the name of the Father, and of the Son, and of the Holy Ghost." Repeating the same words, and applying the water in the same manner, to every individual, he proceeded to baptize the whole number, who kept their seats during the ceremony.

Mahine was not baptized first, because he was the king of the island, but because he was one of the earliest converts, and had been most diligent in his attention, and consistent in deportment. We were careful to avoid giving any preference to rank and station, simply as such; and, on the present occasion, we beheld Hautia, the governor of the island, and others of high rank, sitting by the side of the humble peasants of the land. In reference to civil or political station, we always inculcated the requirements of the gospel, that all should render honour to whom honour is due, invariably presenting a suitable example of the most respectful be haviour to individuals of distinguished rank or station. But in the church of God, and in the participation of the privileges of Christianity, we as invariably taught that all were brethren, that there was no precedence derived from worldly station, that one only was our lord and king, the Saviour himself. This principle we were happy to see recognized by themselves on this occasion, as some of the principal chiefs sat at the lowest end.

The word *tapape*, used in the first instance, was that which appeared the most suitable, as we were anxious to divest the rite of every thing extraordinary or mysterious. The signification of the word is to apply water, without expressing the precise mode of application. They have no word answering to the term *baptize*, as now understood in the English language, though they have distinct words for sprinkling, pouring, bathing, &c., but we considered the simple application of water to approach nearer to the original word *baptisto*, than either of these; and it seemed so appropriate, as to render it unnecessary to introduce any other. Subsequently, however, our opinions changed, and we adopted the original word, which in Tahitian is written *ba-pa-ti-zo*, and used only to signify this sacred rite.

The water was not sprinkled on the face with force, the sign of the cross was not made, nor was water poured on the head from any vessel, but, taking one hand from the vessel containing the water, and holding it over the individual, we allowed so much water as was held in or attached to the hand, to fall upon the crown or forehead of the baptized, pronouncing, at the same time, the name, and the words prescribed in the Gospels.

Some difficulty was experienced with regard to the names, as many of the natives, especially the chiefs, have a number; some of office, others hereditary, and not a few intimately connected with their former idolatry, or its abominable institutions. It was not thought desirable that they should assume a new name on receiving baptism, or that it should interfere with any name of office, station, or hereditary title, that might appertain unto them. But every blasphemously idolatrous or impure name, (and those of some of the Areois and priests were so to a most affecting degree,) we recommended should be discontinued, and that they should select those names, by which, in future, they would wish to be designated. A few of the adults chose foreign, and in general scriptural names, for themselves or their children.

This produced a considerable change in their language. Formerly, all names were descriptive of some event or quality—as Fanauao, day-born, Fanaupo, night-born, Mataara, wakeful or bright-eyed, Matamoa, sleepful or heavy-eyed, Paari, wise, or Matauore, fearless, &c. A number of terms were now introduced, as Adamu Adam, Noa Noah, Davida David, Ieremia Jeremiah, Hezekia Hezekiah, Iacoba James, Ioane John, Petero Peter, &c. with no other signification than being the names of the persons. With regard to infants, we only baptized those whose parents, one or both, were themselves baptized, and desired thus to dedicate their children to God, and engaged to train them in the principles of Christianity; and then we only baptized infants, unless the children of more advanced years understood the nature of the ordinance, and themselves desired to make, by this act, a public profession of their discipleship to Christ, and their wishes to be instructed in his word.

Sometimes the infant was held in the arms of its parent, who stood up while the rite was administered; at other times, and I believe invariably, during subsequent years, we have taken the child in the left arm, and baptized it with the right hand. Whenever any of our own children have been baptized, we have brought them to the chapel, and have performed the ceremony at the same time and in the same way as with the natives; that they might perceive that in this respect there was no difference between us.

The baptism of infants has certainly been among the most inteiesting religious exercises in which we have engaged. It was generally performed after morning service on the Sabbath. We usually addressed a short and affectionate exhortation to the parents, enforcing their responsibility, and duty towards the dear children they were thus offering; not indeed as an innocent child was formerly offered in sacrifice to senseless idols, or to a cruel imaginary deity, but to be trained up in the nurture and admonition of that Divine Parent, who has said, "I love them that love me, and those that seek me early shall find me."

I have been sometimes almost overwhelmed on beholding the intensity of mingled feeling, with which three or four sweet smiling infants have been brought by their respective parents to the rustic baptismal font. I have fancied, in the strongly expressive countenances of the parents, the lively emotions of gratitude, and the bright ray of hope and anticipated joy in the future progress of the child, when it should exhibit the effects of that inward change, of which this was the outward sign.

In strong and distressing contrast with sensations of this hallowed and delightful kind, I have supposed the memory of far different acts, in which, as parents, many of them had been engaged, has remained; I have supposed that recollection has presented the winning look of conscious innocence, which some dear babe has cast upon them, or the plaintive cry which from its lisping tongue first broke upon their ears, but which was unheeded, and they monstrously committed cool, inhuman murder—when they should have cherished the tenderest and softest sensibilities of the human bosom: I believe this has not been in my imagination only. The feeling depicted in the humane and Christian parent's countenance, suffused with tears, has often been an index of no common inward agitation. Subsequent conversation has confirmed the fact; and many have brought their children to present them unto God in baptism, who, while idolaters, had more than once or twice been guilty of the barbarous crime of infant murder. This practice is abolished; and, instead of shameless murder, or pagan sacrifice, the parents now delight to bring their infants to the house of worship, and thus dedicate them to God.

I have been often rather agreeably surprised at the anxiety of the parents to have their children baptized. Without inquiring into the origin of this solicitude, I believe it is not confined to the inhabitants of the South Sea Islands, and is certainly not unpleasant to behold. I recollect at one time the parents of three children came with considerable earnestness, and requested me to baptize their infants, rather earlier than I thought it should be done. It was not at Huahine, and the Missionary, under whose care the station was more particularly placed, was absent; I therefore proposed to defer it till his arrival. They pressed me not to decline; and one of them stated as a reason, that her child had been ill, and she was afraid it should die without having been baptized. "Suppose," I replied, "that it should, you know that the child will not lose thereby. No persons will be admitted to heaven simply because they have been baptized, nor will any be excluded therefrom merely because they have not." "Yes," answered the mother, "I know that; yet I do not feel satisfied now, but when it has been baptized, my mind will become easier." I could not reprove her; I endeavoured, however, to impress upon her mind the conviction, that the ordinance, though a duty, did not itself confer any spiritual benefit, and relieved her mind by informing her, that I would baptize the child at the close of the evening service.

In the preceding detail, I have, perhaps, been more prolix and minute than the importance of the subject may appear to demand; I have been influenced by a desire to give that information relative to our proceedings in this respect, to the friends of Missions in general, and to the patrons of the South Sea Mission in particular, to which, from the interest they have taken, and the support they have afforded, I have considered them justly entitled, and which I cannot but hope will be satisfactory.

Although I have only given the proceedings of one station, I believe that, with the exception of some of the Missionaries baptizing only such adults as they consider to be true Christians, and eligible for church fellowship, the procedure has been uniform in all. With us, those were baptized who made a credible profession of belief in Christ, and a desire to become his disciples, without any immediate view to church fellowship, which we consi dered a subsequent measure.

An address on the nature of baptism, and the duties of those who had received it, was printed after the first administration, and widely circulated, apparently with good effect. The weekly meeting for instructing those who desired baptism, was continued, and the first dispensing of that ordinance produced an astonishing effect upon the people. Multitudes, who had heretofore been indifferent, now appeared in earnest about religion, and the candidates soon amounted to four hundred. Those who had been baptized, also, in general attended the meetings.

A state of religious feeling, such as I never witnessed elsewhere, and equal to any accounts of revivals in America or other parts, of which I ever read, now prevailed, not only in Huahine, but in the other Missionary stations. The schools and meetings were punctually and regularly attended. The inhabitants of remote districts came and took up their abode at the Missionary settlement; and nothing could repress the ardour of the people in what appeared to us their search after the means of obtaining the Divine favour. Often have we been aroused at break of day, by persons coming to inquire what they must do to be saved—how they might obtain the forgiveness of their sins, and the favour of God; expressing their desires to become the people of God, and to renounce every practice contrary to christian consistency.

Many were undoubtedly influenced by a desire of baptism; that had introduced a new distinction, which, notwithstanding our endeavours to prevent, they probably thought must confer some temporal or spiritual advantage on those who received it. But with others it was not so, as the event has satisfactorily proved: many who at this time were awakened to an extraordinary religious concern, have ever since remained stedfast in their principles, and uniform in the practice of every christian virtue. We now felt more than ever the responsibility of our situation, and were afraid lest we should discourage, and throw a stumbling block in the way of those who were sincerely inquiring after God. Yet we felt no less apprehension lest we should be the means of encouraging desires, and cherishing the delusive hopes of such as were either deceiving themselves or others, and, under cover of seeking the favour of God, were actually pursuing that which they imagined would improve their temporal condition, or add to their respectability in society. Some who had been baptized, we found it necessary to admonish, lest they should rest satisfied with the attainments already made, and neglect the more important considerations.

In the interesting and critical duties now devolving upon us, we endeavoured to act with caution, taking the word of God for our directory, and bearing in mind at the same time the peculiar circumstances of the people; avoiding precipitancy in our public measures: so that, if we erred, it might be on the side of extreme carefulness. The everlasting welfare of the people was our only object; this we considered would not eventually suffer, whatever might be the effect of withholding baptism from those who might be proper subjects for it. But by administering this rite to those who sought it from improper motives, should it render them satisfied with the sign, instead of the divine influence signified, we might become accessary to their fatal delusion.

Under the influence of these impressions, we were perhaps led to defer the rite of baptism to those who applied for it, longer than we ought to have done; and I have known many who have been candidates upwards of one or two years. Their views of the doctrine have been in general correct, in their conduct there has been nothing unchristian or immoral, and they have uniformly expressed their desires to become the true disciples of Christ; but during that period we have not baptized them, merely because we have apprehended they did not feel the necessity of that purification of heart of which baptism is only the external sign. When we first administered that ordinance, we had no idea of the natives thronging in such numbers to receive it, and consequently had not deliberated on the term of that probation which we afterwards deemed it desirable to institute.

The same interesting state of the people by which the close of 1819 had been distinguished, marked the commencement of 1820. Never were our direct Missionary labours more arduous and incessant; and yet during no period of our residence there, were they more delightful. We beheld indeed the isles waiting for the laws and institutions of Messiah, and felt that we had been sent to a people emphatically prepared of the Lord, made willing in the day of his power.

The inhabitants of the remote districts which we had periodically visited, were many of them no longer satisfied with an opportunity for conversation on religious subjects once a week, but came and built their houses in the neighbourhood of Fare. We recommended those who remained, to do the same; and soon after the annual meetings in May, they so far complied as to render it unnecessary for us to visit these stations.

Our spacious chapel was opened in the latter end of April, on which occasion I read a translation of the sixth chapter of the second book of Chronicles, and afterwards preached from the sixth verse. Our Missionary meeting was remarkably well attended, and the subscriptions proportionably liberal; they amounted to between three and four thousand gallons of oil, besides cotton, and other trifling articles.

In the midst of this delightful state of things, the stations were visited with a distressing epidemic, which spread through the whole group of islands, and proved fatal to many of the people. It was a kind of influenza, affecting the lungs and throat; many attacked with it lost their voice. We suffered in common with the people, and I was obliged to relinquish all public duty for some weeks. This kind of calamity has been frequently experienced in the islands since they have been the resort of foreign shipping, though we are not aware that it prevailed before. A kind of dysentery appeared after the visit of Vancouver's ship, which called at the islands in 1790, and proved fatal to a vast portion of the population. In the year 1800, the Britannia, a London vessel, anchored at Taiarabu. Two seamen absconded, and a disease followed, less fatal, but very distressing, and more extensive, as scarcely an individual escaped.

These diseases have generally passed through the islands from the east to the west, in the direction of the trade winds. After the above appeared among the people, it was for some months confined to the Windward Islands; and so general was its prevalence, that Pomare one day said to Mr. Nott, "If this had been a fatal or killing disease, like that from Vancouver's ship, no individual would have survived."

As it began to subside, a canoe, called *Hareaino,* arrived from the Leeward Islands, and after remaining a week or two at Tahiti, returned to Huahine. Shortly after this, the people who had been in the canoe were attacked, and the disease ultimately spread as completely through this group, as it had through that at which the foreign vessel touched. Within the last two years, a disorder, in many respects similar to that left by the crew of Vancouver's vessel, has again swept through the islands, and carried off numbers of the people.

CHAP. X.

Former diseases in the islands comparatively few and mild—Priests the general physicians—Native practice of physic—Its intimate connexion with sorcery—Gods of the healing art—The tuabu, or broken back—Insanity—Native warm-bath—Oculists—Surgery—Setting a broken neck and back—The operation of trepan—Native remedies superseded by European medicine—Need of a more abundant supply—Former cruelty towards the sick—Parricide—Present treatment of invalids—Visits to Maeva—Native fisheries—Prohibitions—Enclosures—Salmon and other nets—Use of the spear—Various kinds of hooks and lines—The vaa tira—Fishing by torch-light—Instance of native honesty.—Death of Messrs. Tessier and Bicknell—Dying charge to the people—Missionary responsibility.

THE diseases formerly prevailing among the South Sea Islanders were comparatively few; those from which they now suffer are principally pulmonary, intermittent, and cutaneous. The most fatal are, according to their account, of recent origin. While idolaters, they were accustomed to consider every bodily affliction as the result of the anger of their gods; and the priest was a more important personage, in time of sickness, than the physician. Native practitioners who were almost invariably priests or sorcerers, were accustomed to apply such healing remedies as the islands afforded; and an invocation to some spirit or god attended the administration of every medicine. Tama, Taaroatuihono, Eteate, and Rearea, were the principal gods of physic and surgery. The former, in particular, was invoked for the cure of fractures and bruises.

From the gods the priests pretended to have received the knowledge of the healing art, and to them a part of the fee of the physician was considered to belong. No animal or mineral substances were admitted into their pharmacopœia; vegetable substances alone were used, and these simply pulverized, infused, heated on the fire, or with red-hot stones, and often fermented. Many of their applications, however, were very powerful, especially a species of gourd, or wild cucumber. A preparation, in which the milk of the pulp of the cocoa-nut was a principal ingredient, was sometimes followed by almost instant death. Mr. Barff once took this preparation, at the earnest recommendation of the people; but it nearly cost him his life, although he had not drank more than half the quantity prepared.

Frequently, when some medicines were about to be administered, the friends and relatives of the patient were sent for, that they might be at hand, should the effect be unfavourable. They often expected it would either save or destroy the patient. A number of ceremonies were connected with every remedy applied; and much greater dependence was placed on the efficacy of the prayers, than on the effect of the medicine.

When a person was taken ill, the priest or physician was sent for; as soon as he arrived, a young plantain-tree, procured by some members of the family, was handed to him, as an offering to the god; a present of cloth was also furnished, as his own fee. He began by calling upon the name of his god, beseeching him to abate his anger towards the sufferer, to say what would propitiate him, or what applications would afford relief. Sometimes remedies were applied at the same time, or the relatives sent to fetch certain herbs or roots, but the priest usually went himself to compound the *raau* or medicine: a considerable degree of mystery was attached to their proceedings, and the physicians appeared unwilling that others should know of what their preparations consisted. They pretended to be instructed by their god, as to the herbs they should select, and the manner of combining them. Different *raaus*, or medicines, were used for different diseases; and although they kept the composition of their nostrums a secret, they were not unwilling that the report of their efficacy might be known, in order to their being employed by others. Hence, when a person was afflicted with any particular disease, and the inquiry made as to who should be sent for, it was not unusual to hear it said—"*O ta mea te raau maitai no ia mai,*"—such a one has a good medicine for this disease.

The small-pox, measles, hooping-cough, and a variety of other diseases, to which most European children are subject, are unknown; yet they have a disease called *oniho*, which in its progress, and the effects on the face, corresponds with the small-pox, excepting that it is milder, and the inequalities it leaves on the skin soon disappear. There is another disease, somewhat analogous to this, resembling the species of erysipelas called shingles, for the cure of which the natives apply a mixture of bruised herbs and pulverized charcoal. Inflammatory tumors are prevalent; and the only remedy they apply, is a mixture of herbs bruised with a stone. Asthmatic, and other pulmonary affections also occur, and, with persons about the age of twenty, generally prove fatal.

Among the most prevalent and obstinate diseases to which, as a nation, they are exposed, is one which terminates in a permanent affection of the spine; it usually appears in early life, commencing in the form of an intermittent and remittent. The body is reduced almost to a skeleton; and the disease terminates in death, or a large curvature of the spine, so as considerably to diminish the height of the individual, and cause a very unsightly protrusion of the spine between the shoulders, or a curvature inwards, causing the breast-bones to appear unusually promirent. Multitudes in every one of the Society Islands are to be seen deformed by this disease, which the natives call *tuapu*, literally, projecting; or as we should say, humped-back.

After this curvature has occurred, the patient usually recovers, and, although greatly deformed, does not appear more predisposed to disease than others. Those individuals are often among the most active, intelligent, and ingenious of the people.

Connected with this disease, there are two remarkable circumstances. I am not prepared to say that it is hereditary, but the children of such persons are more frequently the subjects of it than others. It is also singular that it should prevail principally among the lower classes of society, the farmers and the mechanics. I know of no principal chief, and I cannot recollect any one even of secondary rank, thus afflicted: yet their rank and station are hereditary. This single fact renders more striking than it otherwise would be, the difference in appearance between the chiefs and people, and it may certainly warrant the inference that the meagre living of the latter exposes them to maladies, from which a more generous diet and comfortable mode of life exempt their superiors.

Some say this singular complaint was unknown to their ancestors, and has only prevailed since they have been visited by foreign shipping. It does not prevail among the inhabitants of the surrounding islands; but whether it be of recent origin or not in Tahiti, it is very affecting to witness the numbers that have suffered; and we cannot but hope that as industry and civilization advance, and their mode of living improves, it will in an equal ratio disappear from among them.

Blindness is frequently induced by the same disease as that which precedes the spinal curvature. The condition of the blind, when suffered to live, must, under the reign of idolatry, have been truly lamentable—they were generally objects of derision and neglect, if not of wanton cruelty.

Insanity prevailed in a slight degree, but individuals under its influence met with a very different kind of treatment. They were supposed to be inspired or possessed by some god, whom the natives imagined had entered every one suffering under mental aberrations. On this account no control was exercised, but the highest respect was shewn them. They were, however, generally avoided, and their actions were considered as the deeds of the god, rather than the man. Under these circumstances, when the poor wretch became his own destroyer, it was not regarded as an event to be deplored. Deafness was sometimes experienced; and there are a few persons in the islands who can neither speak nor hear distinctly.

In their application to particular diseases, the priests manifested considerable acquaintance with the medicinal properties of the herbs, and their adaptation to the disease, to relieve which they were employed; but their practice must have been very uncertain and ineffectual, though they were held in high esteem by all ranks. Convulsions being sometimes experienced, were considered to result from the direct power of the god. Sudden death was also attributed to the same cause—and an attack so terminating, was called *rima atua:* "hand of god." Those who died suddenly were also said to be *haruhia e te atua,* or *uumehia e te atua:* "seized by the god, or strangled by the god." Indeed, the gods were supposed to send all the diseases with which they were afflicted.

Whatever mystery they might attach to the preparation and use of medicine, their practice of surgery, and application of external remedies, were more simple and straightforward. They did not apply friction in the same manner as the Sandwich Islanders sometimes do, viz. by placing the patient flat on the ground, and rolling a twelve or fourteen pound shot backwards and forwards along the back; but in a far more gentle manner, by rubbing with the hands the muscles of the limbs, and pressing them in the same way as the Indians practise shampooing.

The natives had no method of using the warm-bath, but often seated their patients on a pile of heated stones strewed over with green herbs or leaves, and kept them covered with a thick cloth till the most profuse perspiration was induced. In this state, to our great astonishment, at the most critical seasons of illness, the patient would leave the heap of stones, and plunge into the sea, near which the oven was generally heated. Though the shock must have been very great, they appeared to sustain no injury from this transition.

There were persons among them celebrated as oculists, but their skill principally consisted in removing foreign substances from the eye; and when applied to for this purpose, they, as well as others, received the payment or fee before they commenced their operations;

and if the present did not satisfy them, if they took one splinter, &c. out of the eye, to satisfy the employers, they left another in, that they might be sent for again. Their surgeons were remarkably dexterous in closing a cut or thrust, by drawing the edges carefully together, and applying the pungent juice of the ape, *arum costatum*, to the surface. This, acting like caustic, must have caused great pain.

A fractured limb they set without much trouble: applying splinters of bamboo-cane to the sides, and binding it up till it was healed. A dislocation they usually succeeded in reducing; but the other parts of their surgical practice were marked by a rude promptness, temerity, and barbarism, almost incredible. A man one day fell from a tree, and dislocated some part of his neck. His companions, on perceiving it, instantly took him up: one of them placed his head between his own knees, and held it firmly; while the others, taking hold of his body, twisted the joint into its proper place.

On another occasion, a number of young men, in the district of Fare, were carrying large stones, suspended from each end of a pole across their shoulders, their usual mode of carrying a burden: one of them so injured the vertebræ, as to be almost unable to move; he had, as they expressed it, *fati te tua*, broken the back. His fellow-workmen laid him flat on his face on the grass; one grasped and pulled his shoulders, and the other his legs, while a third actually pressed with both knees his whole weight upon the back, where the bones appeared displaced. It was not far from Mr. Barff's house where the accident occurred, and, observing the people assembled, he went to inquire the cause, and saw them thus engaged. On his asking what they were doing, they coolly replied, that they were only straightening the man's back, which had been broken in with carrying stones. The vertebræ appeared to be replaced; they bound a long girdle repeatedly round his body, led him home, and, without any other treatment, he was in a short time able to resume his employment.

The operation of trepanning they sometimes attempted, and say they have practised with success. It is reported that there are persons living in the island of Borabora on whom it has been performed, or at least an operation very much resembling it: the bones of the skull having been fractured in battle, they have cleared away the skin and coverings, and, having removed the fractured piece of bone, have carefully fitted in a piece of cocoa-nut shell, and replaced the covering and skin; on the healing of which, the man has recovered. I never saw any individual who had undergone this operation, but, from the concurrent testimony of the people, I have no doubt they have performed it.

It is also related, although I confess I can scarcely believe it, that on some occasions, when the brain has been injured as well as the bone, they have opened the skull, taken out the injured portion of the brain, and, having a pig ready, have killed it, taken out the pig's brains, put them in the man's head, and covered them up. They persist in stating that this has been done; but add, that the persons always became furious with madness, and died. They had no idea of phlebotomy as a remedy for disease, but were clever at lancing an abscess, which was generally effected with the thorn from a kind of bramble, or a shark's tooth.

However great the influence of those persons who administered medicine, or practised surgery, might formerly have been, it has entirely ceased since the people have been acquainted with the more certain and efficacious application of English medicine. Like the priests in their temples formerly, the minister of their religion, at every station, is now sought in all cases of sickness, as their physician; and no small portion of our time was occupied in administering medicine, so far as our scanty means would admit.

This is a task necessarily devolving upon the Missionaries, as the only Europeans residing amongst them, either possessing medicine, or knowing how to use it; and it is a claim which we never desired to refuse. It is perfectly compatible with the higher duties of our station—the cure of their spiritual maladies. We have only to regret that we have not possessed better qualifications, and more ample means for its efficient discharge. So long as our family medicine has lasted, we have been ready to share it with those who were in need, and have often been thankful (when afflicted ourselves, and destitute,) to receive the simple remedies they were able to supply.

The Missionary Society has readily furnished us with medical books and instruments; and for our own use, a liberal supply of medicines: but it has generally been inadequate to the wants of the people. Medicine is expensive, and perhaps it would not be considered a just appropriation of the Society's funds, to expend them in providing medicine for those among whom its agents labour; yet it is one of the most affecting sights a Missionary can witness, when visiting his people, to behold them enduring the most painful suffering, pining under the influence of disease, and perhaps sinking into a premature grave, and to know that if he had the means within his reach, he could at least relieve them.

The occurrences are not unfrequent, wherein an anxious parent brings a poor sickly child to his house, with which she is obliged to return unrelieved, not because the disease is remediless, but because the Missionary has not, it may be, a cheap and simple remedy to bestow. The natives would cheerfully purchase so valuable an article as medicine, but they have no means of doing so, by bartering in the islands the produce of their labour. If they send it to England, the return is distant and uncertain; and mistakes, embarrassing to them, are likely to occur. It is to be hoped, however, that as the means of intercommunication become more frequent and regular, these difficulties will be removed. Several generous individuals have laid the people of some of the islands under great obligations, of which they are duly sensible, by sending them out, gratuitously, a liberal supply of the most useful medicines.

It may not be necessary for a Missionary in a civilized nation, where the healing art is cultivated, or going to a country where European colonies are settled, or commercial establishments are formed, to be acquainted with the practice of physic. It is, however, important, and ought to be borne in mind by those who are looking forward to Missionary work, and by those who patronize them, that it would be of the highest advantage for one going to an uncivilized people, to be acquainted with the qualities and use of medicine.

A degree of proficiency that would qualify him to practise in his native country, is not necessary. But so much knowledge as would enable him to be exceedingly serviceable to the people, to win their confidence and affection, and to confer on him an influence the most important and advantageous, in accomplishing the great objects of his mission, might be acquired prior to his departure from England, without in any injurious degree diverting his attention from other pursuits. I speak from painful experience of deficiency of the means for meeting the necessities of my own family, as well as those of the people among whom I have resided. I know they still exist, and therefore express myself more strongly than I should otherwise feel warranted to do.

The introduction of Christianity has been followed by a greater alteration in their general circumstances, than even the medical treatment of the sick. The change has been highly advantageous to the sufferers, who formerly experienced the greatest neglect, and often the most affecting cruelty. As soon as an individual was affected with any disorder, he was considered as under the ban of the gods: by some crime, or the influence of some enemy, he was supposed to have become obnoxious to their anger, of which his malady was the result.

These ideas relative to the origin of diseases, had a powerful tendency to stifle every feeling of sympathy and compassion, and to restrain all from the exercise of those acts of kindness that are so grateful to the afflicted, and afford such alleviation to their sufferings. The attention of the relatives and friends was directed to the gods, and their greatest efforts were made to appease their anger by offerings, and to remove the continuance of its effects by prayers and incantations. The simple medicine administered, was considered more as the vehicle or medium by which the god would act, than as possessing any power itself to arrest the progress of disease.

If their prayers, offerings, and remedies were found unavailing, the gods were considered implacable, and the afflicted person was doomed to perish. Some heinous crime was supposed to have been committed. Whenever a chief of any distinction was afflicted, some neglect or insult was supposed to have been shewn to the gods or the priest, and the most costly offerings were made to avert the effects of their wrath, and secure the recovery of the chieftain. Human victims were sometimes sacrificed, ceremonies performed, and prayers offered. These were not made to the national idol, but to the tutelar god of the family.

They were all, at times, unavailing; and when they imagined, in consequence of the rank or ancestry of the chief, that the deity ought to have been propitious, but they found he was not, and the sufferer did not recover, with a singular promptitude, in powerful contrast with their ordinary conduct towards their gods, they execrated the idol, and banished him from the temple, choosing in his place some other deity that they hoped would be favourable.

The interest manifested in the recovery of their chief would depend much upon his age. If advanced in years, comparatively little concern would be felt for his restoration. Old age was seldom treated with respect, often with contempt and cruelty.

In seasons of illness, especially if protracted, the common people, and the aged, received but little attention. If the malady was not soon relieved by the prayers of the priest, and the remedies he administered, the sufferer was abandoned. Sometimes he was allowed to remain in the house of those with whom he was connected.—But, in general, a small temporary hut was erected with a few cocoa-nut leaves, either near a stream, or at a short distance from the dwelling. Into this, as to the condemned cell, the sick person was removed. For a time, the children or friends would supply a scanty portion of food, but they often grew weary of sending this small alleviation; and it is believed that many have died, as much from hunger, as from disease.

This process was sometimes too slow for those who were connected with the sick, and who desired to share any property they might possess. If they thought there was but little prospect of recovery, they would determine to destroy them at once. Murder was at times perpetrated under these circumstances, with the most heartless and wanton barbarity. The spear or the club was employed, to effect what disease had been too tardy in accomplishing. All the persons in the house when these deeds of horror were performed, were called out; and the friends or companions of the sufferer, armed with spears, prepared for their savage work. It was in vain that the helpless man cried for mercy; instead of attending to his cry, they "would amuse themselves in trying which could take best aim" with the spear they threw; or, rushing upon him with spear in hand, they would exclaim, *Tui i vaho*, pierce through, and thus transfix him to the couch on which he was lying.

Sometimes they buried the sick alive. When this was designed, they dug a pit, and then, perhaps, proposed to the invalid to bathe, offering to carry him to the water, either in their arms, or placed on a board; but, instead of conveying him to the place of bathing, they would carry him to the pit, and throw him in. Here, if any cries were made, they threw down large stones in order to stifle his voice, filled up the grave with earth, and then returned to their dwellings.

The natives once gave me an account of an unhappy sufferer, whom they were conveying to the grave; he perceived it at a short distance before they approached, and, influenced by fear, sprang from the board, and endeavoured to escape. He was pursued, and crippled by a large stone, and thus secured by the murderers. I was acquainted with two persons, who were sawyers, and resided some time in the island of Huahine, who had both been engaged in burying one of their companions. merely because they felt the few attentions required, a burden. One of them, whose name was Papehara, is dead; the other is still living.

It is unnecessary to add to these details. Every friend to humanity will rejoice to know, that since the subversion of that system, under the sanction of which they were practised, they have ceased; and that now, from the influence of Christian principles, although the aged do not receive that veneration which is paid to gray hairs and length of years in some countries, yet that they are treated with kindness.

The sick are also nursed with attention by their relatives and children; and so far from deeming it a burden to attend to them, in Eimeo, Huahine, and, I believe, in some of the other islands, the natives have formed benevolent societies among themselves for the purpose of building houses, supplying with food and clothing those who, in their old age and helpless state, have no friends or children to take care of them. In these dwellings they are lodged, and clothed, and fed. Persons also visit them for the purpose of reading the Scriptures, and praying with them; their present necessities are supplied, the decline of life made easy, and their passage to the grave comparatively tranquil and happy. It is only necessary to contrast this with the former treatment of individuals under similar circumstances, in order to strengthen our conviction of the incalculable diminution of misery which has resulted from their reception of the gospel, and the temporal blessings it has imparted.

Although we discontinued our stated services in Maeva, we visited those who resided there, and only came occasionally to the settlement. They would have taken up their residence in Fare, but the plantations and fisheries required their attention. When proceeding across the beautiful lake, bordered by the shores of Maeva, we often passed the natives employed in taking the fish, with which every part of its waters abound.

Many of the islanders are fishermen by profession, and most of those residing near the shore, derive a great part of their subsistence from the sea. Fish are found on their coasts in great variety and abundance. Cockles are taken in the sands of the lake, and are remarkably sweet. The natives gather them in great quantities, and on public occasions they are baked and mixed with cocoa-nut milk. Muscles are also found in corresponding situations, but not in abundance. In some bays, a small oyster is met with; it is, however, inferior to the oyster found on the coasts of Europe, and is not eaten by the natives, who usually avoid the places where it abounds, on account of the danger to which their feet are exposed from the shells. Wilks and chams are abundant on the reefs, and are a common article of food with the poor people. Many varieties of the *havai*, or echinus, are also found on the reefs, and are roasted and eaten. Among the shell-fish taken for food, the cowrie and large turbo are occasionally included. Land and sea crabs are numerous, but not large. Though there are several kinds of lobsters, they are less plentiful, but very fine. Prawns and shrimps are taken in some parts, and frequently in the inland streams of fresh water.

The finest shell-fish, however, is the turtle; it is occasionally met with in the lagoons of the Society Islands, but is more frequently taken on the low, coralline, sandy reefs in the neighbourhood. Tubai, and other low islands of this kind, a short distance from Borobora, are celebrated for the numbers and quality of the turtle they furnish. This fish was formerly considered sacred, and was cooked within the precincts of the temple, part being offered to the gods, and the remainder eaten only by sacred persons. It is now eaten indifferently by any one, though most of those caught by the people are taken as presents to the chiefs.

The rivers furnish few fresh-water fish; eels are the principal, and they are very fine. Connected with the fresh-water fish, a phenomenon is often observed, for which the natives are puzzled to account. In the hollows of the rocks, and in other places, to which they suppose the sea and the river never gain access, and where the water collected is entirely what falls from the clouds, small but regularly formed fish are sometimes found. The people have frequently expressed their surprise at finding them, and appeared to wonder how they ever came there. They call them *topataua,* literally, rain-drop, supposing they must have fallen from the clouds with the rain.

Eels are great favourites, and are tamed, and fed till they attain an enormous size. Taaroarii had several in different parts of the island. These pets were kept in large holes, two or three feet deep, partially filled with water. On the sides of these pits, the eels formed or found an aperture in a horizontal direction, in which they generally remained, excepting when called by the person who fed them. I have been several times with the young chief, when he has sat down by the side of the hole, and by giving a shrill sort of whistle has brought out an enormous eel, which has moved about the surface of the water, and eaten with confidence out of his master's hand.

The sea-fish are numerous; among the principal is the salmon. The bonito, the flying fish, the operu or herring, the albicores, the sting-ray, the shark, the porpoise, and the dolphin, are caught in the lake or the sea, and are supposed to belong to the owners of the opposite shore. In the latter, any person is in general allowed to use his lines, nets, &c. but if the proprietors of the land on the coast wish to preserve the fish of the adjacent sea, they *rahui*, or restrict, the ground, by fixing up a pole on the reef or shore, with a bunch of bamboo leaves attached to it. By this mark it is understood that the fish are tabu, and fishing prohibited; and no person will trespass on these parts, without the consent of the proprietor.

The native methods of fishing are numerous, some of them rude, others remarkably ingenious. In the shallow parts of the lake they erect singular enclosures of stones for taking a number of small and middling-sized fish. This enclosure they call a *aua ia*, a fish fence.

A circular space, nine or twelve feet in diameter, is enclosed with a stone wall, built up from the bottom of the lake, to the edge of the water. An opening, a foot or two wide, is left in the upper part of the wall, extending four or six inches below the surface. From each side of this opening, a wall of stone is raised to the edge of the water, extending fifty or a hundred yards, and diverging from the aperture, so that the wall leaves a space of water within, of the shape of a wedge, the point of which terminates in the circular enclosure. These walls diverge in a direction from the sea, so that the fish which enter the lake are intercepted only in their return. They are so numerous through the whole extent of the shallow parts of the lake, that it seems hardly possible for a fish to escape. The enclosures are exceedingly valuable; fish are usually found in them every morning, and furnish a means of subsistence to the proprietors, who have no other trouble than simply to take them out with a hand-net. They are also excellent preserves, in which fish may be kept securely till wanted for use. Each enclosure has its distinct owner, whose right to the fish enclosed is always respected. Most of the fish from the lake are taken this way. The net and the spear are occasionally employed, but they rarely use the line here.

They have a singular mode of taking a remarkably timorous fish, which is called *au* or needle, on account of its long sharp head. The fishermen build a number of rafts, which they call *motoi;* each raft is about fifteen or twenty feet long, by six or eight wide, and it is made with the light branches of the hibiscus or purau. At one edge a kind of fence or skreen is raised four or five feet, by fixing the hibiscus poles horizontally, one above the other, and fastening them to upright sticks, placed at short distances along the raft. Twenty or thirty of these rafts are often employed at the same time, the men on the raft go out at a distance from each other, enclosing a large space of water, having the raised part or frame on the outside. They gradually approach each other till the rafts join, and form a connected circle in some shallow part of the lake. One or two persons then go in a small canoe towards the centre of the enclosed space, with long white sticks, which they strike in the water with a great noise, and by this means drive the fish towards the rafts. On approaching these, the fish dart out of the water, and in attempting to spring over the raft, strike against the raised fence on the outer side, and fall on the surface of the horizontal part, when they are gathered into baskets, or canoes, on the outside. In this manner, great numbers of these and other kinds of fish, that are accustomed to spring out of the water when alarmed or pursued, are taken with great facility.

Among the reefs, and near the shore, numbers of fish are seized by preparing an intoxicating mixture from the nuts of the hutu, *betonica splendida*, or the hora, another native plant. When the water is impregnated with these preparations, the fish come from their retreats in great numbers, float on the surface, and are easily caught.

The favour of the gods was formerly considered essential to success in fishing. The gods of fishermen were numerous, though Tahaura and Teraimateti were the principal. Matatine was the deity of those who manufactured nets.

Fishing-nets were various in size and kind; all were remarkably well made and carefully preserved. Their light casting-nets were neatly made, and used with great dexterity, generally as they walked along the beach. When a shoal of small fish appeared, they would throw the net with the right-hand, and enclose sometimes the greatest part of them. The nets used in taking operu, or herrings, were exceedingly large, and generally made of the twisted bark of the hibiscus. Several nets were used at the same time, the meshes of the outside net being very large, and those within smaller, for the purpose of de-taining the fish. This kind of fish visit the coasts in shoals, at one or two seasons of the year only, and as they do not design their nets to last longer than one season, they are not very carefully prepared.

Upea is the common name for net. The *upea ava*, or salmon net, is the longest and most important, and is seldom possessed by any but the principal chiefs; it is sometimes forty fathoms long, and twelve or more feet deep. One of this kind was made by Hautea, the governor of Huahine, soon after our arrival. Although the former pagan ceremonies, and the offerings at the marae, were discontinued, some of the ancient usages were observed, one of which appeared rather singular. As is customary on all occasions of public work, the proprietor of the net required the other chiefs to assist in its preparation. Before he began, two large pigs were killed and baked. When taken from the oven, they were cut up, and the governor's messenger sent with a piece to every chief; on delivery, the quantity was stated which each was desired to prepare towards the projected net. If the piece of pig was received, it was considered as an agreement to furnish it; but to return it, was, in effect, to refuse compliance with the requisition. At this time, however, no one returned the *tarahu*, or price, but all agreed to furnish one or two fathoms of the net. When any other chief wanted a net, he took the same course.

The cord was about a quarter of an inch in diameter, and made with the tough white bark of the mate, *ficus prolixa*, which, next to the romaha, or flax, is considered more durable than any other indigenous vegetable substance. The cord was twisted with the hand across the knee, in two or three strands or threads, and was even and firm. The meshes were about four inches square.

The servants of the chief furnished their quantity of netting, and the needle with which they wrought was not unlike that used by European workmen. As the other parties brought in their portions, the chief and his men joined them together. On entering the house of Hautea, I have found him in a profuse perspiration, toiling in the midst of his men at the manufacture of the net.

The floats were made with short pieces of dry, light, buoyant hibiscus; and the bottom was hung with stones, generally circular and smooth, about three inches in diameter. These were not perforated, but enveloped in pieces of the matted fibre of the cocoa-nut husk, tied together at the ends, and attached to the lower border of the net.

The first wetting of a new net was formerly attended with a number of prayers, offerings, &c. at the temple, and on the beach. I recollect, at Afareaitu, when they were going to take out, for the first time, a large salmon-net, and had put it upon the canoe, the whole party, including the fishermen and chiefs of the district, kneeled down upon a pebbly beach, and offered a prayer to the true God, that they might be successful. This was about day-break; and as the sun rose above the waves, I saw them rowing cheerfully out to sea. Though these nets were called *upea ava,* salmon-nets, a variety of large fish was taken in them; a shark was not unfrequently enclosed, which sometimes made great havock among the fishermen, before they could transfix him with their spears.

This kind of fishing was followed not only as a means of procuring food, but as an amusement. The chiefs were exceedingly fond of it, and often strove to excel. Hautea was celebrated for his skill and strength in taking some kinds of fish. Their country was little adapted for hunting, and the only quadrupeds they ever pursued were the wild hogs in the mountains; but the smoothness and transparency of the sea within the reefs, favoured their aquatic sports; and a chief and his men, furnished with their spears, &c. often set out on their fishing excursions with an exhilaration of spirits equal to that with which a European nobleman pursues the adventures of the chase. The more daring of the young chiefs were generally among the foremost in pursuing the shark, or other dauntless fish; while others, more advanced in years, remained in their canoes at a distance, gratified to behold the sport, and share in some degree the excitement it produced. When the *tautai* or fishing party returned, the nets were hung up on the branches of trees near the shore, as they appear in the view of Fa-re harbour, inserted at page 414, vol. i. Besides the herring, hand, and salmon nets, they had a number of others, adapted to particular places, or kinds of fish.

Next to the net, the spear was most frequently used. It was variously formed, according to the purpose for which it was designed. Since their intercourse with foreigners, the best spears have been made with iron, always barbed, but only on one side. Two or three small spear-heads were occasionally fastened to a single handle. Another kind of spear, in frequent use, was entirely of wood. Nine, ten, or twelve pointed pieces of hard wood, six or eight inches long, were fastened to a handle, from six to eight feet in length. When using this, they generally waded into the sea as high as the waist, and, standing near an opening between the rocks of coral, or near the shore, and watching the passage of the fish, darted the spear, sometimes with one hand, but more frequently with both, and often struck them with great precision.

Their aim with this spear, however, is much less certain than with one headed with iron; which some throw with great dexterity, though others are exceedingly awkward. When fishing on the reefs, they often wear a kind of sandal, made of closely netted cords of the bark of the native *auti*, or cloth-plant. This was designed to preserve their feet from the edges of the shells, the spikes of the echinus, &c. They use the angle or the spear in fishing at the edge of the reef, when the surf is low. I have often, when passing across the bay, stopped to gaze on a group of fishermen standing on a coral reef, or rock, amidst the roar of the billows and the dashing surf and foam, that broke in magnificent splendour around them. With unwavering glance, they have stood, with a little basket in one hand, and a pointed spear in the other, striking with unerring aim such fish as the violence of the wave might force within their reach. They have a curious contrivance for taking several kinds of ray and cuttle-fish, which resort to the holes of the coral rocks, and protrude the arms or feet for the bait, but remain themselves firm within the retreat. The instrument they employ, consists of a straight piece of hard wood, a foot long, round and polished, and not half an inch in diameter. Near one end of this, a number of the most beautiful pieces of the cowrie or tiger-shell are fastened one over another, like the scales of a fish or the plates of a piece of armour, until it is about the size of a turkey's egg, and resembles the cowrie. It is suspended in an horizontal position, by a strong line, and lowered by the fisherman from a small canoe, until it nearly reaches the bottom. The fisherman then gently jerks the line, causing the shell to move as if inhabited by a fish. This jerking motion is called *tootoofe,* the name of the singular contrivance.

The cuttle-fish, attracted, it is supposed, by the appearance of the cowrie, (for no bait is used,) darts out one of its arms or rays, which it winds round the shell, and fastens among the openings between the plates. The fisherman continues jerking the line, and the fish puts forth another and another arm or ray, till it has quite fastened itself to the shells, when it is drawn up into the canoe, and secured.

They use the hook and line both in the smooth water within the reef, and in the open sea; and in different modes display great skill. In this department they seldom have any bait, excepting a small kind of *oobu,* a black fresh-water fish, which they employ when catching albicores and bonitos. Their hooks usually answer the double purpose of hook and bait. Their lines are made with the tough elastic *romaha,* or flax, twisted by the hand.

In no part of the world, perhaps, are the inhabitants better fishermen; and, considering their former entire destitution of iron, their variety of fishing apparatus is astonishing. Their hooks were of every form and size, and made of wood, shell, or bone, frequently human bone. This was considered the most offensive use to which the bones of an enemy could be applied: and one of the most sanguinary modern wars in Tahiti originated in a declaration made by a fisherman of one party, that he had a hook made with the bone of a rival chief who had been slain in a former war.

The hooks made with wood were curious; some were exceedingly small, not more than two or three inches long, but remarkably strong; others were very large. The wooden hooks were never barbed, but simply pointed, usually curved inwards at the point, but sometimes standing out very wide, occasionally armed at the point with a piece of bone. The best were hooks ingeniously made with the small roots of the aito tree, casuarina, or iron wood. In selecting a root for this purpose, they chose one partially exposed, and growing by the side of a bank, preferring such as were free from knots and other excrescences. The root was twisted into the shape they wished the future hook to assume, and allowed to grow till it had reached a size large enough to allow of the outside or soft parts being removed, and a sufficiency remaining to make the hook. Some hooks thus prepared are not much thicker than a quill, and perhaps three or four inches in length. Those used in taking sharks are formidable looking weapons; I have seen some a foot or fifteen inches long, exclusive of the curvatures, and not less than an inch in diameter. They are such frightful things, that no fish, less voracious than a shark, would ever approach them. In some, the marks of the shark's teeth are numerous and deep, and indicate the effect with which they have been used. I do not think the Tahitians take as many sharks as the Sandwich Islanders do: they, however, seldom spare them when they come in their way; and though sharks are not eaten now, the natives formerly feasted on them with great zest.

The shell, or shell and bone hooks, were curious and useful, and always answered the purpose of hook and bait; the small ones are made almost circular, and bent so as to resemble a worm, but the most common kind is the aviti, used in catching dolphins, albicores, and bonitos; the shank of the hook is made with a piece of the mother-of-pearl shell, five or six inches long, and three-quarters of an inch wide, carefully cut, and finely polished, so as to resemble the body of a fish. On the concave side, a barb is fastened by a firm bandage of finely twisted *romaha,* or flax; the barb is usually an inch and a half in length, and is of shell or bone. To the lower part of this, the end of the line is securely fastened, and being braided along the inner or concave side of the shell, is again attached to the upper end. Great care is taken in the manufacture of these pearl-shell hooks, and they are considered much better than any made in Europe.

The line is fastened to the hook in a curious manner, and when taken to sea, is attached to a strong bamboo-cane, about twelve or fifteen feet long; light single canoes are preferred for catching dolphins, bonitos, or albicores. Two or three persons usually proceed to sea, and when they perceive a shoal of these fish, those who angle sit in the stern of the canoe, and hold the rod at such an elevation, as to allow the hook to touch the edge of the water, but not to sink. When the fish approach it, the rowers ply their paddles briskly, and the light bark moves rapidly along, while the fisherman keeps the hook near the surface of the water. The deception of the hook is increased by a number of hairs or bristles being attached to the end of the shell, so as to resemble the tail of a flying-fish. The bonito, &c., darts after, and grasps its prey, and is itself secured. During the season, two men will sometimes take twenty or thirty large fish in this way, in the course of the forenoon.

The most ingenious methods, however, of taking these large fish is by means of what is termed a *tira,* or mast.

A pair of ordinarily sized canoes is usually selected, with a kind of basket-work fixed between them, to contain the fish. To the forepart of the canoes a long curved pole is fastened, branching in opposite directions at the outer end; the foot of this is fixed in a kind of socket, between the two canoes. From each of the projecting branches, lines with pearl-shell hooks are attached, so adjusted as to be kept near the surface of the water. To that part of the pole which is divided into two branches, strong ropes are attached; these extend to the stern of the canoe, where they are held by persons watching the seizure of the hook. The tira, or mast, projects a considerable distance beyond the stem of the canoe, and bunches of feathers are fastened to its extremities. This is done to resemble the aquatic birds which follow the course of the small fish, and often pounce down and divide the prey which the large ones pursue. As it is supposed that the bonitos follow the course of the birds, as much as that of the fishes, when the fishermen perceive the birds, they proceed to the place, and usually find the fish. The undulation of the waves occasions the canoe to rise and sink as they proceed, and this produces a corresponding motion in the hook suspended from the mast; and so complete is the deception, that if the fish once perceives the pearl-shell hook, it seldom fails to dart after it; and if it misses the first time, is almost sure to be caught the second. As soon as the fish is fast, the men in the canoe, by drawing the cord, hoist up the tira, and drag in the fish, suspended as it were from a kind of crane. When the fish is removed, the crane is lowered; and as it projects over the stem of the canoe, the rowers hasten after the shoal with all possible celerity.

Fishing Canoe.

These, and a variety of other methods of fishing, are pursued by day-light; but many kinds of fish are taken by night: sometimes the fishery is carried on by moon-light, occasionally in the dark, but fishing by torch-light is the most picturesque. The torches are bunches of dried reeds firmly tied together. Sometimes they pursue their nocturnal sport on the reef, and hunt the *totara*, or hedge-hog fish. Large parties often go out to the reef; and it is a beautiful sight to behold a long line of rocks illuminated by the flaring torches. These the fishermen hold in one hand, and stand with the poised spear in the other, ready to strike as soon as the fish appears.

In the rivers they also fish by torch-light, especially for eels; and though the scene is different, its impression is not inferior. I have often been struck with the effect of a band of natives walking along the shallow parts of the rocky sides of a river, elevating a torch with one hand, and perhaps a spear in the other; while the glare of their torches was thrown upon the over-hanging boughs, and reflected from the agitated surface of the stream. Their own bronze-coloured and lightly clothed forms, partially illuminated, standing like figures in relief; while the whole scene has appeared in bright contrast with the dark and almost midnight gloom that enveloped every other object.

Since their intercourse with Europeans, English-made steel hooks have been introduced. They like their sharpness at the point, but usually complain of them as too open or wide. For some kinds of fish they are preferred, but for most they find the mother-of-pearl hooks answer much better. Every fisherman, I believe, would rather have a wrought-iron nail three or four inches long, or a piece of iron-wire of the size, and make a hook according to his own mind, than have the best European-made hook that could be given to him. Most of the nails they formerly procured from the shipping were used for this purpose, and highly prized.

Their ideas of the nature of these valuable articles were very singular. Perceiving, in their shape and colour, a resemblance to the young shoots or scions that grow from the roots of the bread-fruit trees, they imagined that they were a hard kind of plant, and procured in the same way. Anxious to secure a more abundant supply, they divided the first parcel of nails ever received, carried part to the temple, and deposited them on the altar; the rest they actually planted in their gardens, and awaited their growth with the highest anticipation. In the manufacture of hooks from nails, they manifested great patience and persevering labour: they had no files, but sharpened the points, and rounded the angles, by rubbing the nail on a stone; they also used a stone, in bending it to the required shape. The use of files, however, has greatly facilitated their operations in the manufacture of fish-hooks.

In connexion with this subject, a striking instance of native simplicity and honesty occurred about the time of our arrival. Two Christian chiefs, Tati and Ahuriro, were walking together by the water-side, when they came to a place where a fisherman had been employed in making or sharpening hooks, and had left a large file, (a valuable article in Tahiti,) lying on the ground. The chiefs picked it up; and, as they were proceeding, one said to the other: "This is not ours. Is not our taking it a species of theft?" "Perhaps it is," replied the other. "Yet, as the true owner is not here, I do not know who has a greater right to it than ourselves." "It is not ours," said the former, "and we had better give it away." After further conversation, they agreed to give it to the first person they met, which they did; telling him they had found it, and requested that if he heard who had lost such a one, he would restore it.

During the year 1820, the Mission in the Windward Islands sustained a heavy bereavement in the decease of Messrs. Bicknell and Tessier. The latter, who was a man of modest and unobtrusive habits, but patient and unremitting industry, in the important work of educating the rising generation, died on the 23d of July. His Christian course had not been splendid or attractive, but it had been undeviating and unsullied. His end was not only peaceful, but triumphant in faith, and glowing in anticipation of the holy and spiritual joys awaiting him in the abodes of blessedness.

Mr. Bicknell, whose health was not firm, followed the remains of his faithful coadjutor to the tomb; and while standing on the edge of the closing grave, and addressing the sorrowing multitude around, felt indisposed from the exposure. This was followed by fever, which terminated his life fourteen days after the death of Mr. Tessier. Though his illness was short, his mind, towards the latter part of it, was tranquil, in reliance on that Saviour who alone can support in the prospect of dissolution.

I have heard he was the first individual who offered his services to the Missionary Society, and was among the first who landed from the Duff in 1796. He remained in Tahiti till the civil war in 1808 drove them from the islands, at which time he visited New South Wales and England. When Pomare invited the Missionaries to return, he was the first to resume his station, which he never abandoned, till called by death from a field, on which he had bestowed upwards of twenty years of patient persevering toil, and from which, though long barren and fruitless, he had ultimately been honoured to reap the first-fruits of a glorious harvest.

In 1818, he removed to the populous district of Papara on the south-west side of Tahiti. This district had, prior to the last war, been the stronghold of idolatry, and was the head-quarters of the pagan army; and the inhabitants, until the death of their chieftain in the memorable battle of Bunaauïa, obstinately opposed the progress of Christianity. Here, under the favourable auspices of Tati, Mr. Bicknell commenced his labours; and while Mr. Tessier daily instructed numbers in the school, Mr. Bicknell collected around him large and attentive congregations, baptized many, and gathered an interesting Christian church.

His latest earthly concern regarded the stedfastness and welfare of his charge. On the last evening of his life, and but a few hours before his departure, he addressed Mr. Crook (who had attended him during his illness, and who was then about to perform divine service among his people,) on the subject. "Tell them," said the dying Missionary, "that my conviction of the truth of those doctrines I have taught, is now stronger than ever. Tell them I am dying, but that these truths are now my support. Tell them to be stedfast." He left, not only a destitute church and afflicted congregation, but a sorrowing widow and five fatherless children, to mourn his departure. Mrs. Bicknell was afterwards united in marriage with Mr. Davies, but she did not long survive, and the children are now orphans in Tahiti. Mr. Caw, who had been sent out to instruct the natives in ship-building, and other arts, but who had been long incapacitated by illness, died about the same time.

CHAP. XI.

General view of a Christian church—Uniformity of procedure in the different stations—Instructions from England—Preparatory instructions—Distinct nature of a Christian church—Qualifications and duties of communicants—The sacrament of the Lord's Supper—Formation of the first church of Christ in the Leeward Islands—Administration of the ordinance—Substitute for bread—Order of the service—Character, experience, and peculiarities of the communicants—Buaiti—Regard to the declarations of scripture—Instances of the power of conscience—Manner of admitting church members—Appointment of deacons—Great attention to religion.

WHILE the Lord of Missions was thus thinning our ranks, he was shewing us that the work in which we were engaged was not ours, but his; that though the agent was removed, the agency under which he had acted was not thereby impeded. The pleasing change we had observed among our people every year, increased during the present in an astonishing manner, and we had the high satisfaction of witnessing the formation and organization of the first church of Christ in the Leeward or Society Islands. It took place early in the month of May, and shortly after the opening of the new chapel.

Although we did not experience that difficulty which, from the peculiar circumstances of the Mission and the people, had attended the first administration of baptism, we regarded it as a matter requiring grave and prayerful deliberation. We felt that our proceedings would influence the views and conduct, not only of those by whom we were surrounded, but perhaps of future generations. A foundation was now to be laid,on which, so far as order and discipline were concerned, the superstructure of the Christian church in that island was to rise in every succeeding age, and by which it would certainly be affected in many important respects. Anxious therefore to begin aright, we sought, and trust we received, Divine guidance, endeavouring to regulate our proceedings altogether by the directions of the sacred volume. It was, however, difficult to divest ourselves entirely of those views of the subject which we had imbibed from the writings of men.

A Christian church we considered to be a society of faithful and holy men, voluntarily associated for the purposes of public worship, mutual edification, the participation of the Lord's supper, and the propagation of Christianity: the Lord Jesus Christ was regarded as its spiritual head; and only such as had given themselves unto the Redeemer, and were spiritually united to him, members. These were our general views. In England we had belonged to different denominations, and however adapted the peculiarities in discipline, of those communions might appear to the circumstances of British Christians, we did not deem it expedient to take any one altogether for our model. It appeared to all more desirable, in the existing state of the people, to divest the churches we might be honoured of God to plant among the Gentiles, of every thing complicated or artificial, that they might be established in the purest simplicity of form, and, as far as possible, according to the directions of revelation. Had any been pertinacious of their peculiarities, they had now the fairest opportunity of acting accordingly.

General good, however, was our object; and that line of procedure, which, as a whole, we could unitedly pursue, in closest accordance with Scripture, and at the same time with greatest advantage to the people, was more desired by every one, than any peculiar views on minor points. I believe it is from the paramount influence of these feelings, more than from any other cause, that such uniformity exists. There was no agreement previously entered into among the Missionaries, but those of each station were left, with the people around who might be brought to a reception of the truth, to assume for themselves such form of constitution and discipline, as should in their views be most accordant with the word of God; and yet I am not aware, that in any material point there is the smallest difference among them.

As the subject had long been one of considerable anxiety, we had written to the Directors of the Society for their advice. They in general referred us to the New Testament. Several persons, however, interested in the progress of truth among the islands, wrote to the Missionaries individually, and also communicated their views to the public through the medium of the Evangelical Magazine. Among others, the Rev. Mr. Greathead, whose views of church government were rather peculiar, wrote very fully. His plans were at first adopted by one or two of the Missionaries; yet the free admission, not only to baptism, but to the ordinance of the Lord's supper, of such persons as sincerely desired to receive the same, without requiring evidence of their being true spiritual converts to Christ, threatened great irregularity and confusion; it was therefore discontinued.

In our public instructions, we inculcated on those who, we had reason to believe, were under the decisive influence of the Spirit of Christ, the duty of commemorating his dying love by that ordinance which he had instituted, and by which his disciples were to shew forth his death till he should come.—Those who had been baptized, now desired to be more particularly informed how, and in what circumstances, they were to observe this injunction of the Lord. We, therefore, proposed to devote one afternoon every week to the instruction of such as had been baptized, and desired to be united in church-fellowship. Fifteen individuals attended the first meeting, and were afterwards joined by others. We met them regularly, and endeavoured to instruct them as fully and familiarly as possible in the duty of partaking of the sacrament; the nature, design, and scriptural constitution of church-fellowship; the discipline to be maintained, the advantages to be anticipated, and the duties resulting therefrom.

Next to the personal piety, which in church-members is considered indispensable, it appeared most important to impress the minds of the people with the distinctness of a Christian church from any political, civil, or other merely human institution. In the system of false religion under which they had lived, and by which their habits of judgment had been formed, the highest civil and sacerdotal offices had been united in one person.—The king was generally chief priest of the national temple; and the high-priesthood of the principal idols was usually held by some member, or near relative, of the reigning family. On many occasions of worship also, the king was the representative of the god. The chiefs and the gods appear always to have exercised a combined influence over the populace. The power of the gods often seemed only exercised to establish the authority of the king, who was by the people regarded as filling his high station by lineal descent from them, while the measures of the government as often appeared to be pursued to inspire fear, and secure acknowledgments for the gods. Hence, when human sacrifices were required, the priest applied to the king, and the king gave orders to provide the victim. Since the kings and chiefs, as well as the people, had embraced the gospel, and many had taken the lead in propagating it, and had uniformly adorned it by their example, and the people sometimes said, that had their chiefs been idolaters or wicked rulers, it would have been improper for them to have interfered in any matters connected with Christianity, but that now they were truly pious, it accorded with their ideas of propriety, that in the Christian church they should, as christian chiefs, be pre-eminent.

We told them they had not imbibed these ideas in a christian, but in a pagan school; that the authority of their kings and chiefs was exerted over their persons, and regarded their outward conduct; that they held their high station under God, for the well-being of society, and were, when influenced by uprightness and humanity, the greatest blessings to the communities over which they presided. We also stated, that in this station every Christian was bound, no less by duty to God than to man, to render obedience to their laws, to respect and maintain their authority, and to pay them every due homage. And we also told them, that in the church of Jesus Christ, which was purely a religious association, so far as distinctions among men, from dignity of station, elevation of office, fame of achievement, or influence of wealth, were concerned, that all members were brethren; and that Christ himself was the only spiritual chief or king; that his influence or reign was not temporal, but, like his authority, spiritual. The only distinction recognized in a Christian church, we informed them, regarded those who acted as officers, and that such distinctions only prevailed in what concerned them as a church, and did not refer to their usual intercourse with the community of which they were members, and in which they were governed by the ordinary regulations established in civilized society.

The duties which those who united in church fellowship were required to perform towards each other, towards those desirous of uniting with them, and to the careless or irreligious, were also fully and frequently brought under their notice, together with the paramount duty of every Christian to endeavour to propagate Christianity, that the Christian church might become a kind of nursery, from which other churches might be planted in the extensive wilderness of paganism around.

Next to this, the institution, nature, design, administration, and uses of the Lord's supper, were familiarly explained, that they might understand, as far as possible, the engagement into which they were desirous to enter, and the observances connected therewith.

The Lord's supper, or sacrament, we regarded as analogous to the passover, symbolical of the death of Christ as an atonement or sacrifice, of which event it was commemorative; that it was designed to perpetuate the remembrance of His death, even to the end of time, and was to be in faith participated by all who build their hopes of admission to the heavenly state on His atonement.

Having been for some months engaged weekly in imparting this kind of instruction to those who had expressed their desire to receive the ordinance of the Lord's supper, the month of May was selected for forming the church. Sixteen individuals, who in the judgment of charity we had every reason to believe were sincere Christians, then met us, and, after imploring the blessing of the great Head of the church, offering a suitable address, and receiving their declaration of faith in Christ, and desire to enjoy the privileges of church-fellowship, a voluntary association was formed, the right-hand of fellowship was given, and they recognized each other as members of the first church of Christ in Huahine.

We did not present any creed or articles of faith for their subscription on this occasion. Sensible of the insufficiency of all mere human writings, however excellent, to restrain the mind, or control the opinions of men, we thought it best to dispense with them, lest the bare assent, or subscription to certain articles of faith, or doctrines of truth, should be substituted, as grounds of confidence, for an experience of the influence of those doctrines on the heart. Their names only were entered in a book kept by the Missionaries for that purpose, and called the Church-book. This little meeting was held in the chapel at Fare, on Friday evening, the 5th of May, 1820: and it is hoped that what was done on earth was confirmed in heaven, and that the union then formed among the disciples of Christ below, though it may be dissolved by death, will be realized in his presence above, and endure through eternity.

On the following Sabbath, May the 7th, an unusual number attended the large place of worship. Mr.

Davies preached in the forenoon, from Luke xxii. 19. In front of the pulpit, a neat table, covered with white native cloth, was fixed, upon which the sacramental vessels were placed. These had been furnished from England. Wheaten bread was an article of diet that we did not very often obtain ourselves, and which the people seldom tasted: we should have preferred it for this ordinance, yet, as we could not, from the irregularity and uncertainty of our supplies at that period, expect always to have it, we deemed it better to employ an article of food as nearly resembling it as possible, and which was at all times procurable. From these considerations, we felt no hesitation in using, on this occasion, the roasted or baked bread-fruit, pieces of which were placed on the proper vessel.

Wine, we were also thankful to possess for this purpose; and although we have sometimes been apprehensive that we might be under the necessity of substituting the juice of the cocoa-nut for that of the grape, or discontinuing the observance of this ordinance, (to which latter painful alternative, some of our brethren have been reduced,) we have been providentially favoured with a sufficiency. Over the elements placed on the table, a beautifully white cloth had been spread, before the accustomed service began. When this was over, although it was intimated that any who wished might retire, no one left the chapel. Mr. Davies, the senior Missionary or pastor of the church, took his station behind the communion-table; Mr. Barff sat at one end; and I took my seat at the other.

When the communicants had seated themselves in a line in front, we sang a hymn. The words of institution, viz. passages of Scripture containing the directions for the observance of this hallowed festival, &c. were read, a blessing implored, and the bread, which was then broken, handed to each individual. The wine was next poured into the cup, a blessing again sought, when the wine was handed to the communicants. After this, another hymn was sung, a short prayer offered, and the service closed.

I have been thus particular in detailing the order observed on this occasion, as affording not only a correct statement of our proceedings at this time, but also a brief general view of the manner of administering this sacred ordinance in the different Missionary stations throughout the islands.

It would be impossible to give any thing like an adequate description of my own emotions, at this truly interesting service. The scene was worth coming from England to witness, and I trust the impression was as salutary as it was powerful and solemn. I am also quite unable to conceive what the feelings of our senior colleague must at this time have been. He had been many years among the people before any change in favour of Christianity took place, and had often beheld them, not only ignorant and wretched, sunk to the lowest state of debasing impurity, and accustomed to the perpetration of the most horrid cruelty, but altogether given to idolatry, and often mad after their idols.

Our joy arose, in a great degree, from the delightful anticipation awakened in connexion with the admission of the anxious multitude, who were waiting to enter into, and we hoped, prepared of God to participate in, all the blessings which this ordinance signified, and in reference to the eternity we hoped to spend with them, when we should join the church triumphant above. His joys, however, in addition to those arising from these sources, must have been powerfully augmented by the recollection of what those individuals once were, and the many hours of apparently cheerless and hopeless toil he had bestowed upon them, now so amply, so astonishingly rewarded.

A state of feeling, almost unearthly, seemed to pervade those who now, for the first time, united with their teachers in commemorating the dying love of Christ. Recollection, perhaps, presented in strong colours the picture of their former state. Their vile abominations, their reckless cruelty, their mad infatuation in idolatry, the frequent, impure, and sanguinary rites in which they had engaged—their darkened minds, and still darker prospects—arose, perhaps, in vivid and rapid succession. At the same time, in striking contrast with their former feelings, their present desire after moral purity, their occupation in the worship of Jehovah, their hopes of pardon and acceptance with him, through the atonement made by the offering of his Son, the boundless and overwhelming effects of his love herein displayed, and the radiant light and hopes of everlasting blessedness and spiritual enjoyment, which, by the event commemorated, they were encouraged to anticipate, were all adapted to awaken, in minds susceptible as theirs, no common train of feelings. Often have we seen the intense emotion of the heart at these seasons, strongly depicted in the countenance, and the face suffused with tears.

The hundreds who remained to witness the scene, were not careless or indifferent spectators. Their deep interest in what was passing, was indicated in their thoughtful and agitated countenances, and the subsequent conduct of many evinced the kind of impression they received. The anxious concern which we had witnessed among the people, since the preceding summer, appeared to increase, and demanded redoubled efforts for their spiritual advantage. Numbers came as candidates for baptism, and regularly attended the meeting for the instruction of such. Others, from among those who had been baptized, desired to be admitted to church fellowship.

Our liveliest affections were awakened on their behalf; but while we had reason to believe many were sincere, we had also reason to fear that others were influenced by less commendable motives. Anxious to afford encouragement or caution, as the circumstances or cha racter of each required, it was not easy to satisfy our own minds as to the best manner of proceeding. We feared to discourage any who were sincerely seeking a more intimate acquaintance with Christ, and who were desirous to be fully instructed in all things concerning his will. On the other hand, we were equally fearful of encouraging the indulgence of improper views, or of admitting to the ordinances of the gospel any who were uninfluenced by those motives which Christ would approve.

There was, however, no part of our charge in whose welfare we now felt so deeply interested, as the little flock, of which the great Shepherd had made us the pastors. So far from considering our work done, with special reference to those whom we had instructed in the nature of a Christian church, and had admitted to this fold, we considered it as only the commencement of a new series of important and interesting duties, arising out of the new relation now subsisting between us. We experienced an attachment binding our hearts to theirs, to which we had before been strangers, and we had reason to believe the feeling was reciprocal.

Their knowledge was but limited, notwithstanding all our efforts to instruct them; their duties increased, their situation became more conspicuous, and their temptations greater. Latent depravity still lurked in their hearts, and it might be expected that their great spiritual adversary would not leave them unmolested. We were also fearful lest the privileges they were raised to enjoy might engender or nourish secret pride, or induce a disposition to rest satisfied with having obtained admission to the outward and visible church of Christ, and thus lead them to neglect that constant seeking after God, and the cultivation of those Christian virtues, by which alone they could sustain, with credit to Christianity, and benefit to their own minds, the situation to which they had been raised. They would naturally become models of imitation to others, and would exert no ordinary influence on the community at large. It was therefore gratifying to behold them humble, prayerful, watchful, and diligent. The weekly meeting with the candidates for communion, whose number was greatly increased, we constantly attended, and recommended the church members not to absent themselves unnecessarily.

At these times we endeavoured to explain the truths in which they were most interested, and, with regard to the members themselves, leaving the first principles of the doctrines of Christ, we endeavoured gradually and gently to lead them on to a more extensive acquaintance with the grand and varied doctrines of the gospel, and the important relative and other duties resulting therefrom.

These meetings were exceedingly interesting, from the simple yet unequivocal evidences often afforded of the operation of the Spirit of the Almighty upon the hearts of the people. Our little church, from time to time, received considerable accessions of such as we had reason to hope were also members of the church of the first-born, whose names are written in heaven.

In the admission of members, we acted with what perhaps many would consider the extreme of caution. Individuals whose moral character has been irreproachable, whose views of divine truth have been clear and scriptural, and whose motives, so far as we could judge, have been pure, have remained two, and sometimes three years as candidates, although we could not prefer any allegation directly against them. The admission of such has been declined, because we feared, that though their knowledge was commendable, and their conduct influenced by the precepts of the gospel, their hearts were not under its decisive influence; in short, that they had not undergone that change of mind, which our Lord himself, in his conversation with Nicodemus, called being "born again," and without which he had declared no man can enter into the kingdom of heaven. In other instances, however, the testimony relative to this change, was so decisive and powerful, that we could not, dared not, hesitate.

The reason the natives have given of their christian hope, has often been not only satisfactory, as it regarded the individual, but important, and in a high degree interesting, as an evidence of the universality of the depravity of man; and also as shewing the effects of Divine truth, under the influence of the Spirit of God, to be the same in every clime, producing the corresponding effects upon men of every diversity in colour, language, and circumstance. Hence, one of the strongest modern evidences in the history of man, of the unequivocal origin of

Christianity, has been afforded, and its perfect unaltered adaptation to the condition of the whole human race.

The same latent enmity to the moral restraints Christianity imposes on the vicious propensities of men, the same unwillingness to admit its uncompromising claims to the surrender of the heart, was experienced here, as is in other parts. The same tendency to suppose the favour of God might be obtained by services which they could perform, and the same unbelief under convictions of sin, and unwillingness to go to the Saviour without a recommendation—that is so often met with in others—was felt by them.

But while, in these respects, the experience of the new converts in the South Sea Islands resembled that of Christians in other parts of the world, there are points in which it has often appeared to us peculiar. We never met with one who doubted the natural depravity, or innate tendency to evil, in the human heart. We never met with any who were inclined to suppose they could, without some procuring cause, be justified in the sight of God. This may perhaps arise from the circumstance of there being no individual among them, whose past life had not been polluted by deeds which even natural conscience told them were wrong, and consequently no arguments were necessary to convince any one that he was guilty before God. They must deny the existence of the Deity, and of all by which the living and true God is distinguished from their own senseless idols, before they could for a moment suppose their past lives appeared otherwise than criminal before Him. Their fearful state, and the consequences of guilt, they never disputed, but were always ready to acknowledge that they must not only appear criminal, but offensive to the Most High, on account of their vices. There were, however, in connexion with these truths, matters associated with the impression upon their minds, that sometimes a little surprised us.

Under declarations of the nature and dreadful consequences of sin, aggravated as theirs had been, the denunciation of the penalties of the law of God, and even under the awakenings of their own consciences to a conviction of sin, we seldom perceived that deep and acute distress of mind, which in circumstances of a similar kind we should have expected. In connexion with this, when such individuals were enabled to exercise faith in the atonement of Christ, and to indulge a hope of exemption from all the fearful effects of sin and guilt, this apprehension has not, in many instances, been attended by that sudden relief, and that ecstatic joy, which is often manifested in other parts of the world, by individuals in corresponding circumstances. Yet, in many instances, we have not doubted the sincerity of their declarations, or the genuineness of their faith in the Redeemer.

We have often tried to account for this apparent anomaly in their Christian character, but have not been altogether satisfied with the causes to which we have sometimes assigned it. It does not appear, generally, that their emotions are so acute as ours, or, that they are equally susceptible of joy and sorrow with persons trained in civilized society. Besides this, though their ideas of the nature and consequences of sin, the blessedness of forgiveness, and the hope of future happiness, were correct so far as they went, yet the varied representations of the punishment and sufferings of the wicked, and the corresponding views of heaven, as the state of the greatest blessedness, being to them partial and new, the impressions were probably vague and indistinct, while with us, from long familiarity, they are at once vivid and powerful. Without pausing to inquire into its cause, it seemed right to mention the fact; better reasons may perhaps hereafter be assigned.

We have often also remarked, that there are but few of what would be called sudden conversions. In general, the process by which their views and feelings have been changed, has been gradual, and almost imperceptible, as to its precise manner of operation, though ultimately most decisive in its nature, and unquestionable in its tendency. Though these gradual transformations are the general means by which, through the Holy Spirit, we hope many have been made partakers of the grace of eternal life, there have been exceptions. Some have been melted under the truth, others have been led to rejoice in the promises of the gospel, and raised to gladness and praise. These facts are adapted to shew, that the Spirit of God is not limited, in the manner of His operations on the human mind, to any one particular kind of order and rule.

The accounts of their views of Divine truth, and their reasons for desiring to join with us, have often been delightful and satisfactory, not only in the Society, but also in the northern isles of the Pacific. One from a native of the latter, although it has appeared in the American Missionary Herald, has not been given to the British public; and its character is so unequivocal, that I cannot deny myself the pleasure of inserting it.

Buaiti, the individual to whom it refers, is between thirty and forty years of age. I believe I had the honour of preaching the gospel in his native islands the first time he ever heard it. It, however, produced no salutary effect then; nor, indeed, until some time after. Since

I left the islands, the preaching and instructions of Mr. Richards have been singularly useful to this individual, as well as to others; he has given every evidence of their having, under the blessing of God, produced an entire and highly beneficial change in his sentiments, feelings, and conduct.

The late queen of the Sandwich Islands, with her usual benevolence, had always treated him with kindness; and the recollection of it is still retained. Buaiti was his native name, but, when he was baptized, he wished to be called Bartimeus; and, in order to preserve the sense of his loss in the death of the queen, he requested that *Lalana*, London, the place of her death, might be added to his name. When he was admitted a member of the Christian church at Lahaina, he was asked by the Missionary, Why do you request to be received into the church? He replied—

Because I love Jesus Christ, I love you, and I desire to dwell with you in the fold of Christ, and to join with you in eating the holy bread, and drinking the holy wine.

What is the holy bread?

It is the body of Christ, which he gave to save sinners.

Do we then eat the body of Christ?

No; but we eat the bread which means his body: and, as we eat bread that our bodies may not die, so our souls love Jesus Christ, and receive him for their Saviour, that they may not die.

What is the holy wine?

It is the blood of Christ, which he poured out on Calvary, in Jerusalem in Judea, to save us sinners.

Do we, then, drink the blood of Christ?

No; but the wine signifies his blood, just as the holy bread signifies his body: and all those who go to Christ, and lean on him, will have their sins washed away by his blood, and their souls saved for ever in heaven.

Why do you think it more suitable that you should join the church than others?

Perhaps it is not, (hesitating.) If it is not proper, you must tell me. But I do greatly desire to dwell with you in the fold of Christ.

Who do you think are the proper persons to be received into the church?

Those who have repented of their sins, and have obtained new hearts.

What is a new heart?

It is one which loves God, and loves the word of God, and does not love sin, or sinful ways.

Do you think you have obtained a new heart?

At one time I think I have, and then again I think I have not. I do not know,—I hope I have a new heart.

What makes you hope you have a new heart?

This is the reason why I hope I have a new heart. The heart I have now is not like the heart I formerly had. The one I have now is very bad, it is unbelieving, and inclined to evil. But it is not like the one I formerly had. Yes; I think I have a new heart.

The satisfaction arising from this simple yet decisive testimony, is increased from a knowledge of the fact stated by Mr. Richards; namely, that these questions and answers were not committed to memory, and merely recited on the occasion, but that they were the undisguised motives and feelings by which he was influenced. He had no knowledge of the questions that would be proposed, until the time when they were publicly asked, and consequently could not have previously framed the replies he gave. The above may be taken as a sample of the kind of declarations made by those who are united in church fellowship; and though it relates to a native of the North Pacific Isles, it resembles in its principal characteristics those of many of the natives of the Southern group. Simplicity is the distinguishing feature in all their religious intercourse of this kind.

There is another very pleasing trait in their christian character, namely, their undoubting reception of the Scriptures, as a Divine revelation. We have plainly and uniformly stated its truths, inculcating among them no opinions or sentiments, on matters of religion, but such as are found in the Bible; declaring that what it taught was essential, and that all the opinions of men, however excellent, are in comparison unimportant. To the Bible we have always referred them, as the authority for what we have taught, and of its declarations we have allowed no evasion. The injunctions of Scripture they have therefore been accustomed to receive implicitly, as they are recorded; and while they exercise their own judgments very freely in matters of human opinion, I never knew one, who professed himself a Christian, inclined to doubt the authority of the Bible. To this standard we have always referred their opinions and their conduct; and by the criterion it furnishes, we always recommended their examining their own condition, rather than comparing their conduct with that of others.

Often, when we have recommended some measure of a religious or general nature, which we have supposed would be advantageous to them, they have inquired, What says the Scripture? Is there any thing about it in the word of God? If, as was sometimes the case, we were under the necessity of stating, that there was nothing in the Scripture directly referring to our recommendation, but that it was according to the general tenour and spirit of the Scriptures, or corresponding with the practice of Christians in England; they would sometimes answer, "That may be very good, but as it is only a matter of opinion with you, we will think about it." On the other hand, so far as those who were members of our churches, or had been baptized, were concerned, I cannot recollect any measure we ever proposed, for which we could refer to the explicit declaration of Scripture as our authority, that they did not at once unhesitatingly adopt. It was much more satisfactory to us that the conduct of their lives should be regulated by principles derived from the Scripture, than by the opinion of their teachers, however highly they might respect them; and we had always rather that they should ask, "What says the word of God?" than, "What say the Missionaries?" The opinions of their teachers may change, or teachers of different opinions may succeed them, but the word of God will endure unalterably the same, being a more sure word, whereunto they do well to take heed.

What the experience of my predecessors in the field may have been, with regard to the manner in which the natives were disposed to admit the claims of the Scriptures to a divine origin, I am not prepared to state with confidence. I believe, however, it was not so much to the divine authority, as to the doctrines of the sacred volume, that they objected. So far as my recollection serves, with regard to the island of Huahine, the inhabitants, though not idolaters, certainly were not Christians, except in name; and in the Sandwich Islands, where, on my first arrival, the people were more opposed than inclined to all that is essential to Christianity, I do not remember to have met with an individual disposed to doubt the origin, or dispute the authority, of revelation. It was to the injunctions and doctrines of the Bible, that humbled their pride, and prohibited their vicious practices, &c. that they objected.

It may be said, that while they believed in idolatry—and revelations from the gods by dreams, or other intimations through the medium of the priests, were acknowledged—that they might suppose the truths of the Bible to be a collection of revelations similar in kind to these, only, as a priest on one occasion stated to me, better preserved, being "made fast upon the paper." But after they had renounced idolatry, and treated with contempt the notions formerly entertained respecting the power of the gods, and regarded all the pretended revelations of them as deceptions of the priests, the claims of the Bible remained undisputed.

The uniform acceptance of the declarations of Scripture as Divine communications to mankind, was not the result of any arguments employed by us. We never attempted to establish by argument what they were not inclined to doubt. Our instructions were, therefore, generally delivered in the simplicity of assertion, or testimony, accompanied with suitable admonition and application to our hearers; taking it as an admitted principle, that the Scriptures contained a declaration of the will of God.

When asked, as we sometimes were, "How do you know the Bible is the word of God?" we did not adduce an infallible church, by which it had been determined what were the canonical books, and by whom they had been preserved; nor did we refer them often to the testimony of history, to prove that the persons, whose names were affixed to the different parts, actually wrote the books ascribed to them, but we referred them to their internal evidence, their harmony or accordance with the works of creation, and the dispensations of Providence, in their display of the Divine character and perfections, their admirable adaptation to the end for which they were given, and the universality of their application to mankind. Next to the agency of that blessed Spirit, under whose influence those Scriptures were first penned, and by which alone they become the means of spiritual illumination to any individual, the internal evidences of the Bible have operated upon the minds of the natives with great force. When they have been asked why they believed the Scriptures to be the word of God, they have answered, "We believe they have a higher than human origin, because they reveal what man could never know; not only in reference to God himself, but our own origin and destinies, and what, when revealed, appears to us true; because its declarations accord with the testimony of our own consciences, as to the moral character of our actions; and because, though written by persons who never saw us, or knew our thoughts, it describes so accurately our inclinations, imaginations, motives, and passions. It must have been dictated by One who knew what man was, better than we know each other, or it could not have displayed our actual state so correctly." These, or declarations to the same effect, if not given in precisely the same words, were the reasons they frequently assigned for believing the divine origin of the Scriptures.

Several remarkable instances of the effect of the word of God, and the power of conscience, occurred about the year 1819. One Sabbath morning, Mr. Nott had been preaching from the words :—"Let him that stole, steal no more." In his discourse, he had refuted the idea they had formerly held, that theft was no crime, but rather an act of merit, if committed with dexterity; and had shewn that the circumstance of detection or escape did not alter the moral quality of the act in the sight of God; that every means employed unjustly to deprive another of his property, was an act of theft, and that restitution ought to be made for past robberies, as well as honesty practised for the future. The next morning, when he arose and opened his door, he saw a number of natives sitting on the ground in the front of his dwelling. Their appearance was rather singular, and the unseasonable time of their assembling led him to inquire the cause. They answered, "We have not been able to sleep all night; we were in the chapel yesterday; we thought, when we were pagans, that it was right to steal when we could do it without being found out. Hiro, the god of thieves, used to assist us. But we heard what you said yesterday from the word of God, that Jehovah had commanded that we should not steal. We have stolen, and all these things that we have brought with us are stolen goods." One then lifted up an axe, a hatchet, or a chisel, and exclaimed, "I stole this from the carpenter of such a ship," naming the vessel, &c. Others held up an umeti, or a saw, or a knife; and, indeed, almost every kind of moveable property was brought and exhibited, with confessions of having been stolen. Mr. Nott said, rather smilingly, "What have you brought them to me for? I do not want them." (The sentiment had often been circulated, that the receiver of stolen goods was as bad as the thief.) "You had better take them home, and, if you have stolen any from your own countrymen, return them; and when the ships come again from which any of the goods have been stolen, take them back, together with a present to the captain or the carpenter, expressive of your desire to make restitution." They all said—"Oh, no, we cannot take them back; we have had no peace ever since we heard it was displeasing to God, and we shall have no peace so long as they remain in our dwellings; we wish you to take them, and give them back to the owners whenever they come." Such was the power of conscience, that although they were even tools, which the natives value more highly than gold, and although Mr. Nott requested them to take them back, he could not persuade one of them to do so; they left them all with him, to be returned to their owners. They went even farther than this: Some had stolen articles from one of the Missionaries at Eimeo. They fitted up the canoe, and with the first fair wind undertook a voyage of upwards of seventy miles, for the purpose of carrying back what they had taken.

In the island of Raiatea, a native walking on one occasion towards the mountains, discovered a hen's nest with a number of eggs in it, at the root of a tree. He eagerly seized the prize, put the eggs in the native cloth he wore, and proceeded with them to his house. On the way, he recollected the commandment—"Thou shalt not steal," and though he had found the nest far from any habitation, in the midst of the woods, and did not know that he had robbed any one except the hen, yet he knew the eggs were not his, and so powerful was the impression of the impropriety of the action, that he returned to the nest, and very carefully replaced the eggs with a light heart.

A similar course was pursued by a native with whom I was once travelling across the island, with regard to a pocket-knife that he had picked up, but afterwards threw down, near the same place, simply because it did not belong to him.

These facts are most pleasing and decisive illustrations of the power of Christian principles. Yet every individual is not influenced by them. These were Christian men; there are others who are such only in name, and who are addicted to the practice of pilfering and theft, especially at those stations near the harbours which are the most frequent resorts of shipping, where the temptations are greatest, and the influence of foreign intercourse most injurious. Nevertheless, when we consider that they were formerly, as every navigator by whom they were visited has testified, almost a nation of thieves—that Hiro, the god of thieves and plunderers, occupied a place in their mythology, and had a temple and priests—we cannot but admire the operation of Christian principles in producing, in such a number of instances, a conscientious regard to justice and honesty. It was, there is reason to believe with many, the result, not of an apprehension of detection, but of a strict regard to moral rectitude, and the declared will of Him who said—"Thou shalt not steal."

The meeting of those who were desirous of uniting with us continued; and from among them who attended, many were added to the church. Besides this meeting, we held one with the communicants only, on the Friday evening preceding the Sabbath when the ordinance was administered, which was the first Sabbath in every month. At these times, new members were proposed by the Missionary, or by any member, to the whole body. Inquiry was made of those present, as to their eligibility, and if any had objections to an individual, he was requested to state them there; if not, one or two of the members were directed to call upon the parties at their habitations, to converse with them, and report the same at the next meeting, for the satisfaction of the church. It was regarded by us a duty, to see these persons more than once during the intervening month.

At the next meeting, these individuals were proposed by name; the recommendation of the persons who had visited them, and of the Missionary, given; and if the members present knew any reasons why they should not be united with them, they were requested to state the same; if not, to signify assent by lifting up the right hand. When the members proposed had been thus individually approved, as they were usually in attendance, they were brought to the chapel, and interrogated singly, as to their reasons for desiring to unite with us. To these questions brief replies were usually rendered; and they were informed that the members of the church, considering them proper persons, were happy to receive them. The right-hand of fellowship was then given by the Missionaries, and subsequently by the members, to those thus received; and the meeting closed with devotional exercises.

We did not require any written confession of faith, nor invariably a verbal account of experience, from the persons admitted. In this latter respect, our procedure was not uniform, but regulated by the peculiar circumstances of the individual.

Towards the close of the year 1820, Mr. Davies left Fare, to supply the station at Papara, in Tahiti, which had been destitute of a Missionary since the decease of Messrs. Tessier and Bicknell. The management of the press, supplying the bocks for the whole of the Leeward Islands, the superintendence of the schools, promoting the civilization of the people, attending the religious meetings, together with our pastoral duties, now pressed so heavily upon us, that we found some assistance requisite. This we necessarily sought among the converts, and were happy to find four persons, members of the church, suitable to act as assistants, whom we proposed to the church to elect as deacons. *Diaconi* is the term by which they are designated; not, however, selected from any strong predilection to the term, or any extraordinary importance attached to it, but because a scriptural term, and one more easily assimilated to the idiom of their language than some others.

On the 15th of February, 1821, they were set apart in the church to this office, by an address from 1 Tim. iii. 10. and prayer for the blessing of God upon them. *Auna, Taua, Pohuetea,* and *Matatore,* were the persons selected, and so long as I continued in the islands, we found them consistent Christians, and valuable coadjutors in managing the temporal concerns of the church, visiting the sick, attending the prayer-meetings, &c.

Religion was now almost the sole business of the people at Fare, and the adjacent districts; and although the meetings were frequent, many continued to visit our dwellings, sometimes by day-break; and often, after we had retired to rest at night, one or two would come knocking gently at our doors or windows, begging us to give them directions, or to answer their inquiries as to the thoughts that distressed their minds. No time, no place, appeared to them unappropriate; and whether they sat in the house, or walked by the way—skimmed the surface of the water in their light canoe, or laboured in the garden—religion was the topic of their conversation. Their motives were various, and probably often of a very mixed character. Some were influenced by a desire to be thought well of by their neighbours; many wished to be baptized without feeling the necessity of, or more earnestly seeking, that spiritual purification which it signified; and others, perhaps, considering church-membership as the highest christian distinction they could gain, desired to be admitted to the communion, as an end of their profession, rather than a means of higher spiritual attainments.

Such individuals, we deemed it, on all occasions, necessary to caution with the greatest simplicity and faithfulness. But while these were the motives by which we have reason to believe many were influenced, there were others who certainly acted from different feelings, who were unable to rest under a sense of guilt and its fearful consequences; who desired to hear more about God, his mercy to sinners, and the love of their Saviour, that their burden of sin might be removed; while some, desirous of expressing their sense of the goodness of God, were anxious to be informed what they might do to promote his praise. I cannot look back upon this period of my Missionary life with indifference; nor can I contemplate the state of the people at this time, without believing that the Spirit of God was powerfully operating upon the minds of many. Of this, their subsequent lives have afforded satisfactory evidence. Instability was one of their prominent traits of character, and did we not believe in a higher agency than their own purposes or principles, we should fear that many would abandon the profession they have made, and return to their former course of life.

Although the advantages resulting from frequent meetings for religious conversation, were too obvious to allow us to withhold every encouragement; and though, under the present circumstances and feelings of the people, they were peculiarly so; yet, as many of the communicants, and several who were desirous of uniting with them, were females, there were many things, in reference to which they needed advice, but which they did not deem suitable to introduce at a public meeting. Mrs. Barff and Mrs. Ellis therefore, being able to converse familiarly in the native language, proposed to meet the female members of the church, and those of their own sex who were desirous of joining them, once a week, for general conversation, and mutual spiritual improvement. This was an interesting meeting; it was held alternately at our respective habitations, Mrs. Barff and Mrs. Ellis both attending. It commenced with singing a hymn; a prayer was offered, and a portion of Scripture read. After this, the most unreserved conversation followed, on religious subjects, the training of their children, and other relative duties connected with the new order of things which Christianity had introduced.

Parental discipline among the people, prior to their reception of Christianity, had been remarkably lax. The children were their own masters as soon as they could act for themselves, and the restraint which the mother was able to impose was trifling indeed. Such was the abundance of provision, that the maintenance of a child was a matter of no anxiety to any one. Hence, if a boy felt offended with his parents, he left them without ceremony, attached himself to another family in an adjacent or remote district, and remained for months without visiting his father's house. To restrain these fugitive habits, and train their children to regular industry, was one of the duties inculcated on Christian parents; yet the children could but ill brook any restraint. I have seen a child, not more than six years old, strike or throw stones at his mother, and the father would oftentimes be scarcely more regarded.

The mothers were now anxious to influence the minds of their children, and gain their respect by kindness. The fathers sometimes had recourse to harsher measures. Hoibu had two sons that were a source of great trouble to him. One of our number went one day into his house, which was a native dwelling, with no other ceiling than the inside of the roof, the ridge-pole extending along the centre, about twenty feet from the floor. After talking some time with the man, the visitor heard something rustling in a long basket of cocoa-nut leaves at the top of the house, and, looking up, saw the legs and arms of a boy protruding from the basket. On inquiring the cause of this, Hoibu said, the boy had been disobedient, and, in order to convince him of his error, he had first talked to him, and then put him into the basket, and, passing a rope over the ridge-pole, had fastened one end of it to the basket, and, pulling the other, had drawn him up there, that he might think on his disobedience, and not be guilty of the same again. He was informed that it was rather a novel mode of punishment, and that it was hoped he would not keep him there long. He said, no, he should lower him before the evening. A similar mode of punishment may, I believe, have been used in some of our public schools, in which a kind of large birdcage has been substituted for a basket; but of this Hoibu had never heard. The invention was his own, and it was scarcely possible to repress a smile at the ludicrous appearance of the suspended boy.

Although the training of their children, and other domestic duties, which the females were now called to discharge, were important matters of inquiry, there were others, more deeply interesting, frequently brought forward at their meetings. Some of these questions regarded the children who were born since the gospel had been introduced, and who they were most anxious should share all its blessings; others frequently referred to such as they had murdered under the influence of idolatry. Sometimes a mother would, in enumerating the crimes of which she had been guilty, recount the number of her innocents she had destroyed, and with anguish relate her struggles of affection, or pangs of remorse, and the distress she now felt; observing, that their images were ever present to her thoughts, and, as it were, constantly haunting her paths, so that she was afraid even to retire to the secret places of the bushes for private prayer, lest their ghosts should rise before her. Often such individuals would say, they feared there was no hope of mercy for them, that they had repeatedly committed the premeditated murder of the innocent, and would perhaps repeat the Scripture declaration, that no murderer hath eternal life abiding in him, and ask, "Ought I to go to Jesus Christ for pardon? were any murderers of their own children ever forgiven?"

While some would ask such questions as these, or state them as the exercises of their own minds, there were others who would speak of the cruelties of which they had been guilty, with a want of feeling that has appeared to border on insensibility to their enormity. Many, however, especially those who were most sensible of the mercy of God through Christ, would on these occasions expatiate on the amazing forbearance of Jehovah, in sparing such merciless creatures as they had been. They would also express their astonishment at the love of Christ in dying for them; and the abundance of his compassion, in continuing to send them the intelligence of his salvation, and, after they had long disregarded it, not only forbearing, but making them willing in the day of his power; melting their hearts, drawing them with cords of affection, and now causing them to rejoice in his love shed abroad in their hearts.

Occasionally they would, in most affecting strains, allude to the anguish which the sight of their neighbours' children produced, by recalling to remembrance those whom they had destroyed. The contrast they often drew between their own childless and desolate condition through their former cruel practice of infant murder—and that of those happy parents—who, under the reign of the Messiah, were surrounded by their children, was touching and painful. These were topics that could not be discussed without emotion, either by those who brought them forward, or by those from whom direction and advice were sought.

There was another matter connected with this, of scarcely inferior interest, and that was, the state of those infants after death. Are their spirits, they would say, in outer darkness, where there is weeping and gnashing of teeth, or are they happy? In reply to this, though opinions were not given with confidence, they were informed, that though they had not sinned, they had suffered death as the effect of Adam's transgression, yet that there was reason to hope and believe they were interested in the covenant of redemption, the condition of which the Lord Jesus Christ had fulfilled, and that therefore they were happy.

It is impossible to conceive the satisfaction of mind which this opinion has inspired in those who had been guilty of the destruction of their offspring, though they were still sensible that the final condition of the murdered infants did not affect the criminal nature of the unnatural deed.

In reference to this point, they would often ask whether they should in heaven know those they had been acquainted with on earth, and especially if there they should recognize the children they had destroyed. In reply, they were informed, that from all that was said on the heavenly state in the Scriptures, there was reason to believe that friends on earth would know each other there, and that it was probable christian mothers would meet their children.

These were not mere speculative inquiries, the parties had a deep personal interest in them; and Mrs. Ellis has been greatly affected in witnessing the emotions with which these discussions have been carried on. I can readily suppose it altogether impossible to conceive of the rapturous expectation with which a christian mother, childless and desolate from her own cruelties, would by faith anticipate meeting in the happy world of spirits the children she had murdered in her days of ignorance on earth, and joining with them to celebrate the praises of Him by whom they had been snatched from the world of sinners ere they had felt its bitter contamination, and she had been brought to share redemption from its curse.

This opinion was not given simply to afford alleviation to the distressed feelings of such unhappy parents, but because it did not seem opposed, but rather favoured, by the word of God, agreeable to the benevolent character of the Deity, and adapted to enlarge our views of His compassion, without affecting His other attributes. We could, therefore, adopt the language and sentiments of the poet, in the belief that,

> "The harp of heaven
> Had lack'd its least, but not its meanest string,
> Had children not been taught to play upon it,
> And sing, from feelings all their own, what men
> Nor angels can conceive of creatures, born
> Under the curse, yet from the curse redeem'd,
> And placed at once beyond the power to fall,—
> Safely, which men nor angels ever knew,
> Till ranks of these, and all of those, had fallen."

The meeting of the females was closed with prayer by one of the natives, who, if a mother, would give the child, she had perhaps been nursing in her lap, to some one sitting by. Their prayers were marked by deep spirituality and strong feeling; and, I believe, these meetings were some of the most affecting seasons of intense and painful, or joyous and hallowed emotion, ever experienced. The individual engaging in the devotional exercise has sometimes, from the strength of feeling, been unable to proceed, and tears alone have afforded relief.

Early in the year 1821, in order to cultivate the most affectionate and profitable intercourse with our people, we proposed, in addition to visits in times of sickness, to pay to each family a pastoral visit, for part of an evening, once a month, or at least once in the course of two months. Mr. Barff and myself, dividing the families between us, were enabled to accomplish this. We were received with kindness by the parties, and it was our study to make these visits advantageous. The time was not spent in useless recital of the passing reports of the day; we addressed ourselves to each individual, when circumstances admitted, directing and encouraging them in their adherence to the Saviour, or inviting them to Him, and concluded our visit by uniting in prayer for the blessing of God upon their household, &c. We trust these domiciliary visits were beneficial; they were often cheering to our own minds. Some of the many happy hours I have been privileged to spend in Missionary occupations, have been those passed in the families of our people on such occasions. Here we sometimes saw the household virtues, the endearments of social and domestic comfort, cherished—shedding their benign, elevating, and purifying influence upon a family, the principal members, and sometimes the greater part of which, were enjoying that blessing which maketh rich, and addeth no sorrow, and were preparing to join the family of the redeemed in the abodes of blessedness.

Associated with these delightful duties, there were others equally needful, but less pleasing, which we were called to discharge, in connexion with the infant church we had been honoured to gather. These were acts of discipline, in the dismissal of those who by their conduct had disgraced the Christian profession. On these occasions, we presented to their consideration the direction of the scriptures, and the duty of the church resulting therefrom; and when it was necessary to dismiss an individual from fellowship, it was always done with solemn prayer and most affecting regret.

We were not called to this painful duty soon or often. One or two instances occurred, before I removed to the Sandwich Islands. They were, however, exceedingly distressing, especially the first, which preyed so constantly upon the mind of the individual, that, though fully convinced of his fault, and the propriety of the proceeding, he never recovered the shock he received. It was exceedingly painful to those, who could no longer, without dishonouring the Christian name, allow him to be identified with them, to separate him. He soon offered every evidence of deep and sincere penitence, and was affectionately invited to return to the bosom of the church: but although he came again among them, a cloud ever after hung over him; and a disease, aggravated by mental anxiety, now attacked his frame, and soon brought him to the grave.

Christian churches were formed upon the same or similar principles in the Windward or Georgian Islands, some months before this was established in Huahine. From the peculiar local circumstances of the people, the churches in Tahiti have been exposed to greater trials than that in Huahine has yet experienced, especially those formed in stations adjacent to the anchorage of shipping. In the vicinity of these, the baneful influence of foreign seamen is most destructive of moral improvement and Christian propriety in the people; and it is probable that there is more immorality among the inhabitants, and more disorder in the churches, at the stations which are the resort of shipping, than in all others throughout the islands. Still the churches there have not been, and are not, without some indication of the Divine care and blessing.

Subsequently, churches were formed in Raiatea, Tahaa, and Borabora, which have in general prospered. As their constitution and proceedings resemble those of Huahine, it is unnecessary to detail their origin or progress. I have selected that in Huahine, not because it is superior to others for its order, or faith, or the piety of its members, but because it was that of which I was, with my esteemed colleague, a pastor, till the providence of God called me to another field of Missionary labour—and because it was planted in the station at which I spent the greater part of the time I resided in the South Sea Islands.

I have also been minute, perhaps too much so, in detailing its nature, order, and discipline. This has not arisen from a desire to give it undue prominency, but because it forms an important epoch in the history of the people, and is a matter of considerable interest with many who are concerned in the extension of the Christian faith throughout the world; I also conceived the patrons of the South Sea Mission entitled to the most ample information on the subject.

It has not been my object to exhibit the plan and order of this, or the other churches in those islands, as models of perfection, nor to claim for them any degree of excellency which others, formed and regulated differently in some minor respects, might not possess; but simply to narrate our own views, and consequent proceedings, in reference to measures which will be regarded with indifference by few, if any, whatever may be their peculiar opinions as to the plan we have pursued. From all, I would ask fervent prayer, that whatever has been contrary to the will of God may be amended, and that what has been agreeable thereto may continue to share his blessing. The church of Christ in Huahine, as well as those in other islands, has had its trials. Some of its members, as might be expected, have departed from the faith and the purity of the gospel. The instances, however, have not been numerous; and I am gratified to know, that a number of young persons, several from the Sunday schools, have joined it; and that, though formed by sixteen individuals in the spring of 1820, it contained, in the autumn of 1827, nearly five hundred members.

CHAP. XII.

Government of the South Sea Islands monarchical and arbitrary—Intimately connected with idolatry—Different ranks in society—Slavery—The proprietors of land—The regal family—Sovereignty hereditary—Abdication of the father in favour of the son—Distinctions of royalty—Modes of travelling—Sacredness of the king's person—Homage of the people—Singular ceremonies attending the inauguration of the king—Language of the Tahitian court—The royal residences—Dress, &c.—Sources of revenue—Tenure of land—Division of the country—National councils—Forfeiture of possessions.

The government of the South Sea Islands, like that which prevails in Hawaii, was an arbitrary monarchy. The supreme authority was vested in the king, and was hereditary in his family. It differed materially from the systems existing among the Marquesians in the east, and the New Zealanders in the south-west. There is no supreme ruler in either of these groups of islands, but the different tribes or clans are governed by their respective chieftains, each of whom is, in general, independent of any other. Regarding the inhabitants of Tahiti, and the adjacent islands, as an uncivilized people, ignorant of letters and the arts, their modes of governing were necessarily rude and irregular. In many respects, however, their institutions indicate great attention to the principles of government, an acquaintance with the means of controlling the conduct of man, and an advancement in the organization of their civil polity, which, under corresponding circumstances, is but rarely attained, and could scarcely have been expected.

Their government, in all its multiplied ramifications, was closely interwoven with their false system of religion, in its abstract theory, and in its practical details. The god and the king were generally supposed to share the authority over the mass of mankind between them. The latter sometimes personated the former, and received the homage and the requests presented by the votaries of the imaginary divinity, and at other times officiated as the head of his people, in rendering their acknowledgments to the gods. The office of high-priest was frequently sustained by the king—who thus united in his person the highest civil and sacerdotal station in the land. The genealogy of the reigning family was usually traced back to the first ages of their traditionary history; and the kings, in some of the islands, were supposed to have descended from the gods. Their persons were always sacred, and their families constituted the highest rank recognized among the people.

The different grades in society were not so distinctly marked in Polynesia, as among the inhabitants of India, where the institution of *caste* exists; nor were they so strongly defined in Tahiti as among the Sandwich Islanders, whose government was perhaps more despotic than that which prevailed in the southern islands. The lines of separation were, nevertheless, sufficiently prominent; the higher orders being remarkably tenacious of their dignity, and jealous of its deterioration by contact with those beneath them.

Society among them was divided into three distinct ranks: the *hui arii*, the royal family and nobility—the *bue Raatira*, the landed proprietors, or gentry and farmers—and the *manahune*, or common people. These three ranks were subdivided into a number of distinct classes; the lowest class included the *titi* and the *teuteu*, the slaves and servants; the former were those who had lost their liberty in battle, or who, in consequence of the defeat of the chieftains to whom they were attached, had become the property of the conquerors. This kind of slavery appears to have existed among them from time immemorial. Individuals captured in actual combat, or who fled to the chief for protection when disarmed or disabled in the field, were considered the slaves of the captor or chief by whom they were protected. The women, children, and others, who remained in the districts of the vanquished, were also regarded as belonging to them; and the lands they occupied, together with their fields and plantations, were distributed among the victors.

We do not know that they ever carried on a traffic in slaves, or sold those whom they had conquered, though a chief might give a captive for a servant to a friend. This is the only kind of slavery that has ever obtained among them, and corresponds with that which has prevailed in most of the nations of the earth in their rude state, or during the earlier periods of their history. This state of slavery among them was in general mild, compared with the affecting cruelty by which it has been distinguished in modern times, among those who support the inhuman system of trafficking in these unhappy beings. If peace continued, the captive frequently regained his liberty after a limited servitude, and was permitted to return to his own land, or remain in voluntary service with his master.

So long, however, as they continued slaves or captives, their lives were in jeopardy. Sometimes they were suddenly murdered, to satiate the latent revenge of their conquerors; at others reserved as human victims, to be offered in sacrifice to their gods. Slavery, in every form, is perfectly consistent with paganism, and was maintained among them as one means of contributing to its support. This kind obtains in most of the islands, but is probably far more oppressive in New Zealand than in the Society Islands. The slaves among the former are treated with the greatest cruelty, and often inhumanly murdered and eaten.

The *manahune* also included the *teuteu*, or servants of the chiefs; all who were destitute of any land, and ignorant of the rude arts of carpentering, building, &c. which were respected among them, and such as were reduced to a state of dependence upon those in higher stations. Although the manahune have always included a large number of the inhabitants, they have not in modern times been so numerous as some other ranks. Since the population has been so greatly diminished, the means of subsistence so abundant, and such vast portions of the country uncultivated, an industrious individual has seldom experienced much difficulty in securing at least the occupancy of a piece of land. The fishermen and artisans (sometimes belonging to this class, but more frequently to that immediately above it,) may be said to have constituted the connecting link between the two.

The *bue raatira*, gentry and farmers, has ever been the most numerous and influential class, constituting at all times the body of the people, and the strength of the nation. They were generally the proprietors and cultivators of the soil, and held their land, not from the gift of the king, but from their ancestors. The petty raatiras frequently possessed from 20 to 100 acres, and generally had more than their necessities required. They resided on their own lands, and enclosed so much as was necessary for their own support. They were the most industrious class of the community, working their own plantations, building their own houses, manufacturing their own cloth and mats, besides furnishing these articles for the king.

The higher class among the raatiras were those who possessed large tracts of land in one place, or a number of smaller sections in different parts. Some of them owned perhaps many hundred acres, parts of which were cultivated by those who lived in a state of dependence upon them, or by those petty raatiras who occupied their plantations on condition of rendering military service to the proprietors, and a portion of the produce. These individuals were a valuable class in the community, and constituted the aristocracy of the country. They were in general more regular, temperate, and industrious in their habits, than the higher ranks, and, in all the measures of government, imposed a considerable restraint upon the extravagance or precipitancy of the king, who, without their co-operation, could carry but few of his measures. In their public national assemblies, the speakers often compared the nation to a ship, of which the king was the mast; and whenever this figure was used, the raatiras were always termed the shrouds, or ropes by which the mast is kept upright. Possessing at all times the most ample stores of native provisions, the number of their dependents, or retainers, was great. The destitute and thoughtless readily attached themselves to their establishments, for the purpose of securing the means of subsistence without care or apprehension of want.

The bue raatira were the middle class in society; forming the most important body in times of peace, and furnishing the strength of their armies in periods of war. Warriors were sometimes found among the attendants on the king or chief; but the principal dependence was upon the raatiras. These, influenced by the noble spirit of independence, accustomed to habits of personal labour, and capable of enduring the fatigues of war, were, probably from interest in the soil, moved by sentiments of patriotism more powerfully than any other portion of the people. The raatiras were frequently the priests in their own family temples; and the priests of the national maraes, excepting those allied by blood to the reigning families, were usually ranked with them.

The hui arii, or highest class, included the king or reigning chieftain in each island, the members of his family, and all who were related to them. This class, though not numerous, was considered the most influential in the state. Being the highest in dignity and rank, its elevation in the estimation of the people was guarded with extreme care; and the individuals of whom it was composed, were exceedingly pertinacious of their distinction, and jealous of the least degradation by the admission of inferiors to their dignity.

Whenever a matrimonial connexion took place between any one of the hui arii with an individual of an inferior order, unless a variety of ceremonies was performed at the temple, by which the inferiority was supposed to be removed, and the parties made equal in dignity, all the offspring of such an union was invariably destroyed, to preserve the distinction of the hui arii, or reigning families.

The king was supreme, and next to him the queen. The brothers of the king, and his parents, were nearest in rank, the other members of the family taking precedence according to their degrees of consanguinity. The regal office is hereditary, and descends from the father to the eldest son: it is not, however, confined to the male sex; these islands have often been governed by a queen. Oberea was the queen of Tahiti when it was discovered by Wallis; and Aimata, the daughter of Pomare II. now exercises the supreme authority in Tahiti and Eimeo. the daughter of the king of Raiatea is also the nominal sovereign of the island of Huahine.

The most singular usage, however, connected with the established law of primogeniture, which obtained in the islands, was the father's abdication of the throne on the birth of his son. This was an invariable, and appears to have been an ancient practice. If the rank of the mother was inferior to that of the father, the children, whether male or female, were destroyed; but if the mother originally belonged to the hui arii, or had been raised to that elevation on her marriage with the king, she was regarded as the queen of the nation. Whatever might be the age of the king, his influence in the state, or the political aspect of affairs in reference to other tribes, as soon as a son was born, the monarch became a subject—and the infant was at once proclaimed the sovereign of the people. The royal name was conferred upon him, and his father was the first to do him homage, by saluting his feet, and declaring him king. The herald of the nation was then despatched round the island with the flag of the infant king. The banner was unfurled, and the young sovereign's name proclaimed in every district. If respected, and allowed to pass, it was considered an acknowledgment by the raatiras and chiefs, of his succession to the government; but if broken, it was regarded as an act of rebellion, or an open declaration of war. Numerous ceremonies were performed at the marae, a splendid establishment was forthwith formed for the young king, and a large train of attendants accompanied him to whatever place he was conveyed.

Every affair important to the internal welfare of the nation, or its foreign relations, continued to be transacted by the father, and those whom he had formerly associated as his counsellors; but every edict was issued in the name and on the behalf of the young ruler; and though the whole of the executive government might remain in the hands of the father, he only acted as regent for his son, and was regarded as such by the nation. The insignia of regal authority, and the homage which the father had been accustomed to receive from the people, were at once transferred to his successor. The lands, and other sources of the king's support, were appropriated to the maintenance of the household establishment of the infant ruler; and the father rendered him those demonstrations of inferiority, which he himself had heretofore required from the people.

This remarkable custom was not confined to the family of the sovereign, but prevailed among the hui arii and the raatiras. In both these classes, the eldest son immediately at his birth received the honours and titles which his father had hitherto borne.

It is not easy to trace the origin or discover the design of a usage so singular, and apparently of such high antiquity, among a people to whom it is almost peculiar.

Its advantages are not very apparent, unless we suppose it was adopted by the father to secure to his son undisputed succession to his dignity and power. If this was the design, the plan was admirably adapted to its accomplishment; for the son was usually firmly fixed in the government before the father's decease, and was sometimes called to act as regent for his own son, before, according to the ordinary institutions, he would himself have been invested with royal dignity.

Considering the inhabitants of the South Sea Islands as but slightly removed from barbarism, we are almost surprised at the homage and respect they paid to their rulers. The difference between them and the common people was, in many respects, far greater than that which prevails between the rulers and the ruled in most civilized countries. Whether, like the sovereigns of the Sandwich Islands, they were supposed to derive their origin by lineal descent from the gods, or not, their persons were regarded as scarcely less sacred than the personifications of their deities.

Every thing in the least degree connected with the king or queen — the cloth they wore, the houses in which they dwelt, the canoes in which they voyaged, the men by whom they were borne when they journeyed by land, became sacred—and even the sounds in the language, composing their names, could no longer be appropriated to ordinary significations. Hence, the original names of most of the objects with which they were familiar, have from time to time undergone considerable alterations. The ground on which they trod, even accidentally, became sacred; and the dwelling under which they might enter, must for ever after be vacated by its proprietors, and could be appropriated only to the use of these sacred personages. No individual was allowed to touch the bodies of the king or queen; and every one who should stand over them, or pass the hand over their heads, would be liable to pay for the sacrilegious act with the forfeiture of his life. It was on account of this supposed sacredness of person that they could never enter any dwelling, excepting those that were specially dedicated to their use, and prohibited to all others; nor might they tread on the ground in any part of the island but their own hereditary districts.

The sovereign and his consort always appeared in public on men's shoulders,* and travelled in this manner wherever they journeyed by land. They were seated on the neck or shoulders of their bearers, who were generally stout athletic men. The persons of the men, in consequence of their office, were regarded as sacred. The individuals thus elevated appeared to sit with ease and security, holding slightly by the head, while their feet hung down on the breast, and were clasped in the arms of the bearer. When they travelled, they proceeded at a tolerably rapid pace, frequently six miles within the hour. A number of attendants ran by the side of the bearers, or followed in their train; and when the men who carried the royal personages grew weary, they were relieved by others.

The king and queen were always accompanied by several pair of sacred men, or bearers, and the transit from the shoulders of one to those of another, at the termination of an ordinary stage, was accompanied with much greater despatch than the horses of a mail-coach are changed, or an equestrian could alight and remount.

* As represented in the engraving, inserted at page 64, Vol. I.

On these occasions, their majesties never suffered their feet to touch the ground; but when they wished to change, what to them answered the purpose of horses, they called two of the men, who were running by their side; and while the man, on whose neck they were sitting, made little more than a momentary halt, the individuals who were to take them onward, fixed their hands upon their thighs, and bent their heads slightly forward: when they had assumed this position, the royal riders, with apparently but little effort, vaulted over the head of the man on whose neck they had been sitting, and, alighting on the shoulders of his successor in office, proceeded on their journey with the shortest possible detention.

This mode of conveyance was called *amo* or *vaha.* It could not have been very comfortable even to the riders, while to the bearers it must have been exceedingly laborious. The men selected for this duty, which was considered the most honourable post next to that of bearers of the gods, were generally exempted from labour, and, as they seldom did any thing else, were not perhaps much incommoded by their office; and although the seat occupied by those they bore was not perhaps the most easy, yet as it was a mark of the highest dignity in the nation, and as none but the king and queen, and occasionally their nearest relatives, were allowed the distinction it exhibited, they felt probably a corresponding satisfaction and complacency in thus appearing before their subjects, whenever they left their hereditary district. The effect must have been somewhat imposing, when, on public occasions, vast multitudes were assembled, and their sovereign, thus elevated above every individual, appeared among them.

In our different journeys and voyages among the islands, where there have been but few means of crossing a stream without fording it, or of landing from a boat or canoe without wading some distance in the water, we have often been glad to be carried, either across a river, or from the boat to the shore. On these occasions they have assisted us to mount in ancient regal style. Though we generally preferred riding on their backs, and throwing our arms round their necks, we have nevertheless, when the river has been deep, seated ourselves upon their shoulders, and in this position have passed the stream, without any other inconvenience than that which has arisen from the apprehension of losing our balance, and falling headlong into the water.—The inhabitants of Rurutu have a singular and less pleasant method of conveying their friends from a boat, &c. to the shore. On the arrival of strangers, every man endeavours to obtain one as a friend, and carry him off to his own habitation, where he is treated with the greatest kindness by the inhabitants of the district; they place him on a high seat, and feed him with abundance of the finest food. After an arrival from a strange island, when a man sees his neighbour carrying a friend or a newcomer on his shoulders, he attacks him—a fight ensues, for the possession of the prize—if the man who formerly possessed it is victorious, he goes home with his man on his shoulders, receives a hearty welcome, and is regarded by the whole district as a brave fellow, and a good man; whereas if he loses the prize, he is looked upon by all his friends as a coward.

I am not aware that the highest rulers in the Society Islands received at any time the same kind of homage which the Hawaiians occasionally paid to those chiefs who were considered to have descended from the gods. When these walked out during the season of tabu, the people prostrated themselves, with their faces touching the ground, as they passed along. A mark of homage, however, equally humiliating to those who rendered it, and probably as flattering to the individuals by whom it was received, was in far more extensive and perpetual use among the Tahitians. This was, the stripping down the upper garments, and uncovering the body as low as the waist, in the presence of the king. This homage was paid to the gods, and also to their temples. In passing these, every individual, either walking on the shore. or sailing in a canoe, removed whatever article of dress he wore upon the shoulders and breast, and passed uncovered the depository of the deities, the site of their altars or the rude temples of their worship.

Whenever the king appeared abroad, or the people approached his presence, this mark of reverence was required from all ranks; his own father and mother were not excepted, but were generally the first to uncover themselves when he approached. The people inhabiting the district through which he passed, uncovered as he approached; and those who sat in the houses by the road-side, as soon as they heard the cry of *te arii, te arii*, "the king, the king," stripped off their upper garments, and did not venture to replace them till he had passed. If by any accident he came upon them unexpectedly, the cloth they wore was instantly rent in pieces, and an atonement offered. Any individual whom he might pass on the road, should he hesitate to remove this part of his dress, would be in danger of losing his life on the spot, or of being marked as a victim of sacrifice to the gods.

This distinguishing mark of respect was not only rendered at all times, and from every individual, to the person of the king, but even to his dwellings, wherever they might be. These houses were considered sacred, and were the only habitations, in any part of the island, where the king could alight, and take refreshment and repose. The ground, for a considerable space on both sides, was in their estimation sacred. A *tii*, or carved image, fixed on a high pedestal, was placed by the road-side, at a short distance from the dwelling, and marked the boundary of the sacred soil. All travellers passing these houses, on approaching the first image, stripped off the upper part of their dress, and, whether the king was residing there or not, walked uncovered to the image at the opposite boundary. After passing this, they replaced their poncho, or kind of mantle, and pursued their journey.

To refuse this homage would have been considered not only as an indication of disaffection towards the king, but as rebellion against the government, and impiety towards the gods, exposing the individuals to the vengeance of the supreme powers in the visible and invisible worlds. Such was the unapproachable elevation to which the superstitions of the people raised the rulers in the South Sea Islands, and such the marked distinction that prevailed between the king and people, from his birth, until he was superseded in title and rank by his own son.

The ceremony of inauguration to the regal office, which took place when the king assumed the government, being one of considerable moment, was celebrated with a rude magnificence, though, like every other observance, it was distinguished by its disgusting abominations, and its horrid cruelty. There was no fixed period of life at which the youth were said to have arrived at years of manhood. Unaccustomed to keep even traditionary accounts of the time of their birth, there were but few whose age was known. The period therefore when the young king was formally invested with the regalia, and introduced to his high office, was regulated by his own character and disposition, the will of his father and guardians, or the exigences of the state; it generally took place some years before he had reached the age of twenty-one.

As it was one of the most important events to the nation, great preparation was made for its being duly celebrated; and whatever could give effect to the pageant was carefully provided. The gods indicated the interest they were supposed to take in the transaction, by the miraculous events that occurred at this time. Among those might be mentioned the sacred *aoa*, a tree resembling the banian of India, that spread over Faa-ape. This was said to have shot forth a new fibrous branch at his birth, and this branch or tendril reached the ground when he was to be made king. Taneua, a bamboo used on the occasion, was said to draw its roots out of the ground at the approach of the ceremony, and to leap into the hand of the person who was sent for it.

The inauguration ceremony, answering to coronation among other nations, consisted in girding the king with the *maro ura*, or sacred girdle of red feathers; which not only raised him to the highest earthly station, but identified him with their gods. The maro or girdle was made with the beaten fibres of the ava; with these a number of *ura*, red feathers, taken from the images of their deities, were interwoven. The maro thus became sacred, even as the person of the gods, the feathers being supposed to retain all the dreadful attributes of vengeance which the idols possessed, and with which it was designed to endow the king. Every part of the proceeding was marked by its absurdity or its wickedness, but the most affecting circumstance was the murderous cruelty attending even the preparation for its celebration.

In order to render the gods propitious to the transmission of this power, a human victim was sacrificed when they commenced the *fatu raa,* or manufacture of this girdle. This unhappy wretch was called the sacrifice for the *mau raa titi,* commencement or fastening on of the sacred maro. Sometimes a human victim was offered for every fresh piece added to the girdle; and when it was finished, another man, called Sacrifice for the *piu raa maro,* was slain; and the girdle was considered as consecrated by the blood of those victims. On the morning of what might be called the coronation day, when the king bathed prior to the commencement of the ceremonies, another human victim was required in the name of the gods.

The pageant, on this occasion, proceeded by land and water. The parties who were to be engaged in the transactions of the day, assembled in the marae of Oro, the tutelar deity of the nation. Certain ceremonies were here performed: the image of Oro, stripped of the sacred cloth in which he usually reposed, and decorated with all the emblems of his divinity, was conveyed to the large court of the temple; the *Papa rahio ruea,* or great bed of Oro, a large curiously formed bench or sofa cut out of a solid piece of timber, was brought out for the throne on which the king was to sit.

When these preliminaries were finished, they proceeded from the temple in the following order.—Tari-moa, one of the priests of the family of Tairi, carried the image of O o. The king followed immediately after the god. Behind him the large bed of Oro was borne by four chiefs. The miro-tahua, or orders of priests, with the great drum from the temple, the trumpets, and other instruments. Each of the priests wore a tapaau, or ornament, on the arm, consisting of the braided leaflets of the cocoa-nut tree. As soon as the image appeared without the temple, the multitude, who were waiting to witness the pageant, retired to a respectful distance on each side, leaving a wide clear space. The priests sounded their trumpets, and beat the sacred drum, as they marched in procession from the temple to the sea-shore, where a fleet of canoes, previously prepared, was waiting for them. The sacred canoe, or state barge of Oro, was distinguished from the rest by the tapaau, or sacred ornaments of platted cocoa-nut leaves, by which it was surrounded, and which were worn by every individual on board.

As soon as the procession reached the beach, Oro was carried on board, and followed by the priests and instruments of music, while the king took his seat upon the sacred sleeping-place of Oro, which was fixed on the shore. The chiefs stood around the king, and the priests around the god, until, upon a signal given, the king arose from his seat, advanced into the sea, and bathed his person. The priest of Oro then descended into the water, bearing in his hand a branch of the sacred mero, plucked from the tree which grew in the precincts of the temple. While the king was bathing, the priest struck him on his back with the sacred branch, and offered up the prescribed ubu, or invocation to Taaroa. The design of this part of the ceremony was to purify the king from all mahuru huru, or defilement and guilt, which he might have contracted, according to their own expression, by his having seized any land, banished any people, committed murder, &c.

When these ablutions were completed, the king and the priest ascended the sacred canoe. Here, in the presence of Oro, he was invested with the maro ura, or sacred girdle, which, the feathers from the idol being interwoven in it, was supposed to impart to the king a power equal to that possessed by Oro. The priest, while employed in girding the king with this emblem of dominion and majesty, pronounced an ubu, commencing with *Faaa tea te arii i tai i motu tabu,* "Extend or spread the influence of the king over the sea to the sacred island," describing also the nature of his girdle, and addressing the king at the close, by saying—*Madua teie a oe ate Arii:* "Parent this, of you O king;" indicating that from the gods all his power was derived.

As soon as the ubu was finished, the multitude on the beach, and in the surrounding canoes, lifted up the right-hand, and greeted the new monarch with loud and universal acclamations of *Maeva arii! maeva arii!* The steersman in the sacred canoe struck his paddle against the side of the vessel, which was the signal to the rowers, who instantly started from the shore towards the reef, having the god, and the king, girded as it were with the deity, on board; the priests beating their large drum, and sounding their trumpets, which were beautiful large turbo, or trumpet-shells. The thronging spectators followed in their canoes, raising their right-hand in the air, and shouting *Maeva arii!*

Having proceeded in this manner for a considerable

distance, to indicate the dominion of the king on the sea, and receive the homage of the powers of the deep, they returned towards the shore.

During this excursion, Tuumao and Tahui, two deified sharks, a sort of demi-gods of the sea, were influenced by Oro to come and congratulate the new king on the assumption of his government. If the monarch was a legitimate ruler, and one elevated to the office with the sanction of the superior powers, these sharks, it was said, always came to pay their respects to him, either while he was bathing in the sea, or during the excursion in the sacred canoe. But it is probable, that when they approached while his majesty was in the water, some of his attendants were stationed round to prevent their coming too near, lest their salutations should have been more direct and personal than would have proved agreeable.

The fleet reaching the shore, the parties landed, when the king was placed on the *papa rahi o ruea,* or sacred couch of Oro, as his throne; but instead of a footstool, the ordinary appendage to a throne, he reclined his head on the *urua Tafeu,* the sacred pillow of Tafeu. This was also cut out of a solid piece of wood, and ornamented with carving.

The procession was now formed as before, and moving towards Tabutabuatea, the great national temple, Tairimoa, bearing the image of Oro, led the way. The king, reclining on his throne, or couch of royalty, followed immediately after. He was borne on the shoulders of four principal nobles connected with the reigning family. The chiefs and priests followed in his train, the latter sounding their trumpets, and beating the large sacred drum, while the spectators shouted *Maeva arii!* as they proceeded to the temple. The multitude followed them into the court of the marae, where the king's couch or throne was fixed upon the elevated stone platform, in the midst of the uru, or carved ornaments of wood erected in honour of the departed chiefs whose bones had been deposited there.

The principal idol Oro, and his son Hiro, were placed by the side of the king, and the gods and the king here received the homage and tribute of allegiance from the people. A veil must be thrown over the vices with which the ceremonies were concluded.

Although this ceremony was one of the least offensive festivities that ever occurred among them, the murderous cruelty with which it commenced, and the wickedness with which it terminated, were adapted to impress the mind with acutest anguish and deepest commiseration. The abominations continued until the blowing of the trumpet on board the canoes required every one to depart from the temple. They now repaired to the banquet or feast provided for the occasion, and passed the remainder of the day in unrestrained indulgence and excess.

The phraseology of the Tahitian court was in perfect accordance with the elevation, and sacred connexion with their divinities, which the binding on the red girdle was designed to recognize and ratify. The preposterous vanity and adulation in language, used in epithets bestowed upon the king of Tahiti and his establishment, fully equal those employed in the most gorgeous establishment of Eastern princes, or the seraglios of Turkish sultans.

It was not only declared that Oro was the father of the king, as was implied by the address of the priest when arraying him in the sacred girdle, and the station occupied by his throne, when placed in the temple by the side of the deities, but it pervaded the terms used in reference to his whole establishment. His houses were called the *aorai*, the clouds of heaven; *anuanua*, the rainbow, was the name of the canoe in which he voyaged; his voice was called thunder; the glare of the torches in his dwelling was denominated lightning; and when the people saw them in the evening, as they passed near his abode, instead of saying the torches were burning in the palace, they would observe that the lightning was flashing in the clouds of heavens. When he passed from one district to another on the shoulders of his bearers, instead of speaking of his travelling from one place to another, they always used the word *mahuta*, which signifies to fly; and hence described his journey by saying, that the king was flying from one district of the island to another.

The establishment and habits of the king often exhibited the most striking contrast; at one time he was seen surrounded by the priests, and invested with the insignia of royalty, and divinity itself; or appeared in public on the shoulders of his bearers, while the people expressed every indication of superstitious reverence and fear. At other times, he might be seen on terms of the greatest familiarity with his attendants and domestics.

He never wore a crown, or any badge of dignity, and, in general, there was no difference between his dress and that of the chiefs by whom he was surrounded, excepting that the fine cloth and matting, called vane, with which he was often arrayed, were more rare and valuable than the dress worn by others. His raiment frequently consisted of the ordinary pareu, or ahu pu, in quality often inferior to that worn by some of the chiefs in attendance upon him.

In some of the islands to the westward, at the ceremonies of the temple, the people, to shew their homage, wound folds of cloth repeatedly round the body of the king, till he was unable to move, and appeared as if it was only a man's head resting on the immense bale of cloth in which he was enclosed. I do not know that the kings of Tahiti ever experienced such treatment from their subjects. The kings of the former were left in this ludicrous and helpless situation, while the people travelled round the island, boxing and wrestling, in honour of their sovereign, throughout every district.

The regal establishment was maintained by the produce of the hereditary districts of the reigning family, and the requisitions made upon the people. Although the authority of the king was supreme, and his power undisputed, yet he does not appear to have been considered as the absolute proprietor of the land, nor do the occupants seem to have been mere tenants at will, as was the fact in the Sandwich Islands.

There were certain districts which constituted the patrimony of the royal family; in these they could walk abroad, as they were sacred lands. The other districts were regarded as belonging to their respective occupants or proprietors, who were generally raatiras, and whose interest in the soil was distinct from that of the king, and often more extensive. These lands they inherited from their ancestors, and bequeathed them to their children, or whomsoever they chose to select as their heirs. At their death the parties to whom land had been thus left, entered into undisturbed possession, as of rightful property.

The practice of *tutuing*, or devising by will, was found to exist among them prior to the arrival of the Missionaries, and was employed not only in reference to land, but to any other kinds of property. Unacquainted with letters, they could not leave a written will, but during a season of illness, those possessing property frequently called together the members of the family, or confidential friends, and to them gave directions for the disposal of their effects after their decease. This was considered a kind of sacred charge, and was usually executed with fidelity.

Every portion of land had its respective owner; and even the distinct trees on the land had sometimes different proprietors, and a tree, and the land it grew on, different owners. The divisions of land were accurately marked by a natural boundary, as a ridge of mountains, or the course of a river, or by artificial means; and frequently a carved image, or tii, denoted the extent of their different possessions. Whether these tiis were designed to intimate that the spirits they represented guarded the borders of their property, or were used as ornaments, I could not learn, but the removal of the ancient landmarks was regarded as a heinous offence.

The produce which the king received from his hereditary estates being seldom sufficient for the maintenance of his household, the deficiency was supplied from the different districts of the islands. The frequency, however, with which the inferior chiefs were required to bring provisions, was neither fixed nor regular, but was governed by the number of the districts, or the necessities of the king's steward. Still there was a sort of tacit agreement between the king and chiefs, as to the times when they should furnish his provision; and the usage among them in this respect, was generally understood.

The provision was usually ready dressed, though occasionally the vegetables and roots were brought uncooked, and the pigs led alive to the king's servants. The pigs, after being presented to the king, were sometimes taken back by the farmer, and fed till required for use. Cloth for the dress of the king's servants, houses for his abode, and canoes not only for himself but also for those of his household, were furnished by the inhabitants of the islands.

Although the king's will was the supreme law, and the government in some respects despotic, it approximated more to a mixed administration, a union of monarchy and aristocracy. The king had usually one confidential chief near his person, who was his adviser in every affair of importance, and was, in fact, the prime minister. Frequently there were two or three who possessed the confidence, and aided the councils, of the king. These ministers were not responsible to any one for the advice they gave. So great, however, was the influence of the raatiras, that a measure of any importance, such as the declaration of war, or the fitting out a fleet, was seldom undertaken without their being first consulted. This was effected by the friends of the king going among them, and proposing the affair in contemplation, or by convening a public council for its consideration.

Their public measures were not distinguished by promptness or decision, excepting when they wreaked vengeance upon the poor and helpless victims of their displeasure. After a meeting of the chiefs had been summoned, it was a long time before all came together, and their meetings were often interrupted by adjournments.

Their councils were usually held in the open air, where the chiefs and others formed a circle, in which the orators of the different parties took their stations opposite to each other. These orators were the principal, but not the only speakers. The king often addressed the assembly. The warriors and the raatiras also delivered their sentiments with boldness and freedom. When a difference of opinion prevailed, and words ran high, the impetuosity of their passions broke through all restraint, and the council terminated in scenes of confusion and bloodshed; or if they ended without open hostility, the chieftains returned to their respective districts, to assemble their tenantry, and prepare for war.

CHAP. XIII.

Power of the chiefs and proprietors of land—Banishment and confiscation—The king's messenger—The niau, an emblem of authority—Ancient usages in reference to crime, &c.—Fatal effects of jealousy—Seizure of property—Punishment of theft—Public works—Supplies for the king—Despotic rapacity—Extortion of the king's servants—Unorganized state of civil polity—Desire a code of Christian laws—Advice and conduct of the Missionaries—Preparation of the laws—Public enactment by the king in a national assembly at Tahiti—Capital punishments—Manner of conducting public trials—Establishment of laws in Raiatea—Preparation of those for Huahine.

Every chief was the sovereign of his own district, though all acknowledged the supremacy of the king. Each island was divided into a number of large portions, or districts, called Mataina, a term also applied to the inhabitants of a district. These mataina had distinct names, and were under the government of a chieftain of rank or dignity belonging to the reigning family, or to the raatiras. This individual was the baron of the domain, or the lord of the manor, and was succeeded in his possessions and his office by his son, or the nearest of his kindred, with a fresh appointment from the king.

For treason, rebellion, or withholding supplies, individuals were liable to banishment, and confiscation of property. The king had the prerogative of nominating his successor, but could not appropriate the lands of the exile to his own use. The removal of a chief of high rank, or of extensive influence, was seldom attempted, unless the measure was approved by the other chiefs. The sovereign was, therefore, more desirous to conciliate their esteem, and engage their co-operation, than to prejudice them against his person or measures. As he had no permanent armed force at his disposal, he could not, on every occasion, accomplish his wishes; and, at times, when he has issued his mandate for the banishment of a raatira, if the other raatiras deemed his expulsion unwarrantable, they have desired him to keep possession of his lands, and then, remonstrating with the king, have declared their determination to maintain the cause of the injured party, even by force of arms. The extent of power possessed by the raatiras, in the number of their tenantry and dependants, was one of the greatest sources of embarrassment to the government, whose measures were only regulated by the will of the ruler, or the exigencies of the state.

In the division of their country, the natives appear to have had a remarkable predilection for the number eight. Almost every island, whatever its size, is divided into eight districts, and the inhabitants into an equal number of mataina, or divisions. In each district the power of the chief was supreme, and greater than that which the king exercised over the whole. This power extended to the persons and lives, as well as the property, of the people.

The inferior chiefs also exercised the same authority over their dependants. The father was magistrate in his own family; the chief in his own district; and the king nominally dispensed law and justice to the whole. The final appeal, in all matters of dispute, was made to the chief ruler; and the parties who resorted to his decision, usually regarded it as binding. The king kept no armed force, neither was there any regular police for the maintenance of public order. The chief of each district was accountable for the conduct of the people under his own jurisdiction. The chieftains who were in attendance on the king, with the servants of his establishment, were the agents usually employed to carry his measures into effect. The servants of the raatiras performed the same duty in their respective localities, and the king often sent his order to the district chief, who employed his own men in its execution.

Notwithstanding the many acts of homage paid to the head and other branches of the reigning family, and their imagined connexion with the gods, the actual influence of the king over the haughty and despotic district chieftains, was neither powerful nor permanent, and he could seldom confide in their fidelity in any project which would not advance their interests as well as his own. Every measure was therefore planned with the most cautious deliberation, the approval and aid of a number of these nobility of the country being essential to carry it into effect; but when the interests of the reigning family and those of the chieftains were opposed, it produced no small embarrassment. These raatiras, who resembled the barons of the feudal system, kept the people under them in a state of the greatest subjection, and received from them not only military service, but a portion of the produce of their lands, and personal labour whenever required.

Whenever a measure affecting the whole of the inhabitants was adopted, the king's *vea,* or messenger, was despatched with a bundle of *niaus,* or leaflets. On entering a district, he repaired to the habitation of the principal chiefs, and, presenting a cocoa-nut leaf, delivered the orders of the king. The acceptance of the leaf was a declaration of their compliance with the requisition, and to decline taking it was regarded as an intimation of their refusal to accede to the measure proposed. Hence the messenger or herald, when he had travelled round the island, reported to the king, who had received and who had refused the *niau*. When the chiefs approved of the message, they sent their own messengers to their respective tenants and dependants, with a cocoa-nut leaf for each, and the orders of the king.

The *niau*, or leaflets of the cocoa-nut tree, was the emblem of authority throughout the whole of the Georgian and Society Islands; and requisitions for property or labour, preparations for war, or the convocation of a national assembly, were formerly made by sending the cocoa-nut leaf to those whose services or attendance was required. To return or refuse the niau was to offer an insult to the king, and to resist his authority.

If the king felt himself strong enough, he would instantly banish such an individual, and send another to take possession of his lands, and occupy his station as chief of the district. Should the offender have been guilty of disobedience to the just demands of the king, though the lands might be his hereditary property, he must leave them, and become, as the people expressed it, a wanderer "upon the road;" but if the king's conduct was considered arbitrary, and the individual justified in his refusal by the other chiefs, they would *tapea*, or detain him, and protest to the king against his removal. The parties generally knew each other's strength and influence, and those who had little hopes of succeeding by an appeal to arms, usually conceded whatever was required.

Personal security, and the rights of private property, were unknown; and the administration of justice by the chiefs in the several districts, and the king over the whole, was regulated more by the relative power and influence of the parties, than by the merits of their cause.

They had no regular code of laws, nor any public courts of justice, and, excepting in offences against the king and chiefs, the rulers were seldom appealed to. The people in general avenged their own injuries. Death or banishment was the punishment usually inflicted by the chiefs, and frequently the objects of their displeasure were marked out as victims for sacrifice.

Destitute, however, as they were of even oral laws or institutes, there were many acts, which by general consent, were considered criminal, and deserving punishment. These were *orure hau*, rebellion, or shaking the government, withholding supplies, or even speaking contemptuously of the king or his administration. So heinous was this offence, that the criminal was not only liable to banishment, or the forfeiture of his life, but a human sacrifice must be offered, to atone for the guilt, and appease the displeasure of the gods against the people of the land in which it had been committed. Lewdness was not regarded as a crime, but adultery was sometimes punished with death. Those among the middle or higher ranks who practised polygamy, allowed their wives other husbands. It is reported that brothers, or members of the same family, sometimes exchanged their wives, while the wife of every individual was also the wife of his *taio* or friend.

Their character in this respect presents a most unnatural mixture of brutal degradation, with infuriated and malignant jealousy; for while their conduct with respect to the taio, &c. exhibits an insensibility to every feeling essential to conjugal happiness, the least familiarity with the wife, unauthorized by the husband, even a word or a look, from a stranger, if the husband was suspicious, or attributed it to improper motives, was followed by instant and deadly revenge.

There is a man now residing in Huahine, whose face and shoulders are frightfully marked with deep scars, inflicted by blows with a carpenter's axe, on this account. A husband and a wife were once sitting together when another man joined the party, and sat down with them. He wore a taupoo or bonnet of platted cocoa-nut leaves; lifting his hand, and taking hold of it by the part that shaded his brows, he waved his hand towards the inland part of the district, in removing his bonnet from his head. The suspicious husband, observing the motion of his hand, considered it as an assignation, that the stranger was to meet his wife there; and without a word, I believe, being spoken by either party, he rose up, took down his spear, which was suspended from the inside of his dwelling, and ran the man through the body, accusing him at the same time of the crime of which he supposed him guilty. Several of the murders of the Europeans, that have been committed in the islands of the Pacific, have originated in this cause.

Theft was practised, but less frequently among themselves than towards their foreign visitors. They supposed it equally criminal, yet they do not in general appear to have attached any moral delinquency to the practice, but they imagined they were more likely to avoid detection when stealing from strangers, than when robbing their own countrymen. Stealing was always considered as a crime among them, and every precaution was taken to guard against it. On this account, their large bales of valuable cloth, and most articles of property not in constant use, were kept suspended from the ridge-pole or rafters of their dwellings; their smaller rolls of cloth were often laid by their pillows; and their pigs were driven under their beds at night, to prevent their being stolen.

This nefarious practice, strange as it may appear, was supported by their false system of religion, and sanctioned by the patronage of the gods, especially by Hiro a son of Oro, who was called the god of thieves. The aid of this god was invoked by those who went on expeditions of plunder, and the priests probably received a portion of the spoils. Chiefs of considerable rank have sometimes been detected in the act of stealing, or have been known to employ their domestics to thieve, receiving the articles stolen, and afterwards sheltering the plunderers. This, however, has generally been practised on the property of foreigners.

Among themselves, if detected, the thief experienced no mercy, but was often murdered on the spot. If detected afterwards, he was sometimes dreadfully wounded or killed. Two very affecting instances of vengeance of this kind are recorded by the early Missionaries. I have also heard that they sometimes bound the thief hand and foot, and, putting him into an old rotten canoe, towed him out to sea, and there left him adrift, to sink in the ocean, or become a prey to the sharks.

The *haru raa*, or seizing all the property of delinquents, was the most frequent retaliation, among the lower class, for this and other crimes. The servants of the chiefs, or injured party, went to the house of the offenders, and took by force whatever they found, carrying away every article worth possessing, and destroying the rest. If the inhabitants of the house received previous intimation of their purpose, they generally removed or secreted their most valuable property, but seldom attempted to resist the seizure, even though every article of food and clothing, and the mats on which they slept, should be taken away.

This mode of retaliation for theft, or other injury, was so generally recognized as just, that, although the party thus plundered might be more powerful than those who plundered them, they would not attempt to prevent the seizure: had they done so, the population of the district would have assisted those, who, according to established custom, were thus punishing the aggressors. Such was the usual method resorted to for punishing the petty thefts committed among themselves. They were generally satisfied with seizing whatever they could find in the houses, yards, or gardens of the offenders; but when it was practised by order of the king or chiefs, the culprit was banished from his house or lands, and reduced to a state of complete destitution.

Great difficulty was often experienced in discovering the thief, or the property stolen; and, on these occasions, they frequently resorted to divination, and employed the sorcerer to discover the offender. The thief, when detected, generally received summary punishment. Mr. Bourne states, that in one of the Harvey Islands, a man found a little boy, about eight years of age, stealing food; the man instantly seized the juvenile delinquent, and, tying a heavy stone to his leg, threw him into the sea. The boy sunk to the bottom, and would soon have paid for his crime with his life, had not one of the native teachers, who saw him thrown into the water, plunged after him, rescued him, and taken him to his own house, where he has ever since resided.

The resources of the government consisted in the personal services of the people, and the produce of the soil. From these the revenue was derived. All public works, such as the erection of national temples, fortifi cations, enclosures from the sea, dwellings for the king, &c. were performed by the whole of the population. In each district, the king had a viceroy, or deputy, to whom his orders were sent with a cocoa-nut leaf. The chiefs sometimes assembled together, and divided the work among themselves. At other times, the king appointed each to his particular share. Every chief then issued his orders to the raatiras under his authority, who prepared the materials, and performed their appointed portion of the work. Canoes for the king's use were furnished in the same way, and also cloth for himself and his household.

Every district brought provisions at stated intervals for the king's use, or for the maintenance of his numerous retinue. Besides what they regularly furnished, orders were often issued for extraordinary supplies, for the entertainment of a distinguished guest, or the celebration of a national festival. No regular system of taxation prevailed, but every kind of property was furnished by the chiefs and people in great abundance, not only for the king, but for the purpose of enriching those who were the objects of his favour.

However abundant the supplies might be which the king received, he was in general more necessitous than many of the chiefs. Applications for food, cloth, canoes, and every other valuable article which he received from the people, were so frequent and importunate, that more than was barely sufficient for his own use seldom remained long in his possession. A present of food was usually accompanied with several hundred yards of native cloth, and a number of fine large double canoes; yet every article was often distributed among the chiefs and favourites on the very day it arrived; and so urgent were the applicants, that they did not wait till the articles were brought, but often extorted from the king a promise that he would give them the first bale of cloth, or double canoe, he might receive. At times they went beyond this; and when a chief, who considered the king under obligations to him, knew that the inhabitants of a district were preparing a present for their sovereign, which would include any articles he wished to possess, he would go to the king, and *tapao,* mark or bespeak it, even before it was finished. A promise given under these circumstances was usually regarded as binding, though it often involved the king in the greatest difficulties, and kept him necessitous.

In the estimation of the people, one of the greatest virtues and highest excellencies of a king, was generosity; and one of the most unpopular dispositions he could cherish, was illiberality. In describing a good chief, or governor, they always spoke of him as one who distributed among his chiefs whatever he received, and never refused any thing for which they asked.

Notwithstanding this generosity on the part of the king, the conduct of the government was often most rapacious and unjust. The stated and regular supplies furnished by the inhabitants, were often inadequate to the maintenance of the numbers, who, attaching themselves to the king's household, passed their time in idleness, and were fed at his table. Whenever there was a deficiency of food for his ordinary followers, or a large party that had arrived as his guests, a number of his servants went out to the settlements of the raatiras, or farmers, and, sometimes without even asking, tied up the pigs that were fed near the dwelling, and plundered the abode, ravaging, like a band of lawless robbers, the plantations or the gardens, and taking away every article of food the poor, oppressed people possessed. Sometimes they launched a fine canoe that might be lying near, and, loading it with their plunder, left the industrious proprietor destitute even of the means of subsistence; and, as they were the king's servants, he durst not complain.

When the king travelled, he was usually attended by a company of Areois, or a worthless train of idlers; and often when they entered a district that was perhaps well supplied with provisions for its inhabitants, if they remained any length of time, by their plundering and wanton destruction, it was often reduced to a state of desolation. Sometimes the king sent his servants to take what they wanted from the fields or gardens of the people; but often, unauthorized by him, they used his name to commit the most lawless and injurious depredations upon the property of the inhabitants; whose lives were endangered, if they offered the least resistance.

Mahamene, a native of Raiatea, gave, at a public meeting in that island, the following account of their lawless plunder. "These teuteu," (servants of the king,) said he, "would enter a house, and commit the greatest depredations. The master of the house would sit as a poor captive, and look on, without daring to say a word.— They would seize his bundle of cloth, kill his largest pigs, pluck the best bread-fruit, take the largest taro, (arum roots,) the finest sugar-cane, the ripest bananas, and even take the posts of his house for fuel to cook them with. Is there not a man present who actually buried his new canoe under the sand, to secure it from these desperate men?"

Nothing fostered tyranny and oppression in the rulers, and reduced the population to a state of wretchedness, so much as these unjust proceedings. Those who, by habits of industry, or desire of comfort for themselves and families, might be induced to cultivate more land than others, were, from this very circumstance, marked out for despoliation. They had no redress for these wrongs, and therefore, rather than expose themselves to the mortifying humiliation of seeing their fields plundered, and the fruits of their labour taken to feed a useless and insulting band that followed the movements of the king, they allowed their lands to remain untilled. They chose to procure a scanty means of subsistence from day to day, rather than suffer the insults to which even their industry exposed them.

So far were these shameless extortions practised, that during the journey of an European through the country, he has been attended by a servant of the king, and when, in return for provisions furnished, or acts of kindness shewn, by the hospitable inhabitants, he has made them even a trifling present, it has been instantly seized by the vassal of the chief, who has followed him for that purpose. The poor people were also allowed to dispose of their produce to the captains or merchants that might visit them for the purpose of barter, but the king or chief frequently requested the greater part, or even the whole, of what they might receive in return for it.

That they should have improved in industry, or advanced in civilization, under such a system, was impossible, and that they should, under such circumstances, have tilled a sufficient quantity of ground to furnish supplies for the shipping, is a matter of greater surprise, than that they should not have cultivated more. The humiliating degradation to which it reduced the farmers, and the constant irritation of feelings to which this wretched system exposed them, were not the only evils that resulted from it. It naturally led the raatiras to regard their chiefs as enemies, and generated disaffection to their administration, while it led the former to consider the latter as inimical to their own interests. It also greatly diminished their resources, for under the discouragements resulting from constant liability to plunder, the people were unable to furnish those supplies, which they would otherwise have found it a satisfaction to render.

This system of civil polity, disjointed and ill adapted as it was to answer any valuable purpose, was closely interwoven with their sanguinary system of idolatry, and sanctioned by the authority of the gods. The king was not only raised to the head of this government, but he was considered as a sort of vicegerent to those supernatural powers presiding over the invisible world. Human sacrifices were offered at his inauguration; and whenever any one, under the influence of the loss he had sustained by plunder, or other injury, spoke disrespectfully of his person and administration, not only was his life in danger, but human victims must be offered, to cleanse the land from the pollution it was supposed to have contracted.

The intimate connexion between the government and their idolatry, occasioned the dissolution of the one, with the abolition of the other; and when the system of pagan worship was subverted, many of their ancient usages perished in its ruins. They remained for some years without any system or form of government, excepting the will of the king, to whom the inhabitants usually furnished liberal supplies of all that was necessary for the maintenance of his household, and the accomplishment of his designs.

The raatiras exercised the supreme authority in the divisions over which the king had placed them. But when circumstances occurred, in which, under idolatry, they would have acted according to their ancient custom, they felt embarrassed. Many of the people, free in a great degree from exposure to seizure, and the more dreadful apprehension of being offered to the gods, evinced a disinclination to render the king the supplies and support he needed.

The sacrificing of human victims to the idols had been one of the most powerful engines in the hands of the government, the requisition for them being always made by the ruler, to whom the priests applied when the gods required them. The king, therefore, sent his herald to the petty chieftain, who selected the victims. An individual who had shewn any marked disaffection towards the government, or incurred the displeasure of the king and chiefs, was usually chosen. The people knew this, and therefore rendered the most unhesitating obedience. Since the subversion of idolatry, this motive has ceased to operate; and many, free from the restraint it had imposed, seemed to refuse almost all lawful obedience and rightful support.

Their government continued in this unsettled state for four or five years; during which, the people brought provisions and supplies to the king, and furnished the accustomed articles for his establishment, either according to arrangements made among themselves, or in obedience to his requisitions. The superior and subordinate rulers over the people, endeavoured to preserve the peace of society, and promote the public welfare, by punishing offenders according to the nature of their crimes, but without any regular or uniform procedure. The only punishment inflicted was banishment, and, in a few instances, seizure for theft. It was, however, evident that another system must be introduced, instead of that which, with the *tabu* idolatry, had been abolished.

It is a fact worthy of note, that although no people in the world could be more vicious than they were prior to their renunciation of paganism, yet such was the moral influence of the precepts of Christianity on the community at large, and consequently on the conduct of many who were Christians only by profession, that for some time crimes affecting the peace of society were but few. Theft, to which ever since their discovery they have been proverbially addicted, was rarely committed. It was not, however, to be expected that this state of things would be permanent; and after a few years, the force of example, and the restraining influence of the preceptive parts of christian truth, began to diminish on the minds of those over whom it had exerted no decisive power, and who, in their altered behaviour, had rather followed popular sentiment and practice, than acted from principles in their own minds. When therefore this class of persons began to act more according to their true character, the chiefs found it necessary to visit their delinquency with punishment; and the welfare of the nation required that measures should be adopted for maintaining the order and peace of the community.

Having as a nation embraced Christianity, they were unanimous in desiring that their civil and judicial proceedings should be in perfect accordance with the spirit and principles of the christian religion. Hence they were led to seek the advice of their teachers, as to the means they should adopt for accomplishing this object. The Missionaries invariably told them that it was no part of their original design to attempt any change in their political and civil institutions, as such; that these matters belonged to the chiefs and governors of the people, and not to the teachers of the religion of Jesus Christ. To this they generally replied, that under the former idolatrous system they should have been prepared to act in any emergency, but they were not familiar with the principles of Christianity in their application to the ordinary relations of life, especially in reference to the punishment of crime.

In compliance with these solicitations, the Missionaries illustrated the general principles of Scripture, that in all the public stations they sustained, they were to do unto others as they would that others should do unto them — that with regard to government, Christianity taught its disciples to fear God, and honour the king—that the power which existed was appointed of God—and that magistrates were for a terror to evil-doors, and a praise to them that do well. These general principles were presented and enforced as the grounds of proceeding in all affairs of a civil or political nature.

The Missionaries, though frequently appealed to, generally left the determination of the matter to their own discretion, declining to identify themselves with either party, in any of their differences. They promised, however, to the chiefs such assistance as they could render in the preparation of their code of laws, and constitution of government, but were exceedingly anxious that it should be the production of the king and chiefs, and not of themselves. They had hitherto avoided interfering with the government and politics of the people, and had never given even their advice, excepting when solicited by the chiefs. When the conduct of petty chiefs or others had affected their own servants, or persons in their employment, if they have taken any steps, it has been as members of the community, and not as ministers of religion.

After the introduction of Christianity, the chiefs were among the first to perceive that the sanguinary modes of punishment to which they had been accustomed were incompatible with the spirit and precepts of the gospel, and earnestly desired to substitute measures that should harmonize with the new order of things. The king applied for assistance in this matter, soon after the general change that took place in 1815. The Missionaries advised him to call a general council of the chiefs, and consult with them on the plans most suitable to be adopted.—Whether his recollection of the unpropitious termination of former councils influenced him, or whether he was unwilling to delegate any of that power to others with which heretofore he had been solely invested, is uncertain; but he objected to the assembling of the chiefs at that time, still requesting advice and counsel from the Missionaries. This they readily afforded, both as to the general principles of the British constitution, the declarations of Scripture, and the practice of Christian nations. Their own sentiments in reference to their duty at this time, will best appear from the following extract of a public letter, bearing date July 2, 1817.—

"During many years of our residence in these islands,

we most carefully avoided meddling with their civil and political affairs, except in a few instances, where we endeavoured to promote peace between contending parties. At present, however, it appears almost impossible for us, in every respect, to follow the same line of conduct. We have told the king and chiefs, that, being strangers, and having come to their country as teachers of the word of the true God, and the way of salvation by Jesus Christ, we will have nothing further to do with their civil concerns, than to give them good advice; and with that view several letters have passed between us and the king. We have advised him to call a general meeting of all the principal chiefs, and, with their assistance and approbation, adopt such laws and regulations as would tend to the good of the community, and the stability of his government; and that in these things, if he desired it, we would give him the best advice in our power, and inform him of what is contained in the word of God, and also of the laws and customs of our own country, and other civilized nations."

The first code of laws was that enacted in Tahiti in the year 1819; it was prepared by the king and a few of the chiefs, with the advice and direction of the Missionaries, especially Mr. Nott, whose prudence and caution cannot be too highly spoken of, and by whom it was almost framed. The code was remarkably simple and brief, including only eighteen articles. It was not altogether such as the Missionaries would have wished the nation to adopt, but it was perhaps better suited to the partial light the people at that time possessed, and to the peculiar disposition of Pomare. He was exceedingly jealous of his rights and prerogatives, and unwilling that the chiefs should assume the least control over his proceedings, or participate in his power. His will still continued to be law, in all matters not included in their code; and with regard to the revenue which the people were required to furnish for his use, he would admit of no rule but his own necessities, and consequently continued to levy exactions, as his ambition or commercial engagements might require.

The Missionaries would have regarded with higher satisfaction an improvement in the principles recognized as the basis of the relation subsisting between the king, chiefs, and people, some division of the power of government—enactments proportioning the produce of the soil to be furnished for the king, and securing the remainder to the cultivators. But having recommended these points to the consideration of the rulers, they did not think it their duty to express any disatisfaction with the code, imperfect as it was.

In the month of May, 1819, the king, and several thousands of the people from Tahiti and Eimeo, assembled at Papaoa, for the purpose of attending the opening of the Royal Missionary Chapel, and the promulgation of the new laws. The anniversary of the Tahitian Missionary Society being held at the same time, the Missionaries from the several stations, in these two islands, were then at Papaoa.

The thirteenth day of the month was appointed for this solemn national transaction; and the spacious chapel which the king had recently erected was chosen as the edifice in which this important event should take place. It was thought no desecration of a building reared for public devotion, and solemnly appropriated to the worship of the Almighty, and other purposes directly connected with the promotion of his praise, that the grave and serious engagements by which the nation agreed to regulate their social intercourse, should be ratified in a spot where they were led to expect a more than ordinary participation of the Divine benediction. During the forenoon, the chiefs and people of Tahiti and Eimeo assembled in the Royal Chapel, and about the middle of the day the king and his attendants entered. The Missionaries were also present; but, regarding it as a civil engagement, attended only as spectators. The king, however, requested Mr. Crook to solicit the Divine blessing on the object of the meeting. He therefore read a suitable portion of the sacred volume, and implored the sanction of the King of kings upon the proceedings that were to follow. Nothing could be more appropriate than thus acknowledging the Power by whom kings reign, and seeking His blessing upon those engagements by which their public conduct was to be regulated. The Divine benediction having been thus sought, the king, who had previously taken his station in the central pulpit, arose, and, after viewing for a few moments the thousands of his subjects that were gathered round him, commenced the interesting proceedings of the day, by addressing Tati, the brother and successor of the late Upufara, who was the leader of the idolatrous and rebel army defeated in November, 1815. "Tati," said the king, "what is your desire? what can I do for you?" Tati, who sat nearly opposite the pulpit, arose and said, "Those are what we want—the papers you hold in your hand—the laws; give them to us, that we may have them in our hands, that we may regard them, and do what is right." The king then addressed himself to Utami, the good chief of Teoropaa, and in an affectionate manner said, "Utami, and what is your desire?" He replied,

"One thing only is desired by us all, that which Tati has expressed—the laws, which you hold in your hand." The king then addressed Arahu, the chief of Eimeo, and Veve, the chief of Taiarabu, nearly in the same manner, and they replied as the others had done. Pomare then proceeded to read and comment upon the laws respecting murder, theft, trespass, stolen property, lost property, sabbath-breaking, rebellion, marriage, adultery, the judges, court-houses, &c., in eighteen articles. After reading and explaining the several particulars, he asked the chiefs if they approved of them. They replied, aloud, "We agree to them—we heartily agree to them." The king then addressed the people, and desired them, if they approved of the laws, to signify the same by holding up their right hands. This was unanimously done, with a remarkable rushing noise, owing to the thousands of arms being lifted at once. When Pomare came to the law on rebellion, stirring up war, &c. he seemed inclined to pass it over, but after a while proceeded. At the conclusion of that article, Tati was not content with signifying his approbation in the usual way only, but, standing up, he called in a spirited manner to all his people to lift up their hands again, even both hands, he setting the example, which was universally followed. Thus all the articles were passed and approved.

The public business of the day was closed by Mr. Henry's offering a prayer unto Him by whom kings reign, and princes decree judgment, and the people retired to their respective dwellings.

Pomare subsequently intimated his intention of appropriating Palmerston's Island as a place of banishment for Tahitian convicts, and proposed to the Missionaries to publish his request that no vessel should remove any who might be thus exiled. The laws which the king read to the people were written by himself, and formed, probably, the first written code that ever existed in the islands; and he afterwards wrote out, in a fair, legible, and excellent hand, a copy for the press. Printed copies were distributed among the people, but the original manuscript, in the king's hand-writing, signed by himself, is in the possession of the London Missionary Society. The laws were printed on a large sheet of paper, and not only sent to every chief and magistrate throughout both islands, but posted up in most of the public places.

The sentence to be passed on individuals who should be found guilty of many of the crimes prohibited by these laws, was left to the discretion of the judge or magistrate; but to several, the penalty of death was annexed; and only a few months after their enactment, the sentence of capital punishment was passed on two individuals, whose names were Papahia and Horopae. They were inhabitants of the district of Atehuru, and were executed on the twenty-fifth of October, 1819, for attempting to overturn the government. Papahia had been a distinguished warrior, and was in the very prime of life. He was a man of a bold and daring character, and of turbulent conduct. He came several times to my house, during our residence at Eimeo; and although, in consequence of his restless and violent behaviour, I was not prepossessed in his favour, my personal acquaintance made me feel additional interest in the melancholy fate of the first malefactor on whom the dreadful sentence of the law was inflicted. The lives of these unhappy men were not taken by thrusting a spear through the body, or beating out the brains with a club, or by decapitation, which were the former modes of punishment, but they were hanged on a cocoa-nut tree, in a conspicuous part of the district.

In the year 1821, a conspiracy was formed to assassinate the king, and two men, who were proceeding to the accomplishment of their murderous purpose, were apprehended, with others concerned in the plot. The names of the two leaders were Pori and Mariri. Sentence of death was passed upon them, and they were hanged on a rude gallows, formed by fastening a pole horizontally between two cocoa-nut trees. These are the only executions that have taken place in the islands. It is not probable that many will be thus punished. The Missionaries interceded on behalf of the culprits, and secured a mitigation of punishment for the rest of the offenders. The judicial proceedings in the different districts of Tahiti, were divested, as much as possible, of all formality; and though some trifling irregularities, and slight embarrassments, as might be expected, were occasionally experienced, among a people totally unaccustomed to act in these matters according to any prescribed form, yet, upon the whole, the administration of justice by the native magistrates, was such as to give general satisfaction. The following account, by an eye-witness* of their proceedings on one of these occasions, will not be uninteresting.

"At the time appointed, a great many people, of both sexes and all ages, assembled under some very fine trees, near the queen's house. A small bench was brought for the two judges; the rest either stood or sat upon the ground, forming something less than a semicircle. We were provided with low seats near the

* Capt. G. C. Gambier, R. N.

judges. The two prisoners were seated cross legged upon the ground, under the shade of a small tree, about twenty paces in front of the judges. They were both ill-looking men, dressed in the graceful tiputa. When all was ready to begin, one of the judges rose, and addressed the prisoners at considerable length, and with a good deal of action—not violent, but firm and gentle motions of the arms. He explained to them the accusation which brought them there, and read to them the law under which, if proved guilty, they would be punished. When he had finished, and called upon them to say whether it was true or not, one of them got up, and answered with great fluency, and good action. He maintained their innocence, and called a witness to confirm it. The witness, very artfully, turned his evidence to the account of the prisoners. Others also, in some way or other, favoured the accused, and the defendants were therefore discharged, from want of evidence."

On the 12th May, 1820, a code of laws was unanimously and publicly adopted in Raiatea, and recognized as the basis of public justice by the chiefs and people of Tahaa, Borabora, and Maupiti. The substance of the Raiatean laws was copied from those enacted by the government of Tahiti during the preceding year. They extended to twenty-five articles, embodying several most valuable enactments omitted by the Tahitian code. The most important of these was the institution of *Trial by Jury*. This was certainly the greatest *civil* blessing the inhabitants of the Pacific had yet received, and future generations will cherish with gratitude the memory of the Missionaries of Raiatea, at whose recommendation, and with whose advice, it was established by law in these islands.

Naturally violent and merciless under a sense of injury, we often found them too severe towards offenders; and while we occasionally interceded on behalf of those whose punishment appeared greater than their crime, we lost no opportunity of conveying just and humane, as well as scriptural ideas on matters of jurisprudence, without, however, interfering with their proceedings, or countenancing the misdeeds of those we might recommend to mercy.

The new laws had now been nearly three years established in Tahiti and Eimeo. Those of Raiatea, Tahaa, and Barabora, had also been for more than twelve months in operation among the inhabitants of these islands.—The chiefs of Huahine had virtually made the latter the basis of their administration of justice, but though acting upon them, no code had yet been officially promulgated.

They had already applied to us for assistance in preparing the laws for the island under their dominion.—This we had cheerfully rendered to the best of our ability, at the same time recommending them still to defer their public enactments until they had deliberately observed the effect of those already in force among the inhabitants of the adjacent islands. It was also proper to obtain the sanction of the queen's sister, then residing at Tahiti, who is nominally the sovereign of Huahine, the government of the island having been formerly presented to her by Mahine, the resident and hereditary chieftain. This grant, which transpired several years before any of the parties embraced Christianity, has often occasioned inconvenience. The internal government of the island has always been maintained by the resident chiefs, but in all matters materially affecting the people, or their relation to the governments of other islands, it has been considered necessary, as a matter of etiquette, or courtesy at least, to consult Teriitaria; and hence it was thought desirable to submit the laws to her inspection, and receive her sanction. Though affecting only the resident chiefs and people, and maintained entirely by the authority of the former, they were to be promulgated in her name, as well as that of Mahine, and the other chiefs of the island. The introduction of new laws being a matter of importance to the nation, it was deemed suitable that a deputation from the chiefs should proceed to Tahiti by the first favourable opportunity, for the purpose of receiving the queen's approval. It was also desirable that Mr. Barff, or myself, should accompany this embassage, that we might make inquiries of Mr. Nott, and others, relative to the adaptation of the laws in force there, to the circumstances of the people, and might alter, if necessary, those prepared for Huahine.

CHAP. XIV.

Pomare's proposed restrictions on barter, rejected by the chiefs of the Leeward Islands—Voyage to Eimeo—Departure for Tahiti—Danger during the night—Arrival at Burder's Point—State of the settlement—Papeete—Mount Hope—Interview with the king—Revision of the laws—Approval of the queen—Arrival of the Hope from England—Influence of letters, &c.—Return to Eimeo—Embarkation for the Leeward Islands—A night at sea—Appearance of the heavens—Astronomy of the natives—Names of the stars—Divisions and computation of time, &c. —Tahitian numerals—Extended calculation—Arrival in Huahine.

EARLY in 1821, the brig which had been purchased in New South Wales for Pomare, arrived in Tahiti. Soon after this, the king sent a messenger to the Leeward Islands, with a bundle of *niaus*, or emblems of royal authority, and a proposal to the chiefs, that they should become joint proprietors of the vessel, and furnish a required quantity of native produce, viz. pigs, arrow-root, and cocoa-nut oil, towards the payment of the Macquarie. The herald left his message and bundle of niaus at Huahine, in the name of Teriitaria, and passed on to Raiatea. In a day or two afterwards we learned that instructions had been sent down to the chiefs, not to dispose of any of the above-mentioned articles, nor to allow the people to barter them to any ship, or even to the Missionaries, but to reserve them all for the cargo of his vessel. We represented to the chiefs the injustice of not allowing every man, provided he paid their just demands, to dispose of the fruits of his own industry, and they assured us that it should be so at Huahine, whatever restrictions might be imposed upon the people of Tahiti. The queen's sister, the nominal ruler of the island, residing at Tahiti, was influenced, they observed, by the advice and measures of Pomare, and often perplexed them by her directions.

On the fourteenth of April, 1821, Pomare's messenger returned from Raiatea. Tamatoa, the king of that island, and the chiefs of those adjacent, had refused to receive the niaus, or to join Pomare in his projected commercial speculations. They had at the same time agreed to unite, and procure a vessel for themselves, in which to trade from the islands to the colony of New South Wales, and had sent up a special messenger, with a letter to the chiefs of Huahine, requesting them to join their enterprise. A public meeting was convened, in which the propositions from Pomare on the one hand, and of Tamatoa on the other, were freely discussed. The result was, that although all were far more disposed to join the Raiatean than the Tahitian chiefs, they declined both for the present, and despatched the respective messengers to their superiors, with declarations to this effect.

The wind, which had set in from the westward on the fourteenth, continued during the whole of the fifteenth and, as it seemed tolerably steady, it was proposed that our boat should be prepared for the voyage to Tahiti. It was also thought best that I should accompany Auna and Matapuupuu on their embassy to the queen's sister. During the evening I waited on the chiefs, and took my leave; the native chieftains did the same; and their final instructions were, to induce, if possible, Teriitaria to come and reside at Huahine; but that if she preferred remaining at Tahiti, she should give up all interference with the government of the island, and delegate it to them, independently of all foreign control.

The wind continuing to blow from the westward through the night, our bark was launched early on the morning of the sixteenth, and we prepared for embarka tion. The boat was rather rude in appearance, being one I had from necessity built, with the assistance of the natives, while visiting in the island of Raiatea, in the early part of 1820. It was about thirty-six feet in length, and capable of carrying forty persons. The breeze increased in strength as the morning began to dawn, and about day-break we sailed from Fare harbour. Auna, Matatore, and Matapuupuu were my companions, and our boat was manned by about ten strong and active natives. As we were bounding over the waves of the harbour, and entering upon the wide-spread bosom of the Pacific, we lost the sprit of one of our matting-sheets in the sea, and could only carry one sail. This circumstance, although it prevented our proceeding so rapidly as we should otherwise have done, contributed perhaps to our safety, for the wind was high and the sea rough. By noon we had entirely lost sight of Huahine, and about sunset we obtained our first distant glance of the lofty peaks of Eimeo. The wind now blew what the natives called a strong *toerau*, or westerly gale, and the agitation of the sea was proportionably increased. The inside of our open boat was, however, perfectly dry, and it appeared to shoot along, as the natives expressed it, upon the tops of the waves, until at length we heard, amid the stillness of the night, the welcome sound of the long heavy surf, rolling in solemn grandeur, and dashing in loud, though distant roar, upon the coral reefs.

This, though adapted to inspire apprehension and terror in the minds of those unaccustomed to navigate among the islands, was a gladdening sound to us, as it indicated an approach to our port of destination. We were several miles distant when we first heard the roaring of the surf upon the reef, but, proceeding rapidly along, we soon came in sight of it. Sailing along in a line parallel with it till we came to an opening, we entered Taloo or Opunohu harbour, and landed near the Missionary settlement shortly after midnight, having sailed a distance of about one hundred miles in the space of twenty hours.

The natives seldom evince much concern about their accommodations, when voyaging or travelling among the islands. Frequently, when landing for the night, they kindle a fire on the sea-beach, and having cooked their bread-fruit, or other provision, which they usually carry with them, they lie down, either in the boat, or on the sand by its side, and, spreading the sails as a tent, or wrapping themselves in them, substitute them for bed and bedding, and sleep comfortably till the morning. Most of those, however, who were my fellow-voyagers on this occasion, had formerly resided at this settlement, and had lived on terms of friendship with many of the inhabitants. To the dwellings of these they repaired, while I pursued my way up the valley to the residence of my friend Mr. Platt, whom I awoke from his midnight repose, and, after receiving from him a kind welcome and some refreshment, I retired to rest till sunrise.

During the forenoon of the 18th, our men went to the mountains, and cut down a new sprit for our sail, and prepared for the prosecution of the voyage. The favourable breeze had, however, been succeeded by a perfect calm, and the rays of the sun were exceedingly oppressive. As it appeared probable that the men would have to row the whole of the way, we agreed to defer our departure till the evening. This afforded me an opportunity of attending public worship with the native Christians of the settlement, and addressing the congregation assembled.

The sun was approaching the western horizon, when we took leave of our friends, and embarked, to prosecute the remainder of our voyage. We sailed across the beautiful bay, which for its size has justly been denominated one of the finest in the world, and passing along within the reefs to Maharepa, we again launched our boat, about eight o'clock in the evening.

The excitement of watching, and fatigue of the preceding part of our voyage, having induced a considerable degree of exhaustion of strength and spirits, we had not advanced far upon the open sea, before I became oppressed with a sensation of drowsiness, which I could not remove. During my voyages among the islands, I have passed many nights at sea with the natives in an open boat, and generally found them watchful and alert during the early hours of night, but wearied and sleepy towards morning; and whenever I have felt rest necessary for myself, have usually taken it before midnight, that I might be more vigilant when my companions should become drowsy. This was my purpose in the present instance. The wind had indeed ceased, but the surface of the sea was agitated with a quick and cross motion; the current was against us; and it was uncertain how soon in the morning we should reach Matavai, our port of destination in the island of Tahiti. I therefore gave Matapuupuu charge of the helm, which I had hitherto kept during the whole of the voyage, and, directing him to awake me in about an hour's time, I wrapped myself in my cloak, lay down upon the seat in the stern of the boat, and, notwithstanding the motion of the sea, and the rattling and shaking occasioned by the movements of the oars, soon fell into a sound sleep.

The refreshing and beneficial effects of my repose were, however, entirely neutralized by the sensations I experienced at its close. I cannot describe my emotions when I awoke, and found it was broad day-light; and, on turning to the helm, saw Matapuupuu fast asleep, with his hands still on the tiller; and then, looking forward along the boat, on beholding every individual motionless; the rowers leaning over their oars, the others stretched along the bottom of the boat, and every one in the most profound sleep. Before I attempted to awake any one, I involuntarily looked for the island we had left. It was still in sight. I then looked on the opposite side, for that to which we were going. It was also in sight, but the lofty mountains rising at the head of Matavai were far to the north, and indicated that the port to which we were bound was many miles behind us. In fact, we appeared to be about midway between Tahiti and Eimeo, drifting to the southward, far away from both, as fast as the current could bear us.

Fully sensible of our critical situation, if the breeze, which just began to ripple the surface of the water, should increase, I instantly awoke my companions, and asked them how they came all to fall asleep together. They looked confused, on beholding the broad light of day beaming upon them, and replied that each had imperceptibly fallen under the influence of sleep, without knowing that any other was under the dominion of the same sensation. Recollecting that I had in the first instance set them the example, I could not much censure their conduct. I therefore directed their attention to the mountains in the vicinity of Matavai and Papeete, on Wilks' Harbour, far in our rear; and as Burder's Point was the nearest part of the coast, urged them to apply with vigour to their oars, that we might reach it before the wind became so strong as to arrest our progress.

The men, refreshed by their slumbers, which had been created by the undulating motion of the boat on the water, and having broken a few cocoa-nuts, and drank the milk, cheerfully grasped their oars, and pulled steadily towards the shore. After about five hours' hard rowing, we reached the beach, and were cordially welcomed by our friends, Messrs. Darling and Bourne, resident Missionaries at Burder's Point. In the afternoon, several of the natives, who had accompanied me to Tahiti, set out for Papara, in order to visit their friends, who had accompanied Mr. Davies from Huahine to that station during the preceding year.

I spent this and the following day at Burder's Point. The respect and affection manifested by the people towards their teachers was gratifying, and the general improvement in the habits of the people, and the appearance of the settlement, highly encouraging. Newly planted gardens and enclosures appeared in every direction; several good houses were finished; a number also were plastered and thatched; and others, though only in frame, and presenting the appearance of mere skeletons of buildings, indicated a state of pleasing and progressive improvement. The public burying-ground, situated on the border of the settlement, was kept remarkably neat.

The outline of the grave was defended by a curb, or border, of large fragments of coral planted in the ground, while the grave itself was covered with small pieces of white coral and shells, brought from the adjacent shore. The school was a good building; and the chapel, erected near the ruins of the ancient marae, which I visited during my stay, was one of the most compact I had seen in the Georgian or Society Islands. The walls were framed and boarded; the roof thatched with *fara*, or palm-leaves. The floor was boarded, the pulpit and appendages remarkably neat, and the whole area of the chapel filled with seats. It was also fitted up with a gallery, the first ever erected in the South Sea Islands; the gallery, and other parts of the interior, which had been finished under the direction and by the assistance of Mr. Darling, were neater, and more European in appearance, than any I had hitherto beheld.

The advancement in civilization had not been so striking or rapid at this station as at some others; the effects of its progress were, however, such as to afford great encouragement, and to warrant the anticipation of its ultimately extending throughout the entire population of a district that had felt the ravages of war, and the demoralization of paganism, as much as any in the South Sea Islands.

About ten in the morning of the 21st, we took leave of our friends at Burder's Point, and, after rowing about four hours between the reefs and the shore, reached Papeete, or Wilks' harbour, where the queen and her sister were residing. On landing, the deputation from the Huahinean chiefs repaired to the abode of Teriitaria, and Matapuupuu delivered their message. She replied, "that she was anxious to remove to Huahine, and would return with them, if Pomare would allow her to leave Tahiti; but said she would see them again, and, before they returned, deliver her final reply."

On the brow of a hill, forming the commencement of a range extending from the vicinity of the shore to the lofty interior mountains, Mr. Crook formerly, at this station, had erected his abode. Having waited on the queen, and other members of the royal family residing with her, I walked up the hill, which Mr. Crook had designated Mount Hope, and was happy to find himself and his family well. The situation he had selected for his abode, though inconvenient on account of its distance from the settlement, and the fatigue induced by the ascent, has nevertheless peculiar advantages; the air is remarkably pure, the temperature generally cooler than on the adjacent lowlands, and the prospect most delightful and extensive.

With his agreeable family I passed the remainder of this day, and the following, which was the Sabbath. The congregation at the public religious services consisted of about five hundred hearers, who were in general attentive; the singing was good, and the voices of the men better than I ever heard elsewhere. The female voices are generally clear and distinct, and they sing well in most of the stations, but the voices of the men are seldom mellow or sonorous.

About ten o'clock on the following day I took leave of the friends at Mount Hope, and, accompanied by the chiefs from Huahine, proceeded to Matavai, where Pomare resided. It was near noon when we arrived, and soon after landing, the messengers waited upon the king, told him they had been sent by the chiefs of Huahine, to request Teriitaria to return and reside there, and expressed their conviction that he would approve of the same. He replied—" *Ua tia ia ia oti ra May e tai ai.* It is agreed, but let May be over, and then go;" alluding to the annual meetings held in the month of May.

I took up my abode with Mr. Nott, and spent the whole of the week in revising, with him and one or two of the chiefs from Huahine, the laws which had been prepared for that island. In this revision we endeavoured to correct what was defective in those already published in Tahiti and Raiatea. This employment occupied a number of hours every day. It was a matter of importance: I was anxious that their laws should be framed with the utmost care, and felt desirous that we should avail ourselves of Mr. Nott's familiar acquaintance with the character of the people, and his observation on the effect of the laws on the inhabitants of Tahiti and Eimeo. I wished also to consult with Mr. Davies, but he was too far off. Mr. Nott stated, that the greatest defects he had observed, arose from the power vested in the hands of the magistrate to punish according to his own discretion those who were found convicted. In consequence of this, the same crime was followed by different punishments, in different parts, or by different magistrates. In order to remedy this, the punishment to be inflicted was annexed to the prohibition of the offence. The laws, it was hoped, would by these means be less uncertain in their influence than they had been.

Another subject of importance was the revenue of the government, and the means of support for the king and chiefs. On this subject, Pomare had refused to make any regulations, preferring to demand supplies from the people as his necessities might require, rather than receive any regular proportion of the produce of the soil. Private property, therefore, was still insecure, and the industrious cultivator of the land was not sure of reaping the fruits of his labour. This was remarkably manifest at the present time, when the king of Tahiti, in his anxiety to pay for the vessel that had been purchased in his name, after making repeated applications to the chiefs for large numbers of pigs, prohibited every individual from selling to a captain or other person any commodity he might have for barter, but required them to bring all to him, in return for which he sometimes gave them articles of the most trifling value. To remedy this defect, several laws were added to those prepared for the people of Huahine, and a certain tax, somewhat resembling a poll-tax, proposed, by which it was fixed what proportion of the produce of the island each individual should furnish for the use of the king, and also of the chief of the district in which he resided. The remainder was to be inviolably his own, for use or disposal. The treatment of offenders between their apprehension and trial, was also regulated. These were the principal additions made to the Huahine code.

The trial by jury had been incorporated in the laws of Raiatea. The alterations were approved of by the chiefs who had come from Huahine, and were by them shewn to Teriitaria, who signified her entire satisfaction in their being adopted as the laws of Huahine. At the same time she informed the chiefs, that after the approaching meetings, she intended to remove to Huahine, but did not wish them on that account to defer the public enactment of the laws, whenever it should appear desirable.

The most important object of our visit being now accomplished, we returned to Papeete, intending to proceed to Eimeo. About noon on the 28th, we embarked in our boat, hoisted our sails, and were on the point of leaving the shore, when a messenger arrived with intelligence that a vessel was approaching Matavai, so that instead of putting out to sea, our course was instantly directed thither. A brig of considerable size was advancing towards the harbour. We hailed her approach with joyful hopes that she would bring us

> ————" News of human kind,
> Of friends and kindred, whom, perhaps, she held
> As visitors, that she might be the link
> Connecting the fond fancy of far friendship."

Meeting the vessel at the entrance of the bay, we found it was the Hope, of London, having Mr. and Mrs. Hayward from England, and Mr. and Mrs. Wilson from New South Wales, on board. As the vessel was under full sail, we could only greet their arrival by signal, and follow them to the harbour. They had, however, scarcely anchored, when we found ourselves alongside, and, ascending the deck, were happy to exchange our mutual congratulations. A number of cattle, some belonging to the passengers, others sent as presents by Mr. Birnie to the chiefs, having suffered much during the voyage, were speedily landed. After this, we accompanied our friends to the shore, elated with the anticipated pleasure of intelligence from home. In this respect we were not disappointed. A few letters which were at hand we received on board, and the rest as soon as the boxes containing them were opened. We broke the seals, skimmed the contents, and glanced at the signatures with no common feelings, reserving a more careful perusal for a season of greater leisure.

No opportunity equally favourable for receiving intelligence from England, had occurred since our arrival. Mr. Hayward had proceeded from the islands to England; he had met our friends and relatives there, and had been enabled to satisfy them in a variety of points, of which, though of confessedly minor importance, they were anxious to be informed. He had left them, and returned direct to us; and the simple fact that we were conversing with one who had traversed scenes long familiar and vividly present to our recollections, and one who had mingled in the society of those dearest on earth to us, appeared to shorten the distance by which we were separated, and to remove the most formidable barriers to intercourse. We had a thousand questions to ask, and the evening was far too short for the answers of half of our inquiries, or the perusal of our letters.

Mingled and intense are the emotions with which a lonely sojourner in a distant and uncivilized part of the world, receives a packet from his native land. This is especially the case when the symbol of mourning appears on the exterior of any of his letters. The unfolded sheet is sometimes put aside, as the eye, in its first glance over the lines, has been arrested by a sentence conveying tidings of the departure of some dear and valued relative or friend.

Notwithstanding the painful sensations occasioned by the knowledge of the fact, that some dear object of the heart's attachment or esteem has been for months consigned to the cheerless grave; epistles from those we have left in our native land, produce emotions more powerful, and satisfaction more elevated, than any other circumstance besides. Letters *sent* home by those in distant climes, may convey all that undiminished affection prompts, but they awaken no recollections connected with the locality, the companions, and the circumstances of those by whom they are written. The scenes and society by which the writers are surrounded, are foreign; and next to the feeling of curiosity, the greatest interest they excite, arises from the connexion with those for whose welfare every concern is felt. Very different are the effects of a letter *from* home, to one in a distant land. Every circumstance connected with it awakens emotion; even the name of the place whence it is dated, recalls a thousand associations of by-gone days. We seem to hear again the familiar voice, and involuntarily feel as if we had mingled once more with the circle which friendship and attachment had often drawn round the domestic hearth; and while perusing letters from home, we have felt the force of the poet's exclamation,

How fleet is a glance of the mind!

Next to the enjoyment of the Divine favour, letters from friends are among the sources of sweetest solace, and most cheering encouragement, to the sojourner in a foreign land. They excite a train of feeling which must be experienced, to be understood. They cheer the spirits often fainting under the effects of an insalubrious clime, the silent prostration of debilitating sickness, or the opposition and the trials of situation. They convey to his mind the gratifying conviction that he is not forgotten by those in whose enjoyments and pursuits he once participated.

This consideration not only revives his spirit, but imparts a fresh impetus to his movements, and adds new energy to all his efforts. Letters from those abroad are gratifying to friends at home; and if so, to those who participate the joys of sincere, enlightened, and glowing friendship, and who are encircled by a thousand sources of enjoyment, how much more welcome must they be to the distant, and often lonely absentee, who, though surrounded by multitudes of human beings, is yet doomed to perfect solitude, in respect to all mutual and reciprocal interchange of sympathy in thought and feeling.

Sure I am, that did the friends of those who have gone to distant, barbarous, and often inhospitable lands, know the alleviation of trials, and the satisfaction of mind, their epistles are adapted to produce, they would not be content with simply answering the letters they may receive, but would avail themselves of every opportunity thus to exchange their sympathies, and impart their joys, to those who are cut off from the many sources of comfort accessible to them.

Did the friends of the exile abroad also know the painful reflections to which a disappointment, in reference to expected intelligence, gives birth, they would endeavour to spare them that distress. In his lonely, distant, and arduous labours, a Missionary requires every solace, assistance, and support that his friends can impart. The communications he receives from his patrons are valuable, but they are frequently too much like letters of business, or treat only of general subjects. His communications from his relatives and friends are of a much more touching and interesting character. These, though they deeply affect, do not engross his soul; he feels connected with, and interested in, the general advancement of the Redeemer's kingdom, and the gigantic energies of those institutions of Christian benevolence and enterprise, which, under God, are changing the world's moral aspect. The reports, &c. of these institutions should be sent, and, in addition to these, a regular correspondence should be kept up with the Auxiliary Missionary Societies with which he may have been connected—the Sabbath-schools in which he may, perhaps, have been a teacher—but especially the Christian church of which he may have been a member. It should not be confined to a bare reply to letters, but should be regular and constant.

Sometimes we have been six, nine, or twelve months on the island of Huahine, and during that, or a longer period, have seen no individual, except our own two families, and the natives. At length, the shout, *E Pahi, e Pahi,* "A ship, a ship," has been heard from some of the lofty mountains around our dwelling. The inhabitants on the shore have caught the spirit-stirring sound, and "A ship, a ship," has been echoed, by stentorian or juvenile voices, from one end of the valley to the other. Numbers flock to the projecting rocks, or the high promontories, others climb the cocoa-nut tree, to obtain a glance of the desired object. On looking out, over the wide-spread ocean, to behold the distant sail, our first attempt has been to discover how many masts she carried; and then, what colours she displayed; and it is impossible to describe the sensations excited on such occasions, when the red British banner has waved in the breeze, as a tall vessel, under all her swelling canvass, has moved towards our isolated abode.

We have seldom remained on shore until a vessel has entered the harbour, but have launched our boat, manned with native rowers, and, proceeding to meet the ship, have generally found ourselves alongside, or on deck, before she has reached the anchorage. At the customary salutations, if we have learned that the vessel was direct from England, and, as was frequently the case, from London, our hopes have been proportionably raised; yet we have scarcely ventured to ask the captain if he has brought us any tidings, lest his reply in the negative should dispel the anticipations his arrival had awakened. If he has continued silent, we have inquired whether he had brought out any supplies; if he has answered No, a pause has ensued; after which, we have inquired whether he had any letters; and if to this, the same reply has been returned, our disappointment was as distressing as our former hopes had been exhilarating. We have remarked, that probably our friends in England did not know of his departure. This has been, we believe, the ordinary cause why so many ships have arrived in the islands from England, without bringing us any intelligence, except what we could gather from two or three odd newspapers that have been lying about the cabin. Though it has been some alleviation to believe, that had our friends known of the conveyance, they would have written; yet the relief thus afforded is but trifling, compared with the pain resulting from the absence of more satisfactory communications. Notwithstanding the length of time we had often been without seeing an individual who spoke our native language, excepting in our own families, we would, in general, rather the vessel had not at that time arrived, than that such arrival should have brought us no intelligence.

No disappointment, however, was experienced on the present occasion. The Hope had brought out a valuable supply of such articles as we needed; and Mr. and Mrs. Hayward, in addition to the letters of which they were the bearers, afforded us much satisfaction by the accounts they gave of those of our friends whom they had seen. The communications from England required the united consideration of the Missionaries; and this, with the distribution of the supplies, detained us a week longer in Matavai.

On the fourth of May, we took our leave. Heavy rains detained us at Papeete until nearly dark, but the weather clearing soon after sunset, we again launched out boat, and, being favoured with a fair wind, arrived in Eimeo before midnight. We were anxious to reach Huahine by the Sabbath, the following being the week in which the Missionary anniversary occurred. Early the next morning, which was Saturday, we arose, and prepared to depart: but the wind being westerly, was contrary, and prevented us. About six in the morning, however, it changed to the north and eastward, and continuing to blow steadily in that direction for an hour or two, we sailed from Eimeo about eight o'clock.

The sea was agitated, and the swell continuing from the westward, after the breeze from that quarter had subsided, was against us. The wind, though favourable, was but light, and our progress consequently slow. Our little bark containing the portion of supplies from the Hope, for the Missionaries in the Leeward Islands, was heavily laden. These amounting to several tons, besides the number of natives on board, not only kept the boat steady, but brought it considerably lower in the water than I had ever seen it before. About mid-day we lost sight of Eimeo. Continuing our course in a north-westerly direction, soon after sun-set, while the radiance of the departed luminary invested the horizon with splendour, we had the high satisfaction to behold the broken summits of what we considered the Huahinean mountains, shewn in beautiful though indistinct contrast with the brightness of the heavens and the sea. The duration of twilight in the tropics is always short; and the rich sunset scene, which the peculiarity of our situation had rendered striking and imposing, was soon followed by the darkness of night, which in much less than an hour veiled the surrounding objects. The glance, however, which we had obtained of the mountains of Huahine, was serviceable and cheering; it convinced us that the current had not swept us aside from our course, and it enabled us to fix satisfactorily the direction in which to steer until morning. Although our rest had been but broken and short during the preceding night, our present situation repressed any desire for repose.

Nothing can exceed the solemn stillness of a night at sea within the tropics, when the wind is light, and the water comparatively smooth. Few periods and situations, amid the diversity of circumstances in human life, are equally adapted to excite contemplation, or to impart more elevated conceptions of the Divine Being, and more just impressions of the insignificancy and dependence of man. In order to avoid the vertical rays of a tropical sun, and the painful effects of the reflection from the water, many of my voyages among the islands of the Georgian and Society groups have been made during the night. At these periods I have often been involuntarily brought under the influence of a train of thought and feeling peculiar to the season and the situation, but never more powerfully so than on the present occasion.

The night was moonless, but not dark. The stars increased in number and variety as the evening advanced, until the whole firmament was overspread with luminaries of every magnitude and brilliancy. The agitation of the sea had subsided, and the waters around us appeared to unite with the indistinct though visible horizon. In the heaven and the ocean, all powers of vision were lost, while the brilliant lights in the one being reflected from the surface of the other, gave a correspondence to the appearance of both, and almost forced the illusion on the mind, that our little bark was suspended in the centre of two united hemispheres.

The perfect quietude that surrounded us was equally impressive. No objects were visible but the lamps of heaven, and the phosphoric fires of the deep. The silence was only broken by the murmurs of the breeze passing through our matting sails, or the dashing of the spray from the bows of our boat, excepting at times, when we heard, or fancied we heard, the blowing of a shoal of porpoises, or the more alarming sounds of a spouting whale.

At a season such as this, when I have reflected on our actual situation, so far removed, in the event of any casualty, from human observation and from human aid, and preserved from certain death only by a few feet of thin board, which my own unskilful hands had nailed together; a sense of the wakeful care of the Almighty has alone afforded composure; and when I have gazed on the magnificent and boundless assemblage of suns and worlds, whose rays have shed their lustre over the scene, and have remembered that they were formed, sustained, and controlled, in all their complex and mighty movements, by Him on whose care I could alone rely, I have almost involuntarily uttered the exclamation of the psalmist, "Lord, what is man, that thou art mindful of him!"

The contemplation of the heavenly bodies, although they exhibit the wisdom and majesty of God, who "bringeth out their host by number, and calleth them all by names, by the greatness of his might," and by whom also the very hairs of the head are all numbered, impressed at the same time the conviction that I was far from home, and those scenes which in memory were associated with starlight nights in my native land.

Many of the stars which I had beheld in England were visible here: the constellations of the zodiac, the splendours of Orion, and the mild twinkling of the Pleiades, were seen; but the northern pole-star, the steady beacon of juvenile astronomical observation, the Great-bear, and much that was peculiar to a northern sky, were wanting. The effect of mental associations, connected with the appearance of the heavens, is singular and impressive. During a voyage which I subsequently made to the Sandwich Islands, many a pleasant hour was spent in watching the rising of those luminaries of heaven which we had been accustomed to behold in our native land, but which for many years had been invisible.—When the polar-star rose above the horizon, and Ursa-major, with other familiar constellations, appeared, we hailed them as long absent friends; and could not but feel that we were nearer England than when we left Tahiti, simply from beholding the stars that had enlivened our evening excursions at home.

But although, in our present voyage, none of these appeared, and the southern hemisphere is perhaps less brilliant than that of the north, it exhibited much to attract attention. The stars in the Fish, the Ship, and the Centaur, the nebulæ or magellanic clouds, and, above all others, Crux, or the "Cross of the South," are all peculiar to this part of the heavens. This latter constellation is one of the most remarkable in the southern hemisphere. The two stars forming the longest part, having nearly the same right ascension, it appears erect when in the zenith, and thus furnishes a nightly index to the flight of time, and a memento to the most sublime feelings of grateful devotion.

With my fellow-voyagers I could enter into nothing like reciprocally interesting conversation on these subjects. Their legends of the nature and origin of the stars were most absurd and fabulous; and my attempts to explain the magnitude, distances, or movements of the heavenly bodies, appeared to them unintelligible—

> Their "souls proud science never taught to stray
> Far as the solar walk or milky-way."

The natives of the islands were, however, accustomed in some degree to notice the appearance and position of the stars, especially at sea. These were their only guides in steering their fragile barks across the deep. When setting out on a voyage, some particular star or constellation was selected as their guide in the night. This they called their *aveia*, and they now designate the compass by the same name, because it answers the same purpose. The Pleiades were a favourite aveia with their sailors, and by them, in the present voyage, we steered during the night. We had, indeed, a lantern and a compass in the boat, but being a light ship's compass, it was of little service.

Although the Polynesians were destitute of all correct knowledge of the sciences, the first principles of which have been recently taught in the academy more regularly than they had heretofore been, they had what might be called a rude system of astronomy. They possessed more than one method of computing time; and their extensive use of numbers is quite astonishing, when we consider that their computations were purely efforts of mind, unassisted by books or figures.

Their ideas, as might naturally be expected, were fabulous and erroneous in the extreme. They imagined that the sea which surrounded their islands was a level plane, and that at the visible horizon, or some distance beyond it, the sky, or *rai*, joined the ocean, enclosing as with an arch, or hollow cone, the islands in the immediate vicinity. They were acquainted with other islands, as Nuuhiva, or the Marquesas, Vaihi, or the Sandwich Islands, Tongatabu, or the Friendly Islands. The names of these recurred in their traditions or songs. Subsequently, too, they had heard of Beritani, or Britain, Paniola, or Spain, &c. but they imagined that each of these had a distinct atmosphere, and was enclosed in the same manner as they thought the heavens surrounded their own islands. Hence they spoke of foreigners as those who came from behind the sky, or from the other side of what they considered the sky of their part of the world.

What their opinions were, as to the material of the heaven to which they gave such definite boundaries, I could never learn; but, according to their mythology, there were a series of celestial strata, or *tua*, teh in number, each stratum being the abode of spirits or gods, whose elevation was regulated by their rank or powers; the tenth, or last heaven, which was perfect darkness, being called a *terai haamama* of tane, and being the abode of the first class only.

We often experienced a degree of confusion in our ideas connected with their use of the term *po*, night or darkness, and its various compounds. They usually, but not invariably, spoke of the region of night as *i raro*, or below. In this instance, in describing the highest heaven as the region of purest light, they spoke of it also as the po. After describing the nine heavens, or stratum of clouds or light, inhabited by the different orders of inferior deities, they describe the tenth, or most remote from the earth, and the abode of the principal gods, as te rai haamama no tane, &c. *te* opening or unfolding to the *po*, or perpetual darkness. From this mode of representation, it appears that the islanders imagined the universe to be chaotic, and that in its vast immensity their islands and ocean, with the sky arching over them, were enclosed, and that below the foundation of the earth, on which they stood, and above the firmament over their heads, this po, or darkness, prevailed.

With respect to the origin of the sun, which they formerly called *ra*, and more recently *mahana*, some of their traditions state that it was the offspring of the gods, and was itself an animated being; others, that it was made by Taaroa. The latter supposed it to be a substance resembling fire. The people imagined that it sank every evening into the sea, and passed during the night, by some submarine passage, from west to east, where it rose again from the sea in the morning. In some of the islands, the expression for the setting of the sun is, the falling of the sun into the sea. On one occasion, when some of the natives were asked where the sun went to, they said, Into the sea. On being asked, further, what prevented its extinction, they said they did not know. It was then inquired, "How do you know that it falls into the sea at all? Did you ever see it?" They said, No, but some people of Borabora, or Maupiti, the most western islands, had once heard the hissing occasioned by its plunging into the ocean.

One of the most singular of their traditions, respecting the sun, deserves attention, from the slight analogy it bears to a fact recorded in Jewish history. It is related that Maui, one of their ancient priests or chiefs, was building a marae, or temple, which it was necessary to finish before the close of the day; but, perceiving the sun was declining, and that it was likely to sink before the work was finished, he seized the sun by his rays, bound them with a cord to the marae, or an adjacent tree, and then prosecuted his work till the marae was completed, the sun remaining stationary during the whole period. I refrain from all comment on this singular tradition; which was almost universally received in the islands.

Their ideas of the moon, which they called *avae* or *marama,* were as fabulous as those they entertained of the sun. Some supposed the moon was the wife of the sun; others, that it was a beautiful country in which the aoa grew. I am not aware that they rendered divine homage either to the sun or moon. Theirs was a far less rational and innocent system than the worship of the host of heaven. They, however, supposed the moon to be subject to the influence of the spiritual beings with whom their mythology taught them to people the visible creation; and to the anger of those spirits, they were accustomed to attribute an eclipse. During an eclipse, the moon is said to be *natua,* bitten or pinched.

The stars, which they call *fetia* or *fetu,* were by some considered as the children of the sun and moon. They are, however, generally supposed to be inhabited by spirits of the departed, or to be the spirits of human beings, several principal stars being designated by the names of distinguished men. The phenomenon called a shooting star, they supposed to be the flight of a spirit. Many of the constellations, and more of the single stars, have distinct names. Mars they call *fetia ura,* red star.

The morning star they call *fetia ao,* star of day; or *horo poipoi,* forerunner of morning. The Pleiades they call *matarii,* small eyes. But one of the most remarkable facts is, that the constellation called the Twins is so named by them; only, instead of denominating the two stars Castor and Pollux, they call them *na ainanu,* the two ainanus; and to distinguish the one from the other, ainanu above, and ainanu below. The nebulæ near the southern pole, called the Magellanic clouds, are denominated *mahu,* mist or vapour, and are distinguished in the same way, one being above, the other below.

Like most uninformed persons, they supposed the earth was stationary, being borne on the shoulders of a god, fixed upon a rock, which they called the rock of foundation supported by pillars, and that the sun, moon, and stars, moved from one side of the arched heavens to the other. When we at first endeavoured to impart to them more correct ideas of astronomy, and exhibited to their view a terrestrial globe, explanatory of the shape of our earth, and illustrative of those of the moon, of the planets, and other heavenly bodies, they were greatly surprised; but when we called their attention to a celestial globe, and represented to them the relative position of the heavenly bodies, and explained the motion of the planets of our system round the sun, they were invariably sceptical at first. It could not possibly be, they said, that the earth went round, as all things remained stationary during the twenty-four hours; which would not be the fact, if the earth on which they stood moved. Frequently they have said, if such was the fact, we should fall from our beds, and all our vessels of food, &c. would be upset. Finding, however, that we persevered in the expression of our sentiments to the contrary, they would sometimes remark—we believe it because you say so, but we cannot understand it. These observations were made only when the subject was first brought under their notice. The intelligent among them now entertain far more consistent views.

Among the Harvey Islands, they worshipped a god of thunder; but he does not appear to have been an object of great terror to any of them. The thunder was supposed to be produced by the clapping of his wings. The ignis fatuus, they considered as one of their most powerful gods, proceeding, in his tutelary visitations, from one marae to another.

The winds were presumed to be under the direction and control of the deities; by whom they were supposed to be kept in a cave, as by Eolus among the ancient pagans. Some contended that there was but one *rua*, hole or cavern, of the winds; others, that there were two, one in the east, the other towards the west, the two quarters whence the winds usually blew. Although they had but one, or at most two caverns, whence they supposed the winds to proceed, they had a distinct name for each wind, designating sometimes both its degree of strength, and its direction. The north wind they called *Haupiti*; the south, *Maraamu*; the east, *Maoai*; the west, *Toerau*. The east, with its variations from north-east to south-east, being the regular trade-wind, is most prevalent, and is seldom unpleasantly violent. Winds from the north are often tempestuous, more so than from the south, whence, although during the season of variable winds they are strong, and continue several days, they are not dangerous. The wind seldom prevails from the west, among the Society Islands, excepting in the months of December, January, and February. At this period they are sometimes violent, usually of short duration, and almost invariably accompanied with rain and heavy unsettled weather.

Though unacquainted with the compass, the islanders have names for the cardinal points. The north they call Apatoa; the south, Apatoerau; the east, Te hitia o te ra, the rising of the sun; and the west, the Tooa o te ra, the falling or sinking of the sun. The climate is warm to an European: the thermometer ranges between 70 and 80, the average height in the shade is 74 degrees.

Their genealogies and chronological traditions do not appear to have been so correctly preserved as those of the Hawaiians, one or two of which I have, that appear, at least for nearly thirty generations, tolerably correct, though they go back one hundred generations. They were, however, as correct in their methods of computing time as their northern neighbours, if not more so. One mode of reckoning time was by *ui's*, or generations; but the most general calculation was by the year, which they call matahiti, and which consisted of twelve or thirteen lunar months, by the tau or matarii, season or half-year, by the month of thirty days, and by the day or night. They had distinct names for each month; and though they all agreed about the length of the year, they were not unanimous as to the beginning of it, or the names of the months, each island having a computation peculiar to itself.

The following is a statement of their divisions of time, copied from a small book on arithmetic, &c. prepared by Mr. Davies, which I printed at Huahine in 1819. It is the method of computation adopted by the late Pomare and the reigning family.

1 Avarehu . . .	The new moon that appears about the summer solstice of Tahiti, and generally answers to the last ten days of December, or the beginning of January.
2. Faaahu	January, and part of February—The season of plenty.
3. Pipiri	February, and part of March.
4. Taaoa	March, and part of April—The season of scarcity.
5. Aununu . .	April, and part of May.
6. Apaapa. . . .	May, and a part of June.
7. Paroro mua . .	June, and a part of July.
8. Paroro muri . .	July, and a part of August.
9. Muriaha . . .	August, and a part of September
10. Hiaia	September, and part of October.
11. Tema	October, and part of November—The season of scarcity.
12. Te-eri	The whole, or a part of, November—The uru, or young bread-fruit, begins to flower.
13. Te-tai	The whole, or a part of, December—The uru, or bread-fruit, nearly ripe.

Their calculations, however, were not very exact. Thirteen moons exceed the duration of the solar year. But, in order to adapt the same moons to the same seasons as they successively occur, the moon generally answering to March, or the one occurring about July, is omitted; and in some years, only twelve moons are enumerated.

Another computation commenced the year at the month Apaapa, about the middle of May, and gave different names to several of the months. They divided the year into two seasons, of the *Matarii*, or Pleiades. The first they called *Matarii i nia*, Pleiades above. It commenced when, in the evening, these stars appeared on or near the horizon; and the half year, during which, immediately after sunset, they were seen above the horizon, was called *Matarii i nia*. The other season commenced when, at sunset, the stars were invisible, and continued until at that hour they appeared again above the horizon. This season was called *Matarii i raro*, Pleiades below.

The islanders had three seasons besides these. The first they called *Tetau*, autumn, or season of plenty, the harvest of bread-fruit. It commenced with the month *Tetae*, December, and continued till Faahu. This is not only the harvest, but the summer of the South Sea Islands. It is also the season of most frequent rain. The next is *Te tau miti rahi*, the season of high sea. This commences with *Tieri*, November, and continues until January. The third is the longest, and is called the *Te tau Poai*, the winter, or season of drought and scarcity. It generally commences in *Paroromua*, July, and continues till *Tema*, October.

The natives have distinct names for each day and each night of the month or moon. They do not, however, reckon time by days, but by nights. Hence, instead of saying, How many days since? they would inquire, *Rui hia aenei?* "How many nights?" The following are the different nights of each moon.

THE NIGHTS OF THE MOON.

1. Ohirohiti.
2. Hoata.
3. Hami-ami-mua.
4. Hami-ami-roto.
5. Hami-ami-muré.
6. Ore-ore-mua.
7. Ore-ore-muri.
8. Tamatea.
9. Ohuna.
10. Oari.
11. Omaharu.
12. Ohua.
13. Omaidu.
14. Ohodu.
15. Omarae.—Te-maramaati, or the moon with a round and full face.
16. Oturu-tea.
17. Raau-mua.
18. Raau-roto.
19. Raau-muri.
20. Ore-ore-mua.
21. Ore-ore-roto.
22. Ore-ore-muri.
23. Taaroa-mua.
24. Taaroa-roto.
25. Taaroa-muri.
26. O-Tane.
27. O-Roomie.
28. O-Roomaori.
29. O-mutu.
30. O-Terieo.—This is the night or day the moon dies, or is changed.

The seventeenth, eighteenth, and nineteenth nights, or nights immediately succeeding the full moon, were considered as seasons when spirits wander more than at any other time; they were also favourable to the depredations of thieves. They do not appear to have divided their months into weeks, or to have had any division between months and days. Totally ignorant of clocks or watches, they could not divide the day into hours. They, however, marked the progress of the day with sufficient exactness, by noticing the position of the sun in the firmament, the appearance of the atmosphere, and the ebbing and flowing of the tide.

Midnight they called the Tui ra po.

One or two in the morning—Maru ao.

Cock-crowing, or about three o'clock in the morning—Aaoa te moa; aaoa being an imitation of the crowing of a cock.

The dawn of day—Tatahita.

Morning twilight—Marao rao.

When the flies begin to stir—Ferao rao.

When a man's face can be known—Itea te mata taata.

The first appearance of the upper part of the sun—Te hatea rao te ra.

Sunrise, or morning—Poi poi.

The sun above the horizon—Ofao tuna te ra.

The sun a little higher, sending his rays on the land—Matiti titi te ra.

About seven o'clock—Tohe pu te ra.

Eight o'clock—Pere tia te ra.

About nine—Ua paare te ra.

Ten or eleven—Ua medua te ra.

Noon-day, or the sun on the meridian—Avatea.

One or two in the afternoon—Taupe te ra.

About three in the afternoon—Tape-tape te ra.

Nearly four—Tahataha te ra.

About five—Hia-hia te ra.

Between five and six—Ua maru maru te ra.

Sun-setting, Ahi, ahi—Evening—Mairi—Te ra, Falling of the sun.

The beginning of darkness—Arehurehu.

Night, or the light quite gone—Po.
When the sea begins to flow towards the land—Pananu te tai.
About eleven at night—Tia rua te rui.

In order to facilitate their commercial transactions, and their intercourse with civilized nations, the English names for the months, and the days of the week, have been introduced. They have also been instructed in our methods of calculating the leap-years, &c.

The English method of mensuration has been introduced, and, with regard to short distances, they begin to understand it. The word *hebedoma*, assimilating easily with the peculiar vowel terminations of their words, and being distinct from any word in use among them, has been introduced to signify a week. It is not, however, so frequently employed by the people, as the word Sabbath. If a native wished to say he had been absent on a voyage or journey six weeks, he would generally say six Sabbaths, or one moon and two Sabbaths.

Considering their uncivilized state, and want of letters, their method of computing time is matter of astonishment, and shews that they must have existed as a nation for many generations, to have rendered it so perfect. It is also an additional proof that they are not deficient in mental capacity.

Their acquaintance with, and extensive use of numbers, under these circumstances, is still more surprising. They did not reckon by forties, after the manner of the Sandwich Islanders, but had a decimal method of calculation. These numerals were,

Atahi, one.
Arua, two.
Atoru, three.
Amaha, four.
Arima, five.
Aono, six.
Ahitu, seven.
Avaru, eight.
Aiva, nine.
Ahuru, ten.

Eleven would be Ahuru matahi, ten and one; and so on to twenty, which was simply Erua ahuru, two tens; twenty-one, two tens and one; and proceeding in this way till ten tens, or one hundred, which they called a *Rau*. The same method was repeated for every successive rau, or hundred, till ten had been enumerated, and these they called one *Mano*, or thousand. They continued in the same way to enumerate the units, ahurus or tens, raus or hundreds, and manos or thousands, until they had counted ten manos, or thousands; this they called a *Manotini*, or ten thousand. Continuing the same process, they counted ten manotinis, which they called a *Rehu*, or one hundred thousand. Advancing still farther, they counted ten rehus, which they called an *Iu*, which was ten hundred thousand, or one million.

They had no higher number than the *iu*, or million: they could, however, by means of the above terms or combinations, enumerate, with facility, tens, hundreds, thousands, tens of thousands, or hundreds of thousands of millions.

The precision, regularity, and extent of their numbers has often astonished me; and how a people, having, comparatively speaking, but little necessity to use calculation and being destitute of a knowledge of figures, should have originated and matured such a system, is still wonderful, and appears, more than any other fact, to favour the opinion that these islands were peopled from a country whose inhabitants were highly civilized.

Many of their numerals are precisely the same as those used by the people of several of the Asiatic islands, and also in the remote and populous island of Madagascar. Occasionally the Islanders double the number, by simply counting two instead of one. This is frequently practised in counting fish, bread-fruit, or cocoa-nuts, and is called double counting, by which all the above terms signify twice as large a number as is now affixed to them.

In counting, they usually employ a piece of the stalk of the cocoa-nut leaf, putting one aside for every ten, and gathering them up, and putting a longer one aside, for every rau, or hundred. The natives of most of the islands, adults and children, appear remarkably fond of figures and calculations, and receive the elements of arithmetic with great facility, and seeming delight.

They estimate the distance of places by the length of time it takes to travel or sail from one to the other. Thus, if we wished to give them an idea of the distance from the islands to England, we should say it was five months; and they would say the distance from Tahiti to Huahine was a night and a day, and from Huahine to Raiatea, from sunrise to nearly noon, &c.

But it is now high time to return from this apparently long digression, which, though somewhat diffuse, has an immediate bearing on the arithmetical calculations, the astronomical knowledge, and the nautical acquirements of these islanders, and bring our voyage to its termination.

The wind being light but fair through the night, and the sea pleasantly smooth, we kept on our course till the light of morning began to appear, and when the day broke had the satisfaction of beholding the island of Huahine at no very great distance, and immediately before us. We approached on the eastern side, but the wind being unfavourable for sailing to the settlement, we stood towards the shore. When we found ourselves within half a mile of the reef, we lowered our sails, and manning the oars, rowed round the northern point of the island. By eight o'clock on the 5th of May, we entered Fare harbour, and on our landing had the happiness to find our families and friends well. It was the Sabbath, and we repaired with gratitude to the house of God, to render our acknowledgments for preservation.

CHAP. XV.

Promulgation of the new code of laws in Huahine—Literal translation of the laws on Murder—Theft—Trespass—Stolen property—Lost property—Barter—Sabbath-breaking—Rebellion—Bigamy, &c.—Divorce, &c.—Marriage--False accusation--Drunkenness—Dogs--Pigs—Conspiracy—Confessions—Revenue for the king and chiefs—Tatauing—Voyaging—Judges and magistrates—Regulations for judges, and trial by jury—Messengers or peace-officers—Manner of conducting public trials—Character of the Huahine code—Reasons for dissuading from capital punishments—Omission of oaths—Remarks on the different enactments—Subsequent amendments and enactments relative to the fisheries—Land-marks—Land rendered freehold property—First Tahitian parliament—Regulations relating to seamen deserting their vessels—Publicity of trials—Beneficial effects of the laws.

THE laws and regulations which had received the sanction of Teriiteria at Tahiti, were approved by the chiefs of Huahine, at a public national assembly held in the month of May, 1822. Mamae, a leading raatira, requesting that the laws might be enacted, his request was acceded to, and, after some slight modifications, were promulgated in Huahine, and Sir Charles Sander's island, under the authority of the queen, governors, and chiefs. They were subsequently printed, and circulated in every part of the islands.

In a letter which Mr. Barff transmitted with a printed copy, speaking of the laws, he remarks, "You will find them, in every material point, the same as when you left the islands: I insert a literal translation of this code, not because it was the last promulgated, nor that I consider it superior in every respect to those by which it was preceded, but because it was adopted by the people with whom I was most intimately connected, and received a greater degree of the attention of my colleague and myself, than any of the others. It might, perhaps, have been abridged, or a mere enumeration of the laws might have furnished all the information that is interesting, yet the first code of laws adopted, written, and printed among a people, who, but a few years before, were ignorant heathen, lawless savages, is a document so important in the history of the people, as to justify its entire insertion. The title is *E Ture na Huahine:* "A Law,* or Code of Laws, for Huahine, caused to grow in the government or reign of Teriiteria, Hautia, and Mahine, subordinate (rulers)" and the Imprint is—"Huahine, printed at the Mission press, 1823."

The following is the Introduction immediately after the names of the queen and two principal chiefs—

"From the favour of God, we have our government. Peace to you (People) of Huahine."

LITERAL TRANSLATION OF THE LAWS OF HUAHINE, AND SIR CHARLES SANDER'S ISLAND.

I. Concerning Murder.

If parents murder their infants, or children unborn, if not the parents but the relatives, if not them, a stranger, or any person who shall wantonly commit murder, shall be punished—shall be transported to a distant land, uninhabited by men—such (a land) as Palmerston's Island. There shall (such criminals) be left until they die, and shall never be brought back.

II. Concerning Theft.

If a man steal one pig, four shall he bring as a recompense, for the owner of the pig two, for the king two. If he have no pigs, two single canoes,

* There is no word in their language for law. The Hebrew word has been introduced, as according with the genius and idiom of Tahitian better than any other.

for the owner of the pig one, for the king one. If (he have) no canoes bales or bundles of native cloth, two of them, if the tusks of the pig were growing up out of its mouth.* Each bale shall contain one hundred fathoms (200 yards) of cloth, four yards wide. If a half-grown pig, five fathoms. If a small pig, twenty fathoms in the bale. For the owner of the pig one half, and for the king the other. If he have no cloth, arrow-root. If the pig stolen was a large one, forty measures.† For a half-grown pig twenty measures, and for a small one ten. For the owner of the pig one part, for the king the other. Let the arrow-root of the king, and the owner of the pig, be equal. If not arrow-root, some other property. Thus let every thing stolen be paid for. Let four-fold be returned as a recompense, double for the king, and double for the owner. If he (the thief) have no property, let him be set to work on the lands of the person he has robbed. If he refuse, his land shall be the king's, and he shall wander on the road ‡ for an unlimited period. If the king restore him, he shall return to his land, if not (thus) restored he shall not return. The magistrates or judges shall award the punishment annexed to this crime in the laws, and that only. The judge shall not demand the value of the property from the relatives of the thief."

To this law, in the revision of the laws which took place in 1826, two or three particulars were added; one increasing the punishment with the repetition of the crime, and then expressly referring to those depredations in which burglary was committed, and a chest or box broken open.

III. RELATING TO PIGS.

If a pig enters a garden, and destroys the produce, let no recompense be required, because of the badness of the fence he entered. If stones are thrown at a pig, and it be bruised, maimed, or killed, the man thus injuring it shall take it, and furnish one equal in size, which he shall take to the owner of the pig killed or injured. If he has no pig, he shall take some other property, as a compensation. For a large pig, twenty measures of arrow-root, and for a smaller one, ten. If not arrow-root, cocoa-nut oil, as many bamboo canes full, as measures of arrow-root would have been required. If not (this) personal labour, for a large pig he shall make

* A full-grown hog, of the largest size, is thus denominated.
† A measure contains five or six pounds weight.
‡ The figurative term for banishment.

twenty fathoms of fencing, for a small one five, for the owner of the pig killed. If it be a good fence, and is broken (through the hunger or obstinacy of the pig) and the produce is destroyed, the pig shall not be killed, but tied up, and the magistrate shall appoint the recompense the proprietor of the garden shall receive. The owner also shall mend the broken fence.

IV. CONCERNING STOLEN GOODS OR PROPERTY.

If a man attempting to steal property obtains it, and sells it to another, and the purchaser knew it to be stolen property which he bought—if he does not make it known, but keeps it a secret, he also is a thief; and as is the thief's, such shall be his punishment. Every person concealing property stolen by another, knowing it to be stolen, is also a thief; and as is the thief's, such shall be his punishment.

V. CONCERNING LOST PROPERTY.

When an article that has been lost is discovered by any one, and the owner is known to the finder, the property shall be taken to the person to whom it belongs. But if such property be concealed, when the finder knew to whom it belonged, and yet hid it, he also is a thief; and that his punishment be equal to that of a thief, is right.

VI. CONCERNING BUYING AND SELLING, OR BARTER.

When a man buys or exchanges goods, let the agreement be deliberately and fairly made. When the bargain is finally and satisfactorily made, if one retains his (the article received,) and the other takes away his (the article given,) but after a short season returns it, the other (person) shall not take it again, unless he desires to do so; if agreeable to him to take it, it is with himself. If it be an article, the damage or defects of which were not perceived at the time of exchanging, but after he had taken it to his house were discovered, it is right that it be returned; but if the defects were known at the time of bartering, and when taken to the house were reconsidered, and then returned, it shall not be received.

VII. CONCERNING THE DISREGARD OF THE SABBATH.

For a man to work on the Sabbath is a great crime before God. Work that cannot be deferred, such as dressing food when a sick person desires warm or fresh food, this it is right to do; but not such work as erecting houses, building canoes, cultivating land, catching fish, and every other employment that can be deferred. Let none travel about to a long distance on the Sabbath. For those who desire to hear a preacher on the day of food (the preceding day) it is proper to travel. If inconvenient to journey on the preceding day, it is proper to travel on the Sabbath (to attend public worship;) but not to wander about to a great distance (to different villages) on the Sabbath. The individual who shall persist in following those prohibited occupations, shall be warned by the magistrates not to do so; but if he will not regard, he shall be set to work, such as making a piece of road fifty fathoms long, and two fathoms wide. If, after this, he work again on the Sabbath, let it be one furlong.

VIII. Concerning Rebellion, or Stirring up War.

The man who shall cause war to grow, shall secretly circulate false reports, shall secretly alienate the affections of the people from their lawful sovereign, or any other means for actually promoting rebellion, the man who acts thus shall be brought to trial; and if convicted of stirring up rebellion, he shall be sent to his own district or island, and if he there again stir up rebellion, his sentence shall be a furlong of road. If he repeat the offence, he shall be banished to some distant island, such as Palmerston's, and shall return only at the will or pleasure of the king.

IX. Regarding Bigamy.

It is not proper that one husband should have two wives, nor that one wife should have two husbands. In reference to the man who may have had two wives from a state of heathenism, let nothing be said, but let it remain; but if one of his wives die and the other remain, he shall not have two again. When a man obstinately persists in taking another wife, the magistrate shall cause his second wife to be separated from him; and shall adjudge, to both, labour as a punishment. The man shall make a piece of road forty fathoms long, and two fathoms wide; and the woman shall make twc floor-mats; if not these, four wearing mats; half for the king, and half for the governor (of the district.)

X. Concerning a Wife formerly Forsaken.

The man who, in a state of heathenism, forsook his wife and married another woman, shall not return to his former wife, neither shall the wife (having married another husband) return to the husband she forsook when in a state of heathenism. The man or the woman that shall persist to return, shall be punished; the punishment adjudged, shall be similar to that which is annexed to the breach of the ninth law.

XI. Concerning Married Women, and Married Men.

This law respects the crime of adultery. [It is unnecessary to give the details of its enactments; it requires pecuniary compensation for the offended party, and prohibits the offenders from marrying during the life of the injured individuals.]

XII. Concerning (Divorce) Putting away Husbands, and Putting away Wives.

That a man should put away his wife, who has not been unfaithful to him, is wrong. The magistrates shall admonish such an one that he receive his wife again. If he will not regard the admonition, let him be punished with labour till the day that he will return to his wife. If he is obstinate, and will not return, then they shall both remain till one of them die; the husband shall not marry another wife. The woman, also, who shall forsake, or put away her husband without cause, the above is the regulation with regard to such. But if a man put away his wife on account of her great anger (violent temper,) and for her bad behaviour, such man put (her) away, the magistrates shall admonish the wife and the husband that they live together; but if they are perverse, they shall remain; the wife shall not take another husband, and the husband shall not take another wife. They shall also be adjudged to labour till they live together again. The husband's work shall be on the road or the plantation. The wife shall perform such work as weaving mats or beating cloth. For the king one part, and for the governor the other part, of the work they shall do.

XIII. Concerning the Not Making Provision for the Support of the Wife.

If a man does not provide food for his own wife, but afflicts her with hunger, the magistrate shall admonish such a husband that he behave not thus; but if he will not hear their counsel, and his wife, on account of this evil treatment, (she) leave, let him be sentenced to labour (if not a weak and sickly man) until the day that he will behave kindly to his wife. The work shall be such as making a road, or erecting a fence, for the king and the governor.

XIV. Concerning Marriage.

Marriage is an agreement between two persons, one man and one woman, that they will be united in marriage. It must not be with a brother or a sister, but with a distant relation or a stranger, that a person may properly (lawfully) marry.

The marriage ceremony shall be performed by a Missionary, or a magistrate. Those purposing to marry shall make the same known unto a Missionary, or to a magistrate; and the Missionary (or magistrate) shall cause it to be known among the people, that the propriety (of it) may be known. Perhaps there may be some cause that would render it improper for them to marry; if not, then they may marry. This is the evil (that would render it unlawful) perhaps the man may have forsaken his wife in some island or country, and may have travelled to another land, deceitfully to marry. So also it may be with the wife, and is also the regulation. Therefore shall the Missionary request the people, if they know of evil conduct in any other land, that would render it unlawful, to make it known to him; then shall the marriage not take place. But if there be no evil (that would render it unlawful) then they may marry. On the day of assembling for worship, the Missionary shall publish this word—he shall then say unto all the people, "Such an one, and such an one, desire or propose to be united in marriage. The people will then seek or inquire, if there be any just cause why they should not live together.

When the day arrives for the celebration of marriage, let persons also come as witnesses. The Missionary shall then direct the man to take the right hand of the woman, when he shall say unto him, "Will (or do) you take this woman to be your lawful wife, will you faithfully regard her alone (as your wife,) until death? Then shall the man answer, "*Yes.*" The Missionary shall now direct the woman to take the man by the right hand, and shall ask her, "Will (or do) you take this man to be your lawful husband, will you be obedient unto him, will you faithfully regard him alone, (as your husband until death?) Then shall the woman answer, "*Yes.*"

After this, the Missionary shall declare unto all the people, "These two persons have become truly (or lawfully) man and wife, in the presence of God and man." The register of the marriage shall be written by the Missionary in the marriage book, and signed by himself, the parties, and the witnesses. Thus shall marriage be solemnized. Let none become man and wife secretly, it is a crime.

XV. Concerning False Accusation.

The man who shall falsely accuse another before a magistrate, with intent to have the accused person brought to trial, or the man who shall falsely come as a witness, it being his intention or purpose, in giving false evidence, that the accused may be convicted or punished; if his accusation or evidence is proved to be false, the penalty that would have been adjudged to the accused, (had he been found guilty,) shall be transferred to such false accuser.

XVI. Unnatural Crime.

This law refers to a crime, for the prohibition of which, perpetual banishment, or incessant hard labour for seven long years, is annexed as the punishment of those who shall be guilty of its perpetration.

The XVIIth regards Seduction—the XVIIIth Rape—and the XIXth Fornication: the punishment annexed to the commission of these crimes is, hard labour for a specified period.

XX. Concerning Drunkenness.

If a man drinks spirits till he becomes intoxicated, (the literal rendering would be poisoned,) and is then troublesome or mischievous, the magistrates shall cause him to be bound or confined; and when the effects of the drink have subsided, shall admonish him not to offend again. But if he be obstinate in drinking spirits, and when intoxicated becomes mischievous, let him be brought before the magistrate, and sentenced to labour, such as road-making, five fathoms in length, and two in breadth. If not with this, with a plantation fence, fifty fathoms long. If it be a woman that is guilty of this crime, she shall plat two large mats, one for the king, and the other for the governor of the district; or make four hibiscus mats, two for the king, and two for the governor; or forty fathoms of native cloth, twenty for the king, and twenty for the governor.

XXI. Damage done by Dogs or Hogs.

Concerning dogs accustomed to steal or bite, and pigs which bite or rend whatever may come in their way. When a dog steals food secretly, and is addicted to the practice of devouring young pigs, kids, or goats, fowls, or any other small kinds of property, the owner of the dog shall make restitution. If a pig has been devoured, a pig shall he return as a recompense; or a fowl, a fowl shall be returned. That which is returned shall be equal to that which has been destroyed. He shall also kill such dog. But if the owner persists in keeping such dog, fourfold shall he return as a remuneration for all it destroys; twofold, (half) for the king, and twofold (half) for the owner of the property destroyed. A dog also addicted to the habit of biting children, shall be killed. The man who knows that he has a savage or biting dog, and refuses to kill it, (after having been by the magistrate requested to do so,) if a child be bitten by such dog, the dog shall be killed, and its owner punished with labour, for persisting in keeping such a mischievous dog. The punishment specified in the XXth law shall be adjudged to him. Hogs also accustomed to devour young or sucking pigs, kids, or fowls, and accustomed to bite or attack children, shall be removed to another place, or killed. If the owner be obstinate, and will neither remove nor kill the pig, after having been admonished by the magistrates, they shall kill the hog, and punish the owner with labour, for obstinately keeping such a dangerous hog. His punishment shall be such as that specified in the twentieth law.

XXII. Concerning Wild or Stray Pigs.

There are no pigs without owners. No one shall hunt pigs on the mountains, or in the valleys, under the pretext that they are without owners. The wild pigs in the woods (or ravines) whose owners are not known, belong to the people of the valley. When the original proprietor is known, though the pigs may have become wild, they are still his. If one of such pigs be destroyed, (or eaten) it shall be paid for; the parties who took it shall make restitution with a pig equal in size to that which has been destroyed. The man who is obstinate in hunting pigs on the mountains or in the valleys, on the pretext that they are pigs without owners, he is the same as a thief; and as is the thief's such also (shall be) his punishment—that (which is) written in the second law.

XXIII. CONCERNING CONSPIRACY.

When one man knows that another man is planning or purposing to murder the king, or is devising to kill any other person, or is planning to steal property, or is purposing to commit any other crime; if he keep such counsel or deed planned in his own heart, and does not reveal it, or, when he is questioned, he conceals, and does not fully declare what he knows, he is like such men, and his punishment shall be equal to that adjudged to those who have engaged in such conspiracy (or criminal design.)

XXIV. CONCERNING THE MAN WHO MAKES KNOWN.

If a number of persons shall form their plans—if two in their plan, then two; if three, then three; if ten, then ten—if, when they have devised the commission of any crime, one of their number shall go to the magistrate, and shall fully disclose unto him the purpose and plan formed (if he be not the foundation of that combination, if he be not the person who [first] devised it,) that man shall not be punished. But those who do not confess shall receive judgment.

XXV. CONCERNING THE UNAUTHORIZED CLIMBING FOR FOOD.

Climb not, unauthorized, another person's tree for food; the man who does this is criminal. To beg, to ask explicitly the owner of the land (is right.) The man who steals food in a garden—or by the side of the house, takes that which is not given by the owner of the land. If the proprietor of the land desire that he may be tried, he shall be tried, and punished with labour. For food stolen from a garden—for the owner of the enclosure he shall perform labour, such as erecting a fence, the length being regulated by the value of the food stolen. But if it was food growing, or unenclosed, he shall make forty fathoms of road, or four fathoms of stone-work.

XXVI. CONCERNING REVENUE FOR THE KING AND GOVERNORS.

Every land that has received the word of God, and those that have not, whose institutions are good, agree that it is right to furnish property for their own king, who holds the government, and for the governors of the districts. It is also a thing frequently exhibited in the word of God, and taught by Jesus, our Lord, when he said, "Render unto Cesar the things that are Cesar's?" Therefore it is right that we do the same. Let every individual contribute towards the revenue of the king. The man of great property must furnish more than the man of less property. Such as governors of districts, shall give two hogs yearly. If not hogs, arrow-root ten measures; if not this, cocoa-nut oil ten bamboos full; they must be good-sized bamboos.

The *raatiras*, farmers, or small landed-proprietors, shall each give one hog annually. If not a hog, arrow-root five measures; if not this, oil five bamboos. Those, also, who do not possess land, but belong to this country—or belonging to another land, but residing here—this shall be their contribution, one pig for one year, (smaller than that furnished by the farmers;) if not a pig, arrow-root three measures, or oil three bamboos.

This is another property that the farmers shall prepare for the king that holds the government: Each district shall prepare every year two mats, ten fathoms long, and two fathoms wide; if not large mats, fine hibiscus mats, one from each (family); if not this, three fathoms of native cloth, each.

This is the property for the governors, which the farmers of their own districts shall furnish for a year: One pig each; if not a pig, arrow-root five measures, or oil five bamboos, good bamboos. And as for the king, two mats shall the inhabitants of the district furnish for their own governor, (the mats) shall resemble, in length and breadth, those for the king. If not large mats, hibiscus mats one each, or cloth three fathoms each. This is the revenue which the districts shall furnish for their governors each year, the inhabitants of each district for their own governor; and this is the property which the governors and people shall provide for the king. The man who, on account of illness, is unable to furnish the property here specified in the year, shall meet with compassion from the king and governors. But if it be from indolence, or any other cause, that he does not, he shall be banished, he shall not be detained by any one. Let the farmers act generously towards their king and governors in furnishing provisions, it is right. Of such as bread-fruit, arum, plantains, yams, and such kinds of food, let a portion be taken to the king and governors; let it be taken undressed. Not like the great feedings; they shall be entirely abolished,—but observing the week or the month it may be brought. There shall be no pigs, but fish, if desired. Thus shall we do well.

XXVII. Concerning Marking with Tatau.

No person shall mark with tatau, it shall be entirely discontinued. It belongs to ancient evil customs. The man or woman that shall mark with tatau, if it be clearly proved, shall be tried and punished. The punishment of the man shall be this—he shall make a piece of road ten fathoms long for the first marking, twenty (fathoms) for the second; or, stone-work, four fathoms long and two wide; if not this, he shall do some other work for the king. This shall be the woman's punishment—she shall make two large mats, one for the king, and one for the governor; or four small mats, for the king two, and for the governor two. If not this, native cloth, twenty fathoms long, and two wide; ten fathoms for the king, and ten for the governor. The man and woman that persist in tatauing themselves successively for four or five times, the figures marked shall be destroyed by blacking them over, and the individuals shall be punished as above written.

XXVIII. Concerning Voyaging in Large Companies.

When a member of the reigning family, or a governor, or other man of rank or influence, shall project a voyage to another land—such as those from Raiatea or Tahiti, visiting Huahine—it is right that he select steady men, such as are of the church, or have been baptized, not immoral and mischievous men, that cease not from crime; such should remain in their (own) land. But if these voyagers continue to bring troublesome persons, when they land upon the shore, the magistrates shall admonish them not to disturb the peace of the place, nor wander about at night. If they do not regard, such disturbers shall be bound with ropes until their masters depart, when they shall be liberated.

XXIX. Concerning the Magistrates or Judges.

When a man is accused of a crime, such a man may, perhaps, take enticing property (a bribe) to the magistrate or judge, that his sentence may be diminished; but the magistrate or judge shall on no account receive such bribe or property. The magistrate or judge who shall receive the property (or present) taken by such individual, is criminal. His office shall be discontinued or taken away; neither shall he ever be eligible to be a magistrate or judge again.

XXX. Concerning New Laws.

If any crime comparatively small should arise, and which is not specified in these laws, it is right that this code be altered. Annually the laws shall be revised or amended. Then shall the prohibition of such crimes as may have been omitted, be inserted, together with the punishment annexed to their commission; that the usages in this land may be straight, or correct.

Regulations for the JUDGES, the JURY, and the MESSENGERS, (or Peace-officers.)

Concerning the Principal Judges.

1. The king, or supreme governors, shall select the chief judges; and when a judge dies, or is interdicted that he may not judge, or when a judge removes to another land, the king and supreme chiefs shall nominate another, to perform the duties thus discontinued.

2. The duties of the chief judges. This is their duty.—When a man is tried, and his guilt fully established, the judge shall pronounce the sentence on his crime. The punishment written in the law, and annexed to his crime (shall be adjudged) and no other sentence.

3. Concerning recording the transactions or proceedings.—The judge shall write the name of the prisoner, with his crime, the names of the parties by whom he was accused, the punishment adjudged for his crime, in a book, for the inspection of the king and the people.

4. Concerning the emolument.—The property or salary of the chief judges shall be given yearly by the king. All fines or confiscations shall belong to the king, or the parties specified in the laws.

Concerning the (subordinate) Judges or Magistrates.

1. The king or supreme chiefs shall select or appoint the magistrates for all the districts.

2. Their duties.—A person accused of any crime, if the principal judge is not at the place, shall be brought before the magistrates of the district, who shall try such individual (in their respective districts;) at other seasons of public trial they shall also assist.

3. When a crime is committed, such as theft, or other similar offence, the person whose property has been stolen shall go to a magistrate and give information of the same. The magistrate shall write the names of the accused and the accuser. If the person whose property has been stolen, or who has been injured, desires that the offender should be prosecuted, he shall be tried; but if not, he shall not at once be brought to trial.

4. The magistrates shall endeavour to extinguish every kind of evil that may appear, especially quarrelling, family broils, obstinate contentions, and fighting, that peace may be preserved. Let not the people treat them with disrespect.

5 When sentence has been pronounced, let the magistrate inspect its execution, and direct the messengers or officers that it be fully attended to.

6. It is with (or it is the duty of) the king to furnish the remuneration for all the magistrates; such remuneration shall be yearly given, for their vigilance in making straight that which was crooked.

Concerning the Jury.

1. No man shall be tried for any great crime without a JURY. There shall always be a jury, excepting in minor offences, quarrels, &c.

2. When a man is tried for any crime, the judge shall select six men to be a jury, men of integrity shall he select; they shall mark or hear attentively the untwisting or investigation (of the matter.) When the evidence and examination are ended, the jury shall confer privately on the statements and evidence they have heard during the trial, the words of the accusers, and the words of the accused, with the evidence or testimony of the witnesses. If they unitedly think the person tried is really guilty, that he committed the crime (there having been the evidence of two credible witnesses,) and if they agree that he is guilty, one of their number shall address the judge, saying, This man is really guilty. Then shall the judge pronounce the sentence upon the criminal; the sentence written in the law shall he pronounce. But if the whole of the jury think the man accused is not guilty, then one of their number shall say, There is no guilt. If it be one of the king's family that is tried, then the jury shall be members of the reigning family, (or individuals also of equal rank:) if a landed proprietor or farmer that is tried, of landed proprietors or farmers, only, shall the jury be composed.

3. If during the trial the jury desire to put any question to the prisoner, or to the witnesses, it is right they should do so.

4. If the accused person observes any one on the jury, whom he knows to be a cruel or evil-minded man, or a man of whom his heart does not approve, it will be right for him to say to the judge, "Remove that man, let him not be on the jury." Then shall the judge seek another man in the place of one so removed, and shall proceed in the trial of the accused. If it be two or three on the jury, of whom the prisoner does not in his heart approve, they shall be removed; but in reference to four, or the whole jury, it will be improper. When two are removed, two others the judge must seek; when three, then must the judge seek to fill the place of those removed, and then judge the person accused.

Concerning the Messengers of the Magistrates.

Their duties.—This is the duty of the messenger, (or peace-officer,) When a man is accused to a magistrate, the magistrate shall send a messenger to the accused person, to bring him before the face of the magistrate, and guard or take the custody of him during the trial. When the trial has terminated, the messenger shall superintend the execution of the sentence pronounced by the magistrate or judge, and, in subordination to the magistrate, shall vigilantly inspect the convicts, till the sentence be accomplished.

2. Concerning their remuneration.—The king shall give annually to the messengers, such property as may be appropriated to them, for their vigilance in preserving the order (and peace) of the land.

DIRECTIONS FOR THE JUDGES AND THE JURY.

1. The Judges and the Jury shall not regard the appearance (circumstances) of men. If a man of influence (be brought before them) let him be a man of influence; if a neighbour, let him be a neighbour; if a relative, let him be a relative; or a friend, let him be a friend; this they shall not regard. That which is written in the laws, and that alone, shall they regard.

2. When an offence is committed, such as theft or adultery; if the injured person desires that the offender should be tried, he shall go himself to the magistrate, and give information. The magistrate shall write his name, and the name of the person accused, that it may be regular in trial. But offences which affect the whole island, such as murder, rebellion, conspiracy, working on the Sabbath, it shall be competent for any person to give information to the magistrate; and the magistrate shall write his name, and the name of the person accused. The magistrate shall not regard or bring to trial on vague reports. It is proper that some individual should go and lodge the information.

3. When a person is brought to trial, and when the magistrates are assembled, the accused and the accuser, and the witnesses, being also present, the magistrate or judge shall publicly declare the crime of which the offender is accused, and shall then ask him if it is a true word or accusation. If the prisoner replies, "Yes, it is a true word," the judge shall pronounce the sentence. But if the person accused replies "It is not, I did not commit the offence:" then shall the judge request the person on whose information he was apprehended, to state his accusation. If there be two witnesses, let there be two; if three, (let there be) three. It is proper that witnesses should have the clearest, strongest evidence. Then shall the judge request the prisoner to declare what he has to say. If there be a person there that knows the accused to be innocent, he shall give his evidence; and if there be two, let there be two; if three, let there be three; they shall deliver all their word or evidence. If the person accused wishes to ask his accuser any questions, it is right for him to do so. He shall inquire of the judge, and the judge shall repeat the question to the accuser.

4. No man shall be confined without cause. When a pig breaks into a garden, the owner of the pig shall not be bound, but information shall be given to the magistrate, and he shall send his messenger to bring the owner

of the pig, that he may be tried according to Law III. The same course shall be adopted in all petty offences: but for murder, theft, rebellion, &c. and all great crimes, it is proper to secure (the offender.) Let not the confinement be long before the person is brought to trial. One, two, or three days will be sufficient. Let it not be longer.

5. When petty offences are committed, the district magistrates shall try the offenders; but in all great crimes, the judges and the jury shall assemble in one place for the trial.

6. When a man is tried by a district magistrate, and sentenced by him, if the person sentenced think that the judge has been irritated with him, and has increased his punishment; if (from these considerations) he shall say—I will go to the chief-judge and the jury to be tried, it is right that he do so. They shall both go before the supreme judge and a jury, to be tried.

7. When a man is tried, convicted, and sentenced by the jury and judge, he shall not be maltreated, as by beating with a stick, piercing with a spear, or enduring any other savage practice. It shall not be practised. The punishment appropriate shall be adjudged.

8. When a man is convicted of any great crime by the judge and the jury, and they unanimously think that he deserves punishment; then the judge shall write on a paper his crime, and his own and the jury's decision on which he has been sentenced. This shall be taken to the king, and if the king approves of their decision, he shall write upon a paper brought by the judge. "It is fully approved," and write his own name underneath, then shall the punishment be inflicted on such offender. If the king desire to mitigate the sentence, he may do so, but cannot increase it.

The names of the judges, magistrates, and messengers, or police officers, for Huahine and Sir Charles Sander's Island, follow this last regulation, and close the first official document issued by the government of these islands—and, next to the sacred writings, the most beneficial ever promulgated among the people.

I have endeavoured to give a correct and even servile translation of this important publication. The idiom and peculiar phraseology of the original, I have almost invariably retained, rather than sacrifice fidelity to improvement of style. In some respects I have wished that several enactments had been otherwise than they are; these parts, however, have not been omitted; and notwithstanding their imperfections, considering the circumstances of the parties by whom they were framed, regarding them also as the first effort of their legislation as a Christian people, and the basis of their future civil institutions, they embody all the great principles, of national security, personal liberty, general order, public morals, and good government. And if no Solon or Lycurgus should appear among them, it is not too much to hope that, amidst the variety of character daily unfolded, and the means of improvement which the introduction of letters imparts, that political economy will not be neglected, but that legislators will arise, whose genius shall model and prepare such improvements in their system of jurisprudence, as shall render it in every respect conducive to general prosperity and individual happiness.

In the Tahitian and Raiatean codes, when first promulgated, the punishment of death was annexed to the crime of murder, and rebellion or treason. But by the laws of Huahine and its dependency, capital punishment was not inflicted for any crime. In the first law prohibiting murder and every species of infanticide, the penalty annexed to its commission, instead of being death, is banishment for life to Palmerston's, or some other uninhabited island. This was in consequence of our particular recommendation. We were convinced, that if, under any circumstances, man is justified in the infliction of death, it is for murder alone; but an examination of those parts of the Bible which are generally supposed to authorize this punishment, did not fix on us the impression that the Almighty had delegated to man the right of deliberately destroying a human being, even for this crime.

In our views of those parts of the sacred writings, we may perhaps have been mistaken. But in reference to the great principles on which public justice is administered, the plan recommended appeared in every respect preferable. Death is not inflicted, even on the murderer, from motives of retaliation or revenge; and if it be considered that his life is forfeited, and is taken to expiate his crime, the satisfaction which the injured party derives from such expiation, must be of a very equivocal kind. At the same time, the very execution of the sentence imparts to the executioner somewhat of the character of an avenger, or excites the apprehension that it is done under the influence of irritated and vindictive feelings.

The great design of capital, and even other punishments, is the security of society, and the prevention of crime. The death of the criminal preserves society from any future injury by his means; and the fatal punishment inflicted, it is presumed, will deter others from the commission of similar offences. The security of the community from all future violation or outrage, is certainly obtained by the death of the criminal; but experience and observation abundantly demonstrate the inadequacy of public executions to restrain from the most appalling deeds. Every repetition of the awful spectacle appears to diminish its horrific character, until those habituated to felony become familiar with its heaviest punishment. The principal end of public executions is thus defeated, the general tone of public feeling lowered, and that which was designed to be the most effectual moral barrier, is at length converted into an occasion, or sought for as an opportunity, for the commission of crime.

By recommending the omission of capital punishments, we avoided this evil, and, regarding the peculiar circumstances of the nation, were in hopes thereby indirectly to elevate the tone of moral feeling, and improve the sensibilities, of a people emerging from a state of barbarism, in which murder, and retaliated murder, had not only been familiar, but committed with brutal satisfaction.

The existence of a number of islands uninhabited, but capable of cultivation, and from the cocoa-nut trees growing on their borders, and the fish to be found near their shores capable of furnishing the means of subsistence, and yet too remote to allow of the convicts returning, or proceeding to another island in any vessel they could construct, appeared to afford the means of answering every end of public justice. The community would be as safe from future injury, as if the offenders had been executed; and we had but little apprehension, that a life of perpetual solitude, and necessary labour, would be regarded by many as more intolerable and appalling than speedy death.

We have always regarded it, as less difficult to render laws, once established, more sanguinary, than lenient afterwards. Another consideration, by which we were also influenced, was, the season to exercise repentance or supplication for mercy, which it would afford the criminal, before he was called to the bar of the Almighty. To the offender this was most important, and as it could be bestowed without danger to the donors, we were always desirous that it should be granted. No opportunity for observing the practical effects of this law has yet occurred, no murder having been committed in any of the islands since its enactment. Within two years after the promulgation of the Tahitian code, four executions for conspiracy and treason took place. The influence of these appeared by no means salutary: and in the revision of the laws of Tahiti in 1826, banishment for life was substituted as the penalty for those crimes to which death had before been annexed. One individual was sentenced to perpetual solitude, and was to have been furnished with a few tools, together with such seeds and roots as it was presumed, would, when cultivated, afford the means of subsistence; but before he was actually transported, circumstances occurred which induced the king to mitigate his sentence. It has never been intended to send any number of felons to the same island: hence distinct and distant islands have been appropriated to the residence of traitors and murderers.

The observations on this article may appear to have been unnecessarily extended: but the important character of the law itself, and the difference in its penalty from that ordinarily inflicted, have induced more lengthened remarks than I should otherwise have offered.

Another distinguishing and important feature in their judicial proceedings, is, the omission of *oaths*, in appointing the Jury, or examining witnesses. No oath is administered on any occasion: the deliberate assertion is received as evidence; and false evidence is regarded as equally criminal with false accusation, and is, I believe, punished accordingly.

The second law is one of those regulations peculiar to particular and local communities. Their swine and their gardens are among their principal sources of maintenance and wealth. The animals are not kept in sties or other enclosures, but range the district at liberty; a great proportion of their food being derived from the cocoa-nuts, bread-fruit, plums, chestnuts, and other fruits that fall from the trees. During the season of fruit, these are abundant, and the pigs feed and sleep in quietness under the shade; but during the other seasons of the year, they are very troublesome. Their materials for fences are not very good; and a large strong and hungry hog will easily force a way into the garden with his tusks or his teeth, and often do great mischief in a very short time. In 1826 this law was revised, and rendered more simple.

The sixth enactment, relating to barter, was very necessary, not only for the exchanges, or trade, carried on among the natives themselves, but for the prevention of misunderstandings and dissatisfaction between them and those foreigners by whom they might be visited for purposes of traffic or barter. They are naturally fickle: and were formerly accustomed to return articles which they exchanged, merely because they desired to repossess whatever they might have given for it; this practice led to frequent altercations, and, when trading with foreigners, to most serious quarrels.

The law which prohibits labour on the Sabbath-day, is perhaps enforced by a penalty disproportioned to the offence. It ought, however, to be observed, that as a nation they were accustomed to pay the strictest regard to this day from religious considerations, before the legal enactment was made, which was principally designed to prevent annoyance to those who were desirous to devote the day to religious services. The road which the offenders were to make, was not much more than a foot-path, with a small trench dug on each side, and raised in the centre, from the sand or earth taken from its borders.

The eighth law, referring to rebellion, is translated from the amended code of 1826, simply because this article was much shorter than in that of 1823. It contains the substance of the former enactment, which had been copied verbatim from the Tahitian code, and was drawn up by Pomare, it fixes the punishment for the third offence to perpetual banishment, instead of leaving it optional with the judges to banish, or sentence to public labour for seven years or for life.

The ninth regulation can only be of temporary application, and the necessity for its introduction arose from the peculiar circumstances of the people, while passing, as it were, from paganism to Christianity. Prior to the subversion of heathenism, polygamy prevailed more or less in all the South Sea Islands: some of the chief women had also a plurality of husbands. This regulation did not require those who had entered into these relations in a heathen state to dissolve them on becoming Christians, and was only designed to prevent any one from making these engagements after they had become such: it is a circumstance which merits notice, that there were very few who did not of their own accord, and by agreement among themselves, disannul this relationship excepting with one individual. They knew that with more than one person, it was inconsistent with the precepts of the Bible; and this consideration induced the discontinuance of their former practice. If their previous habits of life, and the notorious licentiousness of their character, be regarded, their conduct in this respect is a striking illustration of the power of Divine truth upon their minds, and of the attention they considered its injunctions to require. This articlc was amended in 1826, and it was enacted, in the event of a man marrying a second wife, without her knowing that he was already a married man, he should not only be sentenced to public work, but should furnish pecuniary compensation for the female he had thus injured.

The twelfth enactment, which regards the dissolution of the marriage contract, is rather a singular article. The influence of the former institutions appears to require it, or at least something of the kind. Formerly, under whatever circumstances, and with whatever ceremonies, the engagement had been made, nothing could be more brittle than the bond which held together those united in matrimony. The engagement was not regarded as binding any longer than the caprice or inclination of the parties dictated. Accustomed thus to relieve themselves for any unpleasantness in temper, &c. it was to be feared that the separations resulting from them would lead to the arranging of new contracts. To avoid the confusion and inconvenience of this, the present regulation was introduced; and although it was not supposed that hard labour would revive affection in the bosoms of those who, notwithstanding they had solemnly agreed to dwell together for life, had yet become estranged from each other; yet it was presumed, that the admonition from the magistrate, and the consequence of obstinate alienation, might induce the parties to impose a little restraint upon their tempers, and to make an effort to live together in peace and quietness, if not in kindness and in love.

The degradation of the female sex is an invariable accompaniment of paganism; and in addition to the humiliation and slavery to which those in the South Sea

Islands were reduced while the community were heathens, they were often exposed to the sufferings of hunger and want, from the neglect or unkindness of their savage and imperious husbands.

The thirteenth enactment, requiring provision to be made for them, may be regarded as an indication of the light in which the nation at this time viewed their former treatment of the females, or an expression of their determination to prevent its recurrence.

The law concerning marriage is a most important enactment, and may be justly regarded as the basis of all their regulations for domestic comfort, and productive of every household virtue. It was thought that the season of assembling for public lecture during the week, which was on Wednesday evening, would be preferable to the Sabbath, for giving the notice, or, what is termed with us, publishing the banns, but the marriage was not to take place till the following week. Though the law only prescribes the terms in which the contract shall be made, the people usually expect a short address, and prayer for the Divine blessing; and on that account in general, prefer applying to the Missionaries to perform the ceremony. No fees are received by either party for solemnizing the marriage, or entering the record. In the revision of the code in 1826, this law was considerably improved by annexing to the public announcement of the intention of the parties, the reason why such public declaration was made, viz. that any who knew of just cause why the marriage should not take place, might declare the same.

Dogs are numerous in the islands, though not now reared as formerly for food. They are generally remarkably indolent, unsightly, and ill-bred, and are a great nuisance in most of the settlements. Disputes are not frequent among the natives, but they arise as often on account of the depredations of their dogs and hungry pigs, as from any other cause. Neither their dogs nor swine are confined, but prowl about the districts, destroying numbers of fowls, kids, and young pigs. Several instances have occurred, in which children have been attacked and injured by savage and hungry swine. Under such circumstances formerly, redress would have been sought, or vengeance taken, with the club or the spear. To diminish the number of useless animals, and to secure greater care over others, the twenty-first regulation was made, which rendered the owners in some degree responsible for any mischief they might occasion.

Such was the population of the islands formerly, that every bread-fruit and cocoa-nut tree had its respective owner; and a single tree, it is said, had sometimes two proprietors. Subsequently, however, extensive clusters or groves of trees were to be met with, having no other owner than the chief of the district in which they grew. The fruit of these, it was formerly their practice to gather in its season, without asking the consent of any one. The proprietor of the land could at any time appropriate to his own use any number of the trees, by affixing such marks as were indications that they were *rahueïa*, or prohibited. This practice being connected with certain idolatrous ceremonies, was discontinued with the abolition of the system. As the population increased, the people became more careful of their trees, and the practice of gathering promiscuously the fruit from those trees not enclosed, appeared generally undesirable. There are, however, a number of persons at most of the settlements, who have scarcely any other sources whence they can derive a supply. In order to afford them an opportunity of procuring this, and at the same time securing to the proprietors their right to the disposal of the fruit growing on their own lands, the twenty-fifth regulation was framed, and applied to most of the trees whose fruit is used as an article of food.

The government having been hitherto an arbitrary monarchy, the king and chiefs had been accustomed, not only to receive a regular supply of all the articles produced in the islands, but to send their servants to take whatever they required, however abundant the supply furnished might have been. This practice destroyed all security of individual property, and, so long as it continued, was one of the great barriers to the improvement and civilization of the people. It had always appeared to us desirable to introduce such regulations, in reference to this subject, as would procure for the king and chiefs a revenue more ample than the system of extortion and plunder had ever furnished, and at the same time secure inviolate to the people the right of private property.

In proposing any regulation of this kind to the chiefs, we always felt some degree of delicacy, and found the introduction of the measure attended with difficulties. To the chiefs it appeared in some degree depriving them of their power, and rendering them dependent on the donations of the people; and there were others who, connecting the prosperity of the people with the continuance of the monarchical government, were not free from apprehension lest the restraint imposed on the chiefs should diminish their influence in the nation, and destroy the authority of the sovereign. The rulers and chiefs of Huahine, however, readily embraced the plan, and heartily recommending it to the adoption of the people, have found it much more productive than the former system of oppression. To the people, the advantage of this enactment is incalculable. They have already begun to taste the sweets and enjoyments resulting from the secure possession of their property, and the satisfaction of contemplating the produce of their labour as inviolably *their own*. No regulation, before introduced, appears so much adapted to promote extensive agriculture, general industry, and advancement in civilization.

In 1826, this enactment was improved: the proportion of tribute individuals in the several classes of society should furnish, was definitely fixed; and the whole rendered more explicit. Although this regulation has been subsequently introduced, and still further extended, in the codes of some of the other islands, this being the first enactment on the subject promulgated by any of the infant governments of the island, may be regarded as the basis on which the right and security of private property is established.

The remaining articles in the first Huahine code refer to the regulation of their judicial proceedings, and are designed to supply the deficiency that appeared in the laws of other islands previously introduced. In these the power of the judge and magistrate was discretionary, both as to the kind and degree of penalty imposed for several offences. This was found to open a door for the abuse of power, and was frequently very unsatisfactory to the people in general. From these considerations we were led to recommend the chiefs to annex the punishment to the prohibition of the offence, and restrict the magistrate in the infliction of such penalty only as the law enjoined: this plan has appeared in general to give satisfaction, though it is often attended with practical difficulties; these, how-however, the increasing experience of the people will probably enable them to remove.

In 1826, the regulations regarding the local magistrates was improved, and two were appointed to preserve the peace in each district, from whose decision any one could appeal, even in petty cases, to the judges of the island. At the same time, the salaries of the magistrates, as well as those of the judges, which are paid by the king, were definitively fixed.

The law which declared that "No man should be tried for any great offence (viz. one which affected his person, liberty, or possessions) without a jury," we have always regarded as the basis of their public justice. The liberty granted to the prisoner, of objecting to any members of the jury, is a valuable security for the proper and impartial investigation of the case.

In 1826, this enactment was also amended, and it was then enjoined, not only that a chief or raatira should be tried by his equals, but that if a peasant or mechanic were brought to trial, the jury also should be peasants or mechanics. Every friend of liberty, and the natural rights of men, and to the order and good government of society, must rejoice that these infant nations should have enjoyed, so early in their civil existence, the security which the trial of a subject by his peers is adapted to insure. At the same time it was also enacted, that, in all cases, the jury should be unanimous in their verdict.

Besides these regulations, which were included in the first legislative code, established in 1823, and improved in 1826; at this latter period, several important articles were added, and the Huahinian code now contains fifty laws. The first of those introduced at this time, regarded the education or discipline of the children, and was designed to counteract the fugitive habits which the children indulged prior to the establishment of regular schools. Formerly the children were accustomed to resist all parental restraint, and, whenever they chose, to leave their parents' abode, and associate with other children, or take up their residence in any other part of the island.

The facility with which the means of support might in general be obtained, rendered it a matter of little or no inconvenience to the parties to whom such children might, at the age of eight, ten, or twelve years, attach themselves. The person with whose establishment they might unite, exercised no moral or guardian care over them; and their distance from the dwelling of their own parents, removed them from the restraint and superintendence of those on whom naturally devolved the preservation of their morals and the formation of their character. To prevent the sanction and support which children absconding from their homes had been accustomed to receive, and to promote a more general attention to the reciprocal duties of parents and children, this regulation was introduced.

Another enactment prohibited the revival of those amusements and dances which were immoral in their tendencies.

A third prohibited husbands from ill-treating their wives.

The fourth referred to their fisheries, and by fixing the proportion of fish taken which should be given to the king and governors, was adapted to prevent misunderstanding or dissatisfaction.

The most advantageous regulation, however, introduced at this period, for the first time in any of the legal enactments, was that which regarded the boundaries of lands. This law required that all disputes about landmarks should be referred to the judges, or settled by the decision of a jury; and that the boundaries of all the lands, fields, &c. throughout the island, should be carefully ascertained, and, with the dimensions, descriptions of the land, and the names of the owners, should be entered in a book to be called the Book of the Boundaries of Lands. A copy of the boundaries of each land, with the owners, signed by the principal judge, and sealed with the king's seal, was to be prepared, as a document or legal title to the possession of such land in perpetuity.

Many difficulties presented themselves in adjusting the rights of different claimants to the same lands. Prior to the introduction of Christianity, the lands often changed owners during the internal wars that prevailed, and the descendants of those who at this time possessed or occupied the land, preferred their claims. But as those who possessed the lands at the abolition of idolatry, held them either as the fruits of conquest or the gifts of the king, it was decreed that those who possessed them should be considered as their lawful owners, and that no claim referring to a period antecedent to this, should be admitted. This law, by which the lands of the islands were made the freehold property of their possessors, is one of the most important in its influence on the property, that had yet been enacted. The unalienable right in the soil would thus descend from the father to the son, and no man could be deprived of this natural right but by a flagrant violation of the laws of his country.

In the year 1824, when the infant, Pomare III. was recognized by the nation as the successor to his father in the government of Tahiti and Moorea, the Tahitian code was revised and enlarged. At this time, a most important law was introduced; which gave to the nation, for the first time, what might be termed a representative government, and rendered the Tahitian a limited, instead of an absolute monarchy. It was then decreed that members from every district should meet annually, for the purpose of devising and enacting new laws, and amending those already in existence. The duration of the session was to be regulated by the business to be transacted. The inhabitants of the districts were to select their representatives, and fresh deputies or members to be elected every three years. It was at first enacted, that two should be sent from each district; but the same law authorized the body which might be thus convened to increase the number to three or four from each, if it were found desirable. No regulation was to be regarded as a law, but such as had been approved or proposed by them, and had received the sanction of the king; and every regulation made by them, and approved by the king, was to be observed as the law of the land.

The printed report of the session of what may be termed the Tahitian parliament, which assembled in May 1826, contains an alteration of two laws, and four new regulations. The first of these is sufficiently important to justify its translation, it is—

CONCERNING SEAMEN WHO MAY LEAVE THEIR VESSELS.

1. The captain, or master of the vessel, who shall turn one of his crew on shore, without the consent of the governor of the district, is criminal. He shall pay thirty dollars; twenty to the king, six to the governor, and four to the man who shall conduct the seaman back to his ship.

2. The man who shall forsake his ship, and hide himself on shore, shall be immediately apprehended. The man that finds and apprehends (each deserter) shall receive eight dollars, if he was taken near at hand; and fifteen dollars, if brought from a distance.

3. The person who shall entice any man belonging to a ship, so that he abandon his ship, and the man who shall hide or secrete him who shall so abscond, shall be tried, and (if convicted) his sentence shall be to make fifty fathoms of pathway or road, or to build eight yards of stone pier or wall.

4. A seaman who had concealed himself on shore, and who shall be found after his ship has sailed, shall be brought to trial, and his sentence shall be to make fifty fathoms of road.

One of the greatest sources of annoyance to the natives, and inconvenience to foreigners, has been the conduct of seamen who have absconded from their ships, or been turned on shore by the masters of trading vessels. To prevent as much as possible seamen from leaving their ships, this law was enacted; and by subjecting to a punishment with hard labour, both the deserters, and those who may favour their desertion or concealment, is adapted to answer the end proposed.

A copy of this law, with an English translation printed on the same paper, is given, by a person whom the government appoints for that purpose, to the master of every vessel entering any of their harbours. The regulation is so just in its nature, and so salutary is its tendency in regard to those who visit the islands, as well as the community on shore, that the most ready acquiescence in its requirements may be most reasonably expected.

The people have always felt more difficulty in the enforcement of those regulations which refer to subjects of other governments residing among them, than to the natives of their own islands. The sentencing of such sailors as may desert from their ships, or may be found on shore after their vessels have sailed, to hard labour on the public roads or quays, is probably the most effectual plan they could have adopted to deter seamen from the very frequent practice of forsaking their vessels.

The promulgation of an official printed code among the inhabitants of these islands, not only formed an epoch in their history, but introduced a new order of feeling and action in their civil relations, as a community governed by laws which they had deliberately and unitedly adopted. Perspicuity and plainness had been studied in the framing of their laws, and in several instances we should have preferred even greater explicitness. The public administration of justice, under the former system, had been exceedingly unceremonious and simple; and although the change now introduced had rendered it rather more complex, it was not intricate or perplexing. In several of the islands, I believe, court-houses were built. There were none, however, in Huahine when I left, though I have since heard that they were erecting one for the chief judges.

No investigations or trials have ever taken place with "closed doors," but all causes are tried in open court. In some of the islands, the bell-man goes round the district, to give public notice before any trial takes place. Their places of justice have usually been the governor's house, or the open air, frequently the court-yard in front of the chief's dwelling, an open space in the centre of the settlement, or near the sea-beach. A wide-spreading tree, or clump, is usually selected, and under its shade the bench is fixed and the trial attended. The hour of sun-rise is usually chosen, as they prefer the coolness of the morning to the heat of noon.

Important as this change in the civil constitution was to all the great interests of the people, there were doubtless many who were either insensible of the advantages that would accrue to themselves and their posterity, or were unable to appreciate their value. There were others, however, among different ranks in society, who thought and felt differently, and occasionally exhibited the high sense they entertained of natural and acknowledged rights, and the security they expected from the laws they had adopted. Many illustrations of this remark might be adduced, I shall only cite one that occurred in the island in which I resided.

In the autumn of 1822, the queen of Tahiti, the widow of Pomare, visited Huahine. Her attendants, who followed in her train from Tahiti, requiring a piece of timber, she directed them to cut down a bread-fruit tree growing in the garden of a poor man on the opposite side of the bay, near which her own residence stood. Her orders were obeyed, and the tree was carried away. Teuhe, the owner of the spot on which it stood, returning in the evening to his cottage, saw the spoiler had been there; the stump was bleeding, and the boughs lay strewed around, but the stately trunk was gone. Informed by his neighbours that the queen's men had cut it down, he repaired to the magistrate of the district, and lodged a complaint against her majesty the queen. The magistrate directed him to come to the place of public justice the following morning at sun-rise, and substantiate his charge: he afterwards sent his servant to the queen, and invited her attendance at the same hour. The next morning, the Missionary residing there went down to witness the proceeding; and, as the sun rose above the horizon, Ori, the magistrate, was seen sitting in the open air, beneath the spreading branches of a venerable tree: on a finely-woven mat, before him, sat the queen, attended by her train; beside her stood the native peasant; and around them all, what may be termed the police-officers. Turning to Teuhe, the magistrate inquired for what purpose they had been convened. The poor man said, that in his garden there grew a bread-fruit tree, whose shade was grateful to the inmates of his cottage, and whose fruit, with that of those which grew around, supported his family for five or seven months in every year; but that, yesterday, some one had cut it down, as he had been informed, by order of the queen. He knew that they had laws—he had thought those laws protected the poor man's property, as well as that of kings and chiefs; and he wished to know whether it was right that, without his knowledge or consent, the tree should have been cut down.

The magistrate, turning to the queen, asked if she had ordered the tree to be cut down. She answered, 'Yes.'—He then asked if she did not know that they had laws. She said, 'Yes;' but she was not aware they applied to her. The magistrate asked, 'If in those laws (a copy of which he held in his hand) there were any exceptions in favour of chiefs, or kings, or queens.' She answered, 'No;' and despatched one of her attendants to her house, who soon returned with a bag of dollars, which she threw down before the poor man, as a recompense for his loss.—'Stop,' said the justice, 'we have not done yet.' The queen began to weep. 'Do you think it right that you should have cut down the tree without asking the owner's permission?' continued the magistrate. 'It was not right,' said the queen. Then turning to the poor man, he asked, 'What remuneration do you require?' Teuhe answered, If the queen is convinced that it was not right to take a little man's tree without his permission, I am sure she will not do so again. I am satisfied—I require no other recompense.' His disinterestedness was applauded; the assembly dispersed; and afterwards, I think, the queen sent him, privately, a present equal to the value of his tree.

CHAP. XVI.

Visit from the Windward Islands—Opposition to the moral restraints of Christianity—Tatauing prohibited by the chiefs—Account of the dye-instruments, and process, of tatauing—Variety of figures or patterns—The operation painful, and frequently fatal—Revival of the practice—Trial and penalty of the offenders—Rebellion against the laws and government—Public assembly—Address of Taua—Departure of the chiefs and people from the encampment of the king's son—Singularity of their dress and appearance—Interview between the rival parties—Return of Hautia and the captives—Frequency of war in the South Sea Islands—Polynesian war-god—Religious ceremonies and human sacrifices, prior to the commencement of hostilities—National councils—Mustering of forces—Emblems of the gods taken to the war—Strength of their fleets or armies—The battle of Hooroto—Women engaging in war—Martial music—Modes of attack—Single combats, challenges, &c.—The rauti, or orators of battle—Sacrifice of the first prisoner—Use of the sling.

During the year 1821, besides going to Tahiti, I made three voyages to Raiatea, and spent a number of weeks with the Missionaries there. These voyages were not dangerous, although we were often out at sea all night, and sometimes for nights and days together. The Hope, whose arrival at Tahiti in April had afforded us so much satisfaction, called at Huahine on her way to England, with a cargo she had taken at Tahiti. Shortly after this, we were also favoured with a visit from Messrs. Darling and Bourne, who accompanied the captain of the Westmoreland from Tahiti, in the ship's long-boat.

After meeting the Missionaries of the Leeward Islands at Raiatea, they passed some weeks with us in Huahine. Their visit was peculiarly gratifying, being the first we had received from any of the Missionaries in the Windward Islands, though we had been at Fare harbour upwards of three years. The season they spent with us was also distinguished by one or two important circumstances.

Paganism had been renounced in 1816, and a general profession of Christianity followed the commencement of the Mission here; there were, however, a number who felt the restraint Christianity imposed upon their evil propensities, to be exceedingly irksome. These were principally young persons; and though, from the influence of example, or the popularity of religion, they had attached themselves to the Christians, they were probably hoping that a change would take place in the sentiments of the nation more favourable to their wishes, and relax the restriction which the precepts of Scripture had imposed. They did not disturb the tranquillity of the community.

But when the chiefs intimated their intention of governing for the future according to the principles and maxims of the Bible, and that the new code of laws had received the sanction of Pomare-vahine, as well as that of the ruling chiefs on the island, they began to be apprehensive that the existing state of things was likely to be permanent. They then first exhibited a disposition to oppose their application. Several who had transgressed had been by the chiefs admonished and dismissed; the latter, at the same time, firmly declaring their determination to enforce the laws which they had promulgated.

Among other prohibitions, that of tatauing, or staining the body, was included. The simple act of tatauing, or marking the skin, was in itself no breach of the peace, but it was connected with their former idolatry, and always attended with the practice of abominable vices, and on this account was prohibited. The fondness of the Tahitians for these ornaments, as they considered the marks thus impressed, is truly remarkable. It is not confined to them, but pervades the principal groups, and is extensively practised by the Marquesians and New Zealanders. Although practised by all classes, I have not been able to trace its origin. It is by some adopted as a badge of mourning, or memorial of a departed friend; and from the figures we have sometimes seen upon the persons of the natives, and the conversation we have had, we should be induced to think it was designed as a kind of historical record of the principal actions of their lives. But it was, we believe, in modern times adopted by the greater number of the people merely as a personal adornment.

It must have been a painful operation, and was seldom applied to any extent at the same time. There were *tahua*, professors of the art of tatauing, who were regularly employed to perform it, and received a liberal remuneration.

The colouring matter was the kernel of the candle-nut, *aleurites triloba*, called by the natives *tiairi*. This was first baked, and then reduced to charcoal, and afterwards pulverized, and mixed with oil. The instruments were rude, though ingenious, and consisted of the bones of birds or fishes, fastened with fine thread to a small stick. Another stick, somewhat heavier, was also used, to strike the above when the skin was perforated. The figure or pattern to be tataued, was portrayed upon the skin with a piece of charcoal, though at times the operation was guided only by the eye.

A number of idolatrous ceremonies attended its commencement; and when these were finished, the performer, immersing the points of the sharp bone instrument in the colouring matter, which was a beautiful jet black, applied it to the surface of the skin, and, striking it smartly with the elastic stick which he held in his right hand, punctured the skin, and injected the dye at the same time, with as much facility as an adder would bite, and deposit his poison.

So long as the person could endure the pain, the operator continued his work, but it was seldom that a whole figure was completed at once. Hence it proved a tedious process, especially with those who had a variety of patterns, or stained the greater part of their bodies. They usually began to impress these unfading marks upon their persons at an early age, frequently before they had reached the seventh or eighth year. Both sexes were tataued, but the men more than the women.

The tatauing of the Sandwich and Paliser Islanders, though sometimes abundant, is the rudest I have seen; that of the New Zealanders and the Marquesians is very ingenious, though different in its kind. The former consists principally in narrow, circular, or curved lines, on different parts of the face; the lines in the latter were broad and straight, interspersed with animals, and sometimes covered the body so as almost to conceal the original colour of the skin.

The Tahitian tatauing is more simple, and displays greater taste and elegance than either of the others. Though some of the figures are arbitrary, such as stars, circles, lozenges, &c.; the patterns are usually taken from nature, and are often some of the most graceful. A cocoa-nut tree is a favourite object; and I have often admired the taste displayed in the marking of a chiefs' legs, when I have seen a cocoa-nut tree correctly and distinctly drawn, its root spreading at the heel, its elastic stalk pencilled as it were along the tendon, and its waving plume gracefully spread out on the broad part of the calf. Sometimes a couple of stems would be twined up from the heel, and divided on the calf, each bearing a plume of leaves.

The ornaments round the ankle, and upon the instep, make them often appear as if they wore the elegant Eastern sandal. The sides of the legs are sometimes tataued from the ankle upward, which gives the appearance of wearing pantaloons with ornamented seams. From the lower part of the back, a number of straight, waved, or zigzag lines, rise in the direction of the spine, and branch off regularly towards the shoulders. But, of the upper part of the body, the chest is the most tataued. Every variety of figure is to be seen here. Cocoa-nut and bread-fruit trees, with convolvolus wreaths hanging round them, boys gathering the fruit, men engaged in battle, in the manual exercise, triumphing over a fallen foe; or, as I have frequently seen it, they are represented as carrying a human sacrifice to the temple. Every kind of animal—goats, dogs, fowls, and fish—may at times be seen on this part of the body; muskets, swords, pistols, clubs, spears, and other weapons of war, are also stamped upon their arms or chest.

They are not all crowded upon the same person, but each one makes a selection according to his fancy; and I have frequently thought the tatauing on a man's person might serve as an index to his disposition and his character. The neck and throat were sometimes singularly marked. The head and the ears were also tataued, though among the Tahitians this ornament was seldom applied to the face.

The females used the tatau more sparingly than the men, and with greater taste. It was always the custom of the natives to go barefooted, and the feet, to an inch above the ankles, of the chief women, were often neatly tataued; appearing as if they wore a loose kind of sandal, or elegant open-worked boot. The arms were frequently marked with circles, their fingers with rings, and their wrists with bracelets. The thin transparent skin over the black dye, often gave to the tatau a tinge of blue.

The females seldom, if ever, marked their faces; the figures on their feet and hands were all the ornaments they exhibited. Many suffered much from the pain occasioned by the operation, and from the swelling and inflammation that followed, which often continued for a long time, and ultimately proved fatal. This, however, seldom deterred others from attempting to secure this badge of distinction or embellishment of person.

On account of the immoral practices invariably connected with the process of tatauing, the chiefs prohibited it altogether, and, excepting a few foreign seamen, who often evinced as great a desire to have some figure tataued on their arms or hands, as the natives themselves, there had not been an individual marked for some years.

In the month of July, we heard that a number, about forty-six young persons, had been marking themselves. The principal chiefs came to ask our opinion, as to what they should do, there were so many of them. Formerly, they said, the disobedience of so numerous a party to any order of the chiefs, would have been considered equivalent to a declaration of war, and they should have sent armed men after them at once, and either have slain or banished the delinquents; but now, as they had laws, they wished to know whether it would be right that they should all be tried, and, if found guilty, have the sentence annexed to the crime pronounced against them.

We told the chiefs it would not be wrong, and the next morning attended the trial. It was conducted with the greatest candour and forbearance on the part of the magistrates and accusers, and an equal degree of submission on the part of the offenders, though it appeared they had supposed that from their numbers, and the circumstance of one or two young chiefs of distinction being among them, the government would not have noticed their conduct. They were sentenced to build a certain quantity of stone-work on the margin of the sea.

In a day or two afterwards, it was discovered that Taaroarii, the king's son, a youth about eighteen years of age, had also been tataued; and this being considered as an expression of his disapprobation of his father's conduct, and of his determination as to the conduct he designed to pursue, produced a great sensation among the people. His venerable father was deeply affected, and the struggle between affection for his son, and his duty to the people, was evidently strong. The latter prevailed; he directed him to be tried, and attended him to the trial: here he affectionately admonished his son to profit by his experience, and warned the spectators, telling them not to be deceived, and suppose that the laws, by which they had mutually agreed to be governed, would be violated with impunity. Some of the latter observed, If the king's son does not escape, what will become of the common men?

Taaroarii, the chief of Sir Charles Sanders' Island, and the expected successor to his father in the sovereignty of Huahine, was now obliged to dive into the sea for coral, and assist in building the portion of stone-work allotted to him. His friends and attendants performed the greater part of the labour; still there was a feeling of pride, that would not allow him to stand altogether idle. I visited his house one evening, and entered freely into conversation with him on the subject. He observed, that he was sorry for what he had done, but appeared to indicate, that he did not wish it to be thought that the work assigned him was any punishment.

Several unsteady young men and women, who followed the example of the first party, were also tried, and sentenced to similar punishments; and afterwards two principal personages in the island, by having their bodies tataued, joined their party: these two were the son of the king of Raiatea, who was residing at Huahine, and his sister, who had been married to a member of Mahine's family. Their party was now strong, both in point of number and influence, and we expected that the simple circumstance of marking the person with tatau, was only one of the preliminaries of their design; and in this we were not mistaken.

In the month of August, we heard that Taaroarii, with a number of those whom the chiefs had sentenced to labour on the public works, had left their employment, and were gone to Parea, in the northern part of the island. They told the officers of the chief appointed to superintend them, that they intended in a few days to return. The people were greatly attached to the king's son, and the officers, willing to shew him every indulgence, did not oppose his removal; but reports were soon circulated, that he was employing emissaries to invite the disaffected to join him, with the assurance that as soon as they were strong enough, he intended to assume the government of the island, and abolish the laws—that under his reign every one should follow his own inclinations, with regard to those customs which the laws prohibited. His father being absent at Raiatea, he had judged the present a favourable time for making a vigorous effort.

On the evening of the ninth of September, which was the Sabbath, a messenger came from the chiefs while we were engaged at family prayers, informing me that a large party of wild young men had gone to Parea, and that the son of the king of Raiatea was preparing to follow them. I went down to his dwelling; his wife and several of his principal men were persuading him to remain, and not unite himself with those whose designs were evidently unfavourable to the peace of the island. I mingled my entreaties with theirs, but it was of no use. His own men, finding he could not be deterred unless by violence, desisted; while a number of young fellows, like-minded with himself, urging him to depart, he hastened after the party that had gone to Parea. As soon as Hautia, the deputy-governor of the island, heard it, he gave orders for the people to prepare to go, and fetch them back the next day.

On the following morning, accompanied by Messrs. Darling and Bourn, I went down to the settlement about sunrise, to witness the proceedings of an assembly convened to consider the events of the preceding day. It was one of the most interesting of the kind I ever attended. The public council was held in the open air, on the sea-beach, in the shade of several tamanu trees, that grew in front of the governor's house. Hautia sat on a rustic native seat, near the trunk of the principal tree. The chiefs of the different districts, and the magistrates, were assembled near him, while most of the people of the settlement had gathered round to witness their proceedings, full of anxiety for the result.

It appeared from the declarations of several, that the conduct of the young men, and especially the chiefs' sons, had not proceeded from any desire to ornament their persons with tatau, but from an impatience of the restraint the laws imposed; that they had merely selected that as a means of shewing their hostility to those laws, and their determination not to regard them; that if they might be allowed, without molestation, to follow their own inclinations, no disturbance of the present sort would be attempted; but that if the restraints of the laws were imposed, and its penalties enforced, they were determined to withstand them. It was also reported that they were armed, and intended to resist all attempts to enforce their obedience.

After a short declaration, it was proposed to go and address them first with kindness, but firmness, inviting them to return; that if they accepted the invitation, well; if not, that they should attempt to bind them, and bring them back; that if they resisted, to use force, but by no means to have recourse to arms unless they should be first assaulted. This was acceded to by all present. The men repaired for their arms, and in half an hour the greater part of the inhabitants of the district assembled in front of the chief's house, ready to set out as soon as he should lead them.

Before they started, Taua, a tall well-made chief, who had formerly been a warrior and a priest, and who was one of their orators, stood up in the midst, and addressed the assembly. His person was rather commanding, his features masculine, his head uncovered, and his hair short, black, and slightly curled. A light mantle of finely braided bark was thrown loosely over his shoulders, and his loins were girded with a purau, and in his hand he held a light spear.

He spoke with considerable judgment and effect. They might as well, he said, leave their weapons at home, as to any use which he expected they would be required to make of them, but that still it was perhaps best to go prepared, and to shew these misguided young men, especially the king's two sons, that it was their determination to make the laws, to which they had openly agreed, the rule of public conduct, to maintain them as they were, and not to bend them to the views of those whose object was to introduce disorder and to foster crime; to let them know at once, that though they were chiefs, they, as well as their subjects, must respect the laws, or sustain the consequences. We think they will submit, (he added,) but perhaps we are mistaken, and the issue of this day is not altogether certain. God, who overrules all events, and sometimes uses the wicked to accomplish his purposes, may, perhaps, design by them to punish and to humble us, and to give them a temporary ascendency; we ought therefore to look to Him.

I do not vouch for the accuracy of the language, but these are the sentiments he expressed.

Drawing to a close, he turned towards us, as we were sitting on a rustic rail near the outside of the assembly, and observed, that though he apprehended there was no danger, it would be well to be prepared; for should they be overcome, although the young chiefs might be inclined to favour us, they could not restrain their followers; that our property would be a temptation; and that as we were supposed to have facilitated the introduction and enforced the observance of the laws, it might be necessary, in order to our safety, that we should leave the island, even before sun-set. A degree of excited animation, attended with a lively and impressive action and an impassioned feeling, which greatly affected us, breathed through the whole of his harangue, and during the latter part we could not refrain from tears.

Shortly after Taua closed, Hautia, who was clad in a loose parau round his loins, a light and beautifully fringed purau mat thrown like a mantle loosely over his shoulders, and holding a light spear in his hand, arose, and came and took leave of us, and then set off towards Parea, surrounded by the chiefs, and followed by their adherents.

When he rose, and gave with his spear the signal to move onward, there was an evident indication of strong excitement, which continued till they had left the court-yard, not only among those who were going, but among the women, children, and others, who were spectators. Hautia's wife walked on by the side of her husband; many of the other women also went to see the issue of the rencounter. We remained till all had departed.

The chiefs and their people did not proceed in one unbroken column, but, after the departure of Hautia and his companions, followed in small detached parties, consisting of a chief, and three, four, or five of his dependants. Their appearance, equipment, and dress presented the most singular spectacle I ever witnessed. The symmetry of form, well-shaped and finely turned limbs and graceful steps of some, together with their tasteful, cumberless dress, the light spear in their hand, and the excitement of their countenance, presented a figure that could not be contemplated without admiration; and the only feelings of a different order, on beholding such an individual, were those of regret at the errand on which he was going.

There were others, however, very different in appearance, which made the contrast the more striking; some exceedingly corpulent and heavy, others spare in habit, all arrayed in a different kind of dress from that they ordinarily wore, and some presenting frightful figures. Many wore a kind of turban, others a bandage of human hair, across their forehead, and round the back of the head.

The most singular head-dress was that worn by Buhia, one of the chiefs of Maeva. It was a kind of wig, consisting of long and yellow beards, fastened in a sort of net-work fitted to the head. Whether they were the beards of vanquished enemies, that had been taken as trophies by his ancestors, as the Americans are accustomed to preserve the scalps of their prisoners, I did not learn. The singularity of his appearance was greatly increased by two or three small whales' teeth, the roots of which were fixed to the net-work, while the points projected through the hair like very short horns: one was placed over each eye, and, I think, one over one of the ears. The other parts of his dress were altogether those of an ancient warrior; and his appearance was so singular, that I could not forbear stopping him a moment, to examine his head-dress, and inquire about it. He informed me that the hair was the beards of men, and that the design of it was to excite terror. On my inquiring what the horny appearances were, I was informed that they were the *neho* or *tara* of *taehae tahito*, teeth or horns of ancient cannibals or wild men. I informed him they were young whales' teeth; but he seemed inclined to doubt it. I could not but think, as I looked at him, that he certainly had succeeded tolerably well in rendering himself a terrible object. One of his attendants, Maro, a plump-bodied, round-faced, good-natured looking man appeared in perfect contrast with his chief, and it was impossible to behold him without a smile. His person was rather stout and short, his hair was cut close to his head, the upper part of his body was uncovered, but round his waist he wore a pareu reaching to his knees. He had a drummer's jacket on, highly ornamented, and scarlet-coloured; it was, however, too small for him to get it on his back, or to pass his muscular arms through the sleeves; it was therefore fixed on the outside of his pareu, the body of the jacket hanging down in place of the skirts of a coat, while the sleeves were passed round his waist, and tied in a knot in front. His equipment was in perfect accordance with his uniform, for the only weapon that he had was a short brass-barrelled blunderbuss, called by the natives *vaha rahi*, or great-mouth.

Although the events of the morning had been such as were adapted to awaken very different feelings, yet when he turned round his good-natured face to bid me farewell, I could not forbear smiling. His person, dress, arms, and a habit of leaning forwards, which, as he hastened by, exhibited very fully the scarlet jacket, rendered him altogether most ludicrous in his appearance.

When the parties had all started, we returned to the valley to breakfast, but were surprised, as we passed through the settlement, to behold almost every house deserted. We inferred that those women and children who had not accompanied the men to Parea, had retired to places of greater security, or better observation. After breakfast, we spent some time in prayer that no blood might be shed, but that the issue of the interview between the rival parties might be conciliatory. We then launched our boat, fixed our masts and rudder, twisted up our matting sails, and waited, not without anxiety, the arrival of intelligence.

The chiefs, before they left, had appointed the following signals. If there was no resistance made by the young chiefs and their adherents, all would remain quiet till they returned. If they had to fight, they would send a man to fire a musket so near the valley that we might hear it. If the rival party was numerous, and there was danger, two would be fired.

We remained in a state of great suspense during the forenoon, and hardly saw an individual in the settlement. About twelve o'clock we heard one musket fired, and very shortly afterwards another. This only increased our embarrassment, for although two had been fired, they had not been fired together, and, judging from the report, we inferred that one was much nearer than the other. We, therefore, determined to wait farther intimation, before we took any measures for our personal security. In this state of uncertainty we continued, supposing a conflict had certainly commenced; and that the two shots we had heard, had, perhaps, occasioned an equal loss of lives.

At two o'clock in the afternoon, however, our anxiety was relieved by the arrival of Tauira, whom the chiefs had sent to inform us that all was peace; that Moeore, the son of the king of Raiatea, and his adherents, had surrendered on the arrival of Hautia, and that the parties were retiring to the settlement. The messenger was almost breathless with speed; and while resting, he united with us in rendering grateful acknowledgments for the agreeable tidings. In an hour or two, Taauroa, one of our people, arrived, and told us the reports we heard were only random shots, fired as expressions of joy, and that it had been done without the knowledge of the chiefs.

Towards sunset we walked to the adjoining district of Haapape, where we were happy to meet Hautia and his friends returning; the young chief, who was about six-and-twenty years of age, with his adherents, following in their train as captives. We mingled our congratulations for the issue of the events of the day. We were also thankful to learn, that although one individual had a very narrow escape, yet no life had been lost, and no person injured.

Two days afterwards we attended the trial of the rebels, at a special court, held in the open air. The conduct of each was candidly and impartially examined; and, as many, it was found, had gone merely to accompany the chief, or to procure food, without any intention of joining in the rebellion, they were liberated. The others, who had not only designed but commenced hostilities, by plundering the plantations, killing and eating the hogs of the party favourable to the laws, were sentenced to public labour, and were set to work in small parties, with police-officers to attend them. Although they were repeatedly interrogated as to the reasons for their conduct, they said but little in reply. In the evening of the same day, great numbers of the people attended our weekly service, when I endeavoured to improve the recent painful events, from the history of Absalom's rebellion.

There have been several slight insurrections in Tahiti since the promulgation of the laws, but they have affected only a small number. Disturbances at Raiatea and Borabora, similar to the above, have occurred; at one time, Mr. Williams very narrowly escaped with his life, a plan having been formed for his assassination. They have not been recent, and the laws seem firmly established; but there are many, in all the islands, who find them an irksome restraint, and would most willingly, if an opportunity offered, abrogate them. Such individuals desire the return of the time when there was no law, and every one followed his own inclinations. In Huahine, though they have been frequently violated, I do not think any attempt has been made to disannul them, since the one above alluded to.

The South Sea Islanders are greatly addicted to war. It occurred very frequently, prior to the introduction of Christianity. During the fifteen years Mr. Nott spent in the islands, while the people were pagans, the island of Tahiti was involved in actual war ten different times. The Missionaries were painfully familiar with it. It surrounded their dwelling; and the wounded in battle have often, with the wounds fresh and bleeding, repaired to their houses for relief. This, however, was the only time that I saw any thing like it, though we often heard its rumours.

Oro was the principal war god, but he was not the only deity whose influence was important on these occasions. Tairi, Maahiti, Tetuahuruhuru, and Rimaroa, "long hand, or arm," the ancient gods of war, were all deities of the first rank, having been created by Taaroa, according to their fabulous traditions, before Oro existed.

In modern times, Oro's influence has been principally sought in every war. This they imagined was the chief object of his attention; and when it proceeded in its bloodiest forms, it was supposed to afford him the highest satisfaction. Somewhat of his imagined character may be inferred from the fact of his priest requiring every victim offered in sacrifice, to be covered with its own blood, in order to his acceptance.

When war was in agitation, the first human sacrifice that was offered to Oro was the Matea: this was called fetching the god to preside over the *nuu* or army. The god was brought out; when the victim was offered, a red feather was taken from his person, and given to the party, who bore it to their companions, and considered it as the symbol of his presence and sanction, during their subsequent preparations. Another human sacrifice was now taken, called the Maui faatere, "the throwing or darting" equivalent to the public declaration of war, and such it was also considered by the opposing party. In 1808, when the late Pomare heard that Taute, his former chief minister, and the most celebrated warrior in the nation, had joined the rebel chiefs, and that the Maui faatere had been offered, and the sanction of the gods thus implored, he was so affected that he wept; and it was in vain that one of his orators, in alluding to this event subsequently, exelaimed, Who is Taute? He is a man, and not a god, his head reaches not to the skies. Who is Taute? The king's spirits and courage never revived.

If it was a naval expedition, canoes were now collected and equipped, and the weapons put in order, the spears and clubs cleaned with a boar's tusk, pointed with bones of the sting-ray, and having been carefully polished, the handle of every weapon was covered with the resinous gum of the bread-fruit, that it might adhere to the warrior's hand, and render his grasp firm.

When the implements of destruction were ready, and this seldom occupied many days, another human sacrifice was offered, called the *haea mati*—the tearing of the mati wreath of peace. This was immediately before the expedition started; and if accepted, Oro generally inspired one of his prophets, who declared that the fleet or army should be victorious. On all these occasions, human sacrifices, covered with their own blood, were offered to Oro, in numbers proportioned to the magnitude of the undertaking, or the force of the parties confederated.

While these ceremonies were proceeding, national councils were held. Peace, or war, was usually determined by a few leading individuals, including the king, priests, and the principal chiefs. The prayers and sacrifices offered, oracles consulted, responses received, and councils held, were only parts of the external machinery by which, as it regarded the mass of the population, these movements were regulated. This, however, was not always the case, and peace or war was often the result of the impressions produced by the popular orators on the general assemblies. These harangues were specimens of the most impassioned natural eloquence, bold and varied in its figures, and impressive in its effects.

I never had an opportunity of attending one of their national councils when the question of war was debated, under all the imposing influence imparted by their mythology, whereby they imagined the contention between the gods of the rivals was as great as that sustained by the parties themselves. A number of the figures and expressions used on these occasions are familiar, but, detached and translated, they lose their force. From what I have beheld in their public speeches, in force of sentiment, beauty of metaphor, and effect of action, I can imagine that the impression of an eloquent harangue, delivered by an ardent warrior, armed perhaps for combat, and aided by the influence of highly excited feeling, could produce no ordinary effect; and I have repeatedly heard Mr. Nott declare, (and no one can better appreciate native eloquence,) he would at any time go thirty miles to listen to an address impassioned as those he has sometimes heard on these occasions.

When war was determined, the king's *vea*, or herald, was sent round the island, or through the districts dependent on the parties, and all were required to arm, and repair to the appointed rendezvous. Sometimes the king's flag was carried round. The women, the children, and the aged, called the *ohua*, were either left in the village, or lodged in some place of security, and the men hastened to the field.

Their arms were kept with great care, in high preservation. In some of the houses, on our arrival in the Leeward Islands, especially in the dwelling of Fenuapeho, the chief of Tahaa, every kind of weapon was in such order, and so carefully fixed against the sides of the house, that the dwelling appeared more like an armoury than a domestic abode. Many a one, whom the summons from the chief has found destitute in the morning, has been known to cut down a tall cocoa-nut tree, finish his lance or his spear, and join the warriors at the close of the same day. The chief of each district led his own tenantry to the war—reported, on his arrival, the number of men he had brought—and then formed his *buhapa*, or encampment, with the rest of the forces.

A number of ceremonies still remained to be observed. The priests were important personages in every expedition, their influence with the gods was considered the means of victory, and they received a proportionate share of consideration. The first service of this kind was called the *taamu raa ra*—the binding of the sacredness or supernatural influence; and while the chiefs and warriors had been employed in the preliminaries of war, the priests had been unremitting in their prayers that the *ra atua*, &c. the influence of the gods, &c. might be turned against their enemies, or that the gods would leave them defenceless. When their prayers were successful, it was supposed that the gods of their enemies left them, and came to the party by whom they were thus implored, and, entering the canoes, clubs, spears, and other weapons of their army, insured its triumph. As a compensation for this important service, the chiefs assembled; a quantity of cloth, mats, and perhaps a canoe, was spread before them, surmounted by a branch of the sacred *mero*, and a few red feathers, emblematical of the tutelar gods. The priests were then sent for, and the whole presented from the heads of the army by an orator, the burden of whose address was—"This is the recompense for your fatigue in imploring the aid of the gods by night and by day."

A second ceremony followed, called *fairaro:* a large

quantity of cloth, mats, &c. were given to the priests, that they might persevere in their labours. This was succeeded by a third, of the same kind, called the *haameii*, in which, in addition to the other kinds of property, a number of fine pigs, each distinguished by a distinct name, were given to the priests, that they might redouble their vigilance to induce their own gods to keep with them, and the gods of their enemies to forsake those enemies, and, by means of the weapons of those who now sought their favour, to exert their power against the parties they had formerly aided.

The *atoa fareia Manaha*—the building of the house of Manaha—was the most singular ceremony. It was designed for the abode of the gods and spirits, who they supposed fought with them, and whom they desired to have near at hand. In order to propitiate the gods, a human sacrifice was offered. The work was begun, and the house must be finished in one day, on which day every individual must abstain from all kinds of food. Into this house the *toos*, or images of the spirits, were sometimes taken; but although the priest always offered his prayer here, the gods were usually left in their sacred temples, and only a feather was taken from their images, which they supposed to be endowed with all their power.

The last religious ceremony, prior to the commencement of conflict, was the *haumanava*. Slight temples were erected in the sacred canoes of Oro, and the other gods. In these, the red feathers taken from the idols were deposited; they were called *manutahi no Tane*, &c. or single bird of Tane; all the gods were supposed to be present, having been brought from their elysian abodes by the prayers of the priests. There was a kind of intermediate race of beings, between men and gods, who were employed as messengers, to fetch the latter in cases of emergency; each god had his own messenger, hovering about the habitations of men, in the shape of a bird or a shark. When the priest by prayers sought the aid of these gods, they imagined that the messenger set off to the place of the god's abode, somewhere in *fare papa*, near "the foundation of the world," and made the usual declaration—*Mai haere i te ao e tamae ti te ao*, "Come to the world, or state of light, there is war in the world."

The sacred feathers being deposited in the temporary maraes erected in the canoes, a large number of the finest hogs they could procure were killed, and baked in the temple on shore, the heads cut off, and placed on a small altar in the canoe, before the symbol of the idol's presence. The remaining part of the body was eaten by the priests, and those who feasted on the sacrifices. Whether they fought by sea or on shore, as their principal engagements were near the shore, a fleet usually accompanied the army, and on board the canoes the principal idols were generally kept. The arrangements being now completed, with the symbols of their gods, and the offerings they made, they speedily set out for the combat, confident of victory.

Nuu and *papaupea* were the terms usually employed to designate an army, though it is probable the former was applied principally to an army, or fleet, filled with fighting men, and the latter to an army on shore, together with the multitude that followed for the purposes of plunder. Their armies must formerly have been large: when Captain Cook was there in 1774, he supposed the fleet to consist of not fewer than 1700 canoes, each carrying forty men; making altogether 6000 fighting men. I think, however, there must have been some mistake in his calculation. In the last war but one, in which the people of Huahine were engaged with those of Raiatea, at the battle of Hooroto, in the latter island, according to the testimony of Mahine, the present king of Huahine, who was there, and whose father was the general of the forces, the fleet consisted of ninety ships, or war-canoes, each about one hundred feet long, filled with men, who, besides their ordinary arms, possessed the two guns left with Mai by Captain Cook, from the use of which they expected an easy victory. This was one of the most sanguinary conflicts that had occurred for many years. Tenamia, the king of Huahine, went down to avenge the cause of Ohunehaapaa, whose son is still living in Raiatea. Ohunehaapaa had been banished by the Raiatean chiefs, and the chiefs and people of Huahine undertook to reinstate him. The Windward fleet anchored at Tipaemau, when the Raiateans fled to Tahaa. The Huahinean chief sent to demand from Tapaa the surrender of the land. This was refused, and both parties prepared for battle. Next day the hostile fleets met near Hooroto, and a most bloody and obstinate engagement ensued; both parties lost so many, that when piled up, on the day after the battle, the dead bodies are said to have formed a heap as high as the young cocoa-nut trees. They still determined to persevere till one party should be destroyed. Mauai, a native of Borabora, inspired by Oro, intimated his will that they should desist. An armistice was concluded; the warriors of two districts of Huahine, Faretou, and Fareihi, being comparatively uninjured, sailed over to Tahaa, for the purpose of plunder. They, however, met with a more determined resistance than they had expected, and were not only repulsed, but almost cut off. Mato, the father of the present king of Huahine, and general of the army, was slain. The survivors were glad to return to their own island, and the Raiateans were too much enfeebled to prevent them.

In this war, the greater part of the chiefs and warriors of the Leeward or Society Islands were destroyed. The island of Huahine never recovered from the shock of this murderous conflict.

The slaughter of the routed army was continued till the evening closed on the scene of murder and of blood, or until the fugitives had either reached their fortifications and strongholds in the mountains, or had eluded the pursuit of their enemies.—When the men went to battle, the women generally remained; but some of them fearlessly attended their husbands to the field, and either followed in the rear, or fought in the midst of the ranks. They carried the same kind of weapons as the men, but frequently used only their nails and their hands. Many were slain in the field, or during the retreat.

The flags of the gods, or the emblems of the idols, were carried to the battle, to inspirit the combatants, but the martial banners they employed were hoisted on board the different fleets. Rude and harsh kinds of music were sometimes heard in their canoes, but I do not know that they were used by the armies on shore.

It is a singular fact, that although they left their images in their respective temples, no offerings were presented after the haumanava had been performed, and no sacrifice was deposited on the altars of any of the temples, lest the gods should hereby be induced to forsake the army, and remain behind.

Tamai is the general term for war, in all its diversified forms; the same word is also used to denote quarrelling; *aro* is the term for battle. The modes of attack and defence were various, and regulated by circumstances: there was either the *aro viri*, skirmishing of advanced portions of each army, or an *arota*, close engagement, when they fought hand to hand. The general divisions of their army having been stated in the account of the battle of Bunaauïa, it is needless to repeat them; in addition to these, there was the corps of reserve.

The forces were marshalled for the fight by the principal leader, who was said to *tarai te aro*, shape or form the battle; when this was accomplished, the signal was given, and with deafening shouts and imprecations they rushed with bold and menacing impetuosity to *u*, or join in combat. Sometimes their attacks were made by night, but then they generally bore a *rama*, or torch. To ambuscades they seldom had recourse, though they occasionally adopted what was called the *aro nee*, or attack by stealth.

When their modes of attack were deliberate, the celebrated warriors of each army marched forward beyond the first line of the body to which they belonged, and, on approaching the ranks of the enemy, sat down on the sand or the grass. Two or three from one of these parties would then rise, and advancing a few yards towards their opponents, boastfully challenge them to the combat. When the challenge was accepted, which was often with the utmost promptitude, the combatants advanced with intimidating menaces.

These often addressed each other by recounting their names, the names and deeds of their ancestors, their own achievements in combat, the prowess of their arms, and the augmented fame they should acquire by the addition of their present foes to the number of those they had already slain; in conclusion, inviting them to advance, that they might be devoted to their god, who was hovering by to receive the sacrifice. With taunting scorn the antagonist would reply much in the same strain, sometimes mingling affected pity with his denunciations. When they had finished their harangue, the *omoreaa,* club of insult, or insulting spear, was raised, and the onset commenced. Sometimes it was a single combat, fought in the space between two armies, and in sight of both.

At other times, several men engaged on both sides, when those not engaged, though fully armed and equipped, kept their seat on the ground. If a single combat, when one was disabled or slain, the victor would challenge another; and seldom thought of retreating, so long as one remained. When a number were engaged, and one fell, a warrior from his own party rose, and maintained the struggle; when either party retreated, the ranks of the army to which it belonged rushed forward to sustain it; this brought the opposing army on, and from a single combat or a skirmish, it became a general engagement. The conflict was carried on with the most savage fury, such as barbarous warriors might be expected to evince—who imagined the gods, on whom their destinies depended, had actually entered into their weapons, giving precision and force to their blows, direction to their missiles, and imparting to the whole a supernatural fatality.

The din and clamour of the deadly fury were greatly augmented by the efforts of the Rauti. These were the orators of battle. They were usually men of commanding person and military prowess, arrayed only in a girdle of the leaves of the ti-plant round their waist; sometimes carrying a light spear in the left, but always a small bunch of green ti-leaves in the right hand. In this bunch of leaves the principal weapon, a small, sharp, serrated, and barbed *airo fai*, (bone of the sting-ray,) was concealed, which they were reported to use dexterously when in contact with the enemy. The principal object of these Rautis was, to animate the troops by recounting the deeds of their forefathers, the fame of their tribe or island, and the interests involved in the contest. In the discharge of their duties they were indefatigable, and by night and day went through the camp rousing the ardour of the warriors. On the day of battle they marched with the army to the onset, mingled in the fury, and hurried to and fro among the combatants, cheering them with the recital of heroic deeds, or stimulating them to achievements of daring and valour.

Any attempt at translating their expressions would convey so inadequate an idea of their original force, as to destroy their effect. "Roll onward like the billows—break on them with *te haruru o te tai*, the ocean's foam and roar when bursting on the reefs — hang on them as *te uira mau tai*, the forked lightning plays above the frothing surf—give out the vigilance, give out the strength, give out the anger, the anger of the devouring wild dog,—till their line is broken, till they flow back like the receding tide." These were the expressions sometimes used, and the recollection of their spirit-stirring harangues is still vivid in the recollection of many who, when any thing is forcibly urged upon them, often involuntarily exclaim, "*Tini Rauti teia*," this is equal to a Rauti.

If the battle continued for several successive days, the labours of the Rautis were so incessant by night through the camp, and by day amid the ranks in the field, that they have been known to expire from exhaustion and fatigue. The priests were not exempted from the battle, they bore arms, and marched with the warriors to the combat.

The combatants did not use much science in the action, nor scarcely aim to parry their enemy's weapons; they used no shield or target, and, believing the gods directed and sped their weapons with more than human force upon their assailants, they depended on strength more than art for success. Their clubs were invariably aimed at the head, and often, with the lozenge-shaped weapon, they would *tapai*, or cleave, the skulls of their opponents. Their spears they directed against the body, and the *maui* was often a deadly thrust, piercing through the heart.

When the first warrior fell on either side, a horrid shout of exultation and of triumph was raised by the victors, which echoed along the line, striking a panic through the ranks of their antagonists, it being considered an intimation of the favour of the gods towards the victorious parties. Around the body the struggle became dreadful; and if the victors bore him away, he was despoiled of his ornaments, and then seized by the priests, or left to be offered to the gods at the close of the battle.

The first man seized alive was offered in sacrifice, and called *te mataahaetumu Taaroa*—the first rending of the root. The victim was not taken to the temple, but laid alive upon a number of spears, and thus borne on men's shoulders along the ranks, in the rear of the army, the priest of Oro walking by the side, offering his prayer to the god, and watching the writhings and involuntary agitation of the dying man. If these agonies were deemed favourable, he pronounced victory as certain. Such indications were considered most encouraging, as earnests of the god's co-operation.

When a distinguished chief or warrior fell, the party, to which he belonged, retired a short distance, collected some of their bravest men, and then, in a body, with fury and revenge rushed upon their antagonists, to *vare toto*, clear away the blood. The shock was terrific when they met the opposing ranks, and numbers frequently fell on both sides.

During the engagement, the parties often retreated, so that there was a considerable space between the ranks in these seasons, as when advancing to the onset. The slingers were then employed, and they often advanced in front of the ranks to which they belonged, and with boasting threats warned their enemies to fly or fall. The most dangerous missile was the *uriti* or stone, from the *ma* or sling. The latter was prepared with great care, and made with finely braided fibres of the cocoa-nut husk, or filaments of the native flax, having a loop to fasten it to the hand at one end, and a wide receptacle for the stone in the centre. The sling was held in the right hand, and, armed with the stone, was hung over the right shoulder, and caught by the left hand on the left side of the back. When thrown, the sling after being stretched across the back, was whirled round over the head, and the stone discharged with great force.

The most expert slingers were celebrated through the islands, as well as the most renowned among the warriors; and when one of these presented himself, a cry ran through the opposite ranks. Beware, or be vigilant, *e ofai mau o mea*—an adhering stone is such a one; or *e ofai tano e ofai buai*—a sure or a powerful stone is such an one. The stones, which were usually about the size of a hen's egg, were either smooth, being polished by friction in the bed of a river, or sharp, angular, and rugged; these were called *ofai ara*—faced or edged stones. When thrown with any degree of elevation, they were seen and avoided, but they were generally thrown horizontally four or five feet from the ground, when they were with difficulty seen, and often did much execution. The slingers were powerful and expert marksmen.

CHAP. XVII.

Singular custom of the chiefs in marching to battle—Sanguinary and exterminating character of their engagements—Desolation of the country —Estimation in which fighting men were held—Weapons—Dress—Ornaments—Various kinds of helmet, &c.—Ancient arms, &c. superseded by the introduction of fire-arms—Former ideas respecting the musket, &c.—Divination or augury—Savage and merciless conduct of the victors —Existence of wild men in the mountains—Account of one at Bunaauïa who had fled from the field of battle—Treatment of the captives and the slain—Division of the spoil, and appropriation of the country—Maritime warfare—Encampments—Fortifications—Instance of patriotism—Methods of concluding peace—Religious ceremonies and festivities that followed—Present sentiments of the people in reference to war—Triumph of the principles of peace—Incident at Rurutu.

The custom of the warriors sitting on the ground to wait for the combat, was not the only singular practice of the Tahitians in proceeding to battle. There was another, which they called *pito*. When two leading chiefs marched together to the onset, they not only walked side by side, but arm in arm. In this manner, Pomare-vahine, and Mahine, the chiefs of Huahine and Eimeo, marched to the battle of Narii. This was designed to shew their union, and that they would conquer or fall together. When a single chief led on his own men, he also walked in *pito* with his principal *aito* or warriors, two on each side, the nearest to him having hold of his arms. On approaching the enemy they separated, but fought near the person of their chieftain, whose life it was considered their special duty to defend, at the exposure of their own.

The battle sometimes terminated by both parties retreating, to recover, and prepare for a fresh campaign, but it was more frequently continued till the flight of one party left the other master of the field.

The carnage and destruction which followed the *fati* or breaking, and *hea* or flying, of one of the armies, was dreadful. It was called *tahaea*, and in it the gods were supposed to engage as well as the men. Those who were *vi*, or beaten, fled to their canoes, or to their *paris* or fastnesses in the mountains, while the victors, who were called *upoatia*, erect heads, pursued them with reckless slaughter. A prostrate warrior, as he lay at the feet of his antagonist, wounded or disarmed, would perhaps supplicate mercy, exclaiming *Tahitia iau ia ora wau*—Spare me, may I live. If the name of the king or chief, of the victor, was invoked, the request was often granted; but frequently a reproach or taunt, and a deadly blow or thrust, was the only reply.

Such was the implacable rage with which they carried on the work of destruction, that victory on the field did not satisfy them; they repaired to the villages and other places, where the wives, children, &c. of the vanquished had been left for security, and satiated their rage with most affrighting cruelty.

By whatever considerations civilized and enlightened nations may be influenced in the practice of war, and upon whatever principles they may desire to conduct it, war, barbarous, murderous unrelenting war, is the delight of savages; and among no portion of the most cruel and warlike of the human race has it perhaps prevailed more extensively, or proved a greater scourge, than among the interesting inhabitants of the islands of the Pacific. With the Society and Sandwich Islanders, it has, since the introduction of Christianity, ceased. In the Friendly, Figi, and other groups, it still prevails: in the Marquesas, and New Zealand, it rages with unabated violence, and spreads devastation and wretchedness among the infatuated and hapless people.

Among the Society Islanders, in consequence of the influence of the climate, luxurious mode of living, and effeminacy of character, induced thereby, the obstinacy and the continuance of actual combat were not equal to that which obtained in other tribes; yet we learn from the frequency of its occurrence, and the deadly hatred which was cherished, that the passion for war was not less powerful with them than with the New Zealander or the Marquesian; and its consequent cruelties and demoralization were perhaps unequalled in any other part of the world. Their wars were most merciless and destructive. Invention itself was tortured to find out new or varied modes of inflicting suffering; and the total extermination of their enemies, with the desolation of a country, was often the avowed object of the war. This design, horrid as it is, has been literally accomplished: every inhabitant of an island, excepting the few that may have escaped by flight in their canoes, has been slaughtered; the bread-fruit trees have been cut down, and left to rot; the cocoa-nut trees have been killed by cutting off their tops or crown, and leaving the stems in desolate leafless ranks, as if they had been shivered by the lightning.

Their wars were not only sanguinary, but frequent; yet from a variety of ceremonies, which preceded the expeditions, they were seldom prompt in commencing hostilities. What they were prior to the first visits of foreigners, we have not the means of correctly ascertaining, but since that time, the only period during which correct dates can be affixed to events in their history, the short and simple annals of Tahiti are principally filled with notices of destructive wars; and the effects of desolation still visible, prove that they have not been less frequent in the other islands.

The occasions of hostility were also at times remarkably trivial, though not so their consequences. The removal of a boundary mark; the pulling down of the king's flag; the refusing to acknowledge the king's son as their future sovereign; speaking disrespectfully of the gods, the king, or the chiefs; the slightest insult to the king, chiefs, or any in alliance or friendship with them; with a variety of other more insignificant causes—were sufficient to justify an appeal to arms, or an invasion of the offender's territory with fire and spear. Although there were no standing armies or regular troops in the South Sea Islands, nor any class of men exclusively trained and kept for military purposes, war was followed as a profession as much as any other, and considered by many as one to which every other should be rendered subservient.

Provision for war was attended to when every other consideration was disregarded. In the perpetration of the unnatural crime of infanticide, boys were more frequently spared than female children, solely with a view to their becoming warriors. In all our schools, we were surprised at the disproportion between the boys and the girls that attended, and at the small number of women in the adult population; and on inquiring the cause, were invariably told that more girls than boys were destroyed, because they would, if spared, be comparatively useless in war. War therefore, being esteemed by the majority as the most important end of life, every kind of training for battle was held in the highest repute.

In times of war, all capable of bearing arms were called upon to join the forces of the chieftain to whom they belonged, and the farmers, who held their land partly by feudal tenure, were obliged to render military service whenever their landlord required it. There were, besides these, a number of men celebrated for their valour, strength, or address in war, who were called *aito,* fighting-men or warriors. This title was the result of achievements in battle; it was highly respected, and proportionably sought by the courageous and ambitious. It was not, like the chieftainship and other prevailing distinctions, confined to any class, but open to all; and many from the lower ranks have risen, as warriors, to a high station in the community.

Originally their weapons were simple, and formed of wood; they consisted of the spear, which the natives called *patia* or *tao,* made with the wood of the cocoa-nut tree, or of the *aito,* iron-wood, or casuarina. It was twelve or eighteen feet long, and about an inch or an inch and a half in diameter at the middle or the lower end, but tapering off to a point at the other. The spears of the inhabitants of Rurutu, and other of the Austral Islands, are remarkable for their great length, and elegant shape, as well as for the high polish with which they are finished.

The *omore,* or club, was another weapon used by them; it was always made of the *aito,* or iron-wood, and was principally of two kinds, either short and heavy like a bludgeon, for the purpose of close combat, or long, and furnished with a broad lozenge-shaped blade. The Tahitians did not often carve or ornament their weapons, but by the inhabitants of the Southern Islands they were frequently very neatly, though partially, carved. The inhabitants of the Marquesas carve their spears, and ornament them with human hair;* and the natives of the Harvey Islands, with the Friendly and Figiian islanders, construct their weapons with taste, and carve them with remarkable ingenuity.

The *paeho* was a terrific sort of weapon, although it was principally used at the *heva*, or seasons of mourning. It resembled, in some degree, a club; but having the inner side armed with large shark's teeth, it was morc frequently drawn across the body, where it acted like a saw, than used for striking a blow. Another weapon of the same kind resembled a short sword, but, instead of one blade it had three, four, or five. It was usually made of a forked *aito* branch; the central and exterior branches, after having been pointed and polished, were armed along the outside with a thick line of sharks' teeth, very firmly fixed in the wood. This was only used in close combat, and, when applied to the naked bodies of the combatants, must have been a terrific weapon. The bowels or lower parts of the body were attacked with it, not as a dagger is used, but drawn across like a saw.

They do not use the *patia*, or dagger, of the Sandwich Islands, but substitute an equally fatal weapon, the *aero fai*, or back-bone of the stinging ray, which being ser-

* This practice corresponds with that of the Malayans, among whom Dr. Buchanan saw a chief, the top of whose spear was ornamented with a tuft of hair, which he had taken from a vanquished foe, as he lay dying or dead at his feet.

rated on the edges, and barbed towards the point, is very destructive in a dexterous hand. Some of the natives of the Palliser Islands used the *ihi*, javelin or short spear, while fighting at a distance.

The dress and ornaments of the warriors of Tahiti, and the adjacent islands, were singular, and unlike those of savage nations, being often remarkably cumbersome. Their helmets, though less elegant and imposing than the fine Grecian-formed helmet of the Hawaiians, were adapted to produce considerable effect. Some of the Tahitians wore only a fillet or bandage round the temples, but many had a quantity of cloth bound round in the form of a high turban, which not only tended to increase their apparent stature, but broke the force of a blow from a club, or a thrust from a spear.

The most elegant head-dresses, however, were those worn by the inhabitants of the Austral Islands, Tubuai, Rurutu, &c. Their helmets were considerably diversified in form, some resembling a tight round cap, fitted closely to the head, with a light plume waving on the summit. Those used by the natives of Tubuai, and High Island, resembled an officer's cocked hat, worn with the ends projecting over each shoulder, the front beautifully ornamented with the green and red wing and tail feathers of a species of paroquet. The Rurutuan helmet* is graceful in appearance, and useful in the protection it affords to the head of the wearer. It was a cap fitted to the head, and reaching to the ears, made with thick stiff native cloth, or a cane frame-work. The lower part of the front is ornamented with bunches of beautiful red and green

* A Rurutuan helmet, a number of spears, a paeho, and many of the implements of war here described, have been deposited in the Missionary Museum, Austin Friars, London.

feathers, tastefully arranged, and above these a line of the long slender tail-feathers of the tropic, or man-of-war bird, is fixed on a wicker frame; the hinder part of the cap is covered with long flowing human hair, of a light brown or tawny colour, said to be human beard; this is fastened to a slight net-work attached to the crown of the helmet, and, being detached from any other part, often floats wildly in the wind, and increases the agitated appearance of the wearer.

On each side, immediately above the ears, numerous pieces of mother-of-pearl, and other shells, are fastened, not as plates, or scales, but depending in a bunch, and attached to the helmet by a small strong cord, similar to those passing under the chin, by which the helmet is fastened to the head. These shells, when shaken by the movements of the wearer's head, produce a rattling noise, which heightens the din of savage warfare.

The Rurutuan helmet, though more complete and useful, was far less imposing than the *fau* worn by the Georgian and Society Islanders. This was also a cap fitted closely to the head, surrounded by a cylindrical structure of cane-work, ornamented with the dark glossy feathers of aquatic birds. The hollow crown frequently towered two or three feet above the head, and, being curved at the top, appeared to nod or bend with every movement of the wearer.

This was a head-dress in high esteem, and worn only by distinguished men, who were generally sought out by the warriors in the opposing army. To subdue or kill a man who wore a fau, was one of the greatest feats. I have been often told, by a gigantic man who resided some time in my house, and was one of the warriors of Eimeo, that when the army of the enemy has come in sight, they used to look out for the fau rising above the rest of the army, and when they have seen one, pointing to it, animate each other by the exclamation, "The man with the fau; ha! whosoever shall obtain him, it will be enough." Bùt, however imposing in appearance these high helmets may have been, they afforded no defence; and although formed only of cane-work and feathers, must have been cumbersome.

The slingers, and the most light and agile among the fighting men, wore, in battle, only a maro, a loose mantle, or ahubu.

Some of the fighting men wore a kind of armour of net-work, formed by small cords, wound round the body and limbs, so tight, as merely to allow of the unencumbered exercise of the legs and arms, and not to impede the circulation of the blood. This kind of defence was principally serviceable in guarding from the blows of a club, or force of a stone, but was liable to be pierced by a spear. In general, however, the dress of the Tahitian warriors must have been exceedingly inconvenient. To make an imposing appearance, and defend their persons, seem to have been the only ends at which they aimed; differing greatly in this respect from the Hawaiians, who seldom thought of guarding themselves, but adopted a dress that would least impede their movements.

The Tahitians went to battle in their best clothes, and often had the head, not only guarded by an immense turban, but the body enveloped in folds of cloth, until the covering was many inches in thickness, extending from the body almost to the elbows, where the whole was bound round the waist with a finely braided sash or girdle. On the breast they wore a handsome military gorget, ingeniously wrought with mother-of-pearl shells, feathers, dog's hair, white and coloured.

Their ancient dresses and weapons have, since their intercourse with Europeans, been superseded in a great degree by the introduction of common fire-arms, the bayonet, and the sword. *Pupuhi* is the general name for gun. *Puhi* signifies to blow with the mouth, *pupuhi* to blow repeatedly, and this name has been given to a musket, from the circumstance of the foreigners, whom the natives first saw firing, bending down the head on one side to take aim, and bringing the mouth nearly in contact with the piece, into or through the barrel of which they supposed the person blew, and thus produced the explosion; hence it is called the blower.

They imagined that the first ships they saw were islands; their inhabitants supernatural, vindictive, and revengeful beings. The flag of one of the first vessels hanging from the ship into the water, a native approached, and took a piece of it away; this being perceived, he was fired at, and wounded, as they all supposed, by the thunder.

When we consider this, we shall not be surprised at their ideas of the source of motion in the ball. The opinion of its being blown from the mouth of the musketeer, has long been corrected; still the name is retained, and a cannon is called *pupuhi fenna*, to blow land, or country, from its contents spreading over a wide tract of country; the musket they call *pupuhi roa*, long gun; the blunderbuss *vaharahi*, wide or great mouth; the pistol, *pupuhi teuumu;* a swivel, *pupuhi tioi*, turning gun; the bullets or balls they call *ofai*, or stones. Arms, ammunition, and ardent spirits, were formerly the principal articles in demand by all classes; and being the most valuable kinds of barter, they maintained a high price. Ten or twelve hogs, worth at least from one to two pounds a head, was, for a long time, the regular price of a musket; and one hundred pigs have been paid for a cannon. I have seen upwards of seventy tied up on the beach, at Fare, as the price of a single old cannon, which had been preserved from the wreck of an English vessel, at another island. These articles have, however, long ceased to be in demand among the Tahitians.

It does not appear that their wars were more sanguinary and cruel when they fought at a distance with muskets, than when they grappled hand to hand with club and spear. The numbers killed might be greater, but fewer were wounded. Although familiar with the musket during their last wars, they are by no means expert marksmen: they understand little about taking aim, and often fire without placing the but-end of the musket against the shoulder, or presenting their piece. They grasp it in the most awkward manner, holding it above the head, or by the side, and in this singular position fire it off. I was once with a party of natives, when one of them fired at a bullock but a few yards distant, and missed it.

War was seldom proclaimed or commenced with promptitude, being always considered as one of the most important matters in which the nation could engage. Hence the preparatory deliberations were frequent and protracted.

The greatest importance was always attached to the will of the gods: if they were favourable, conquest was regarded as sure; but if they were unfavourable, defeat, if not death, was as certain. Divination, or enchantment, was employed for the purpose of knowing their ultimate decision, and at these times they always pretended to follow implicitly supernatural intimation, though all this juggling and contrivance was designed only to deceive the people into a persuasion that the god sanctioned the views of the king and government. The divinations were connected with the offerings, and the success or failure of the expedition was often chiefly augured from the muscular action in the heart or liver of the animal offered, or the involuntary acts and writhing contortions of the limbs of the human sacrifice in the agonies of death.

When the murder and destruction of actual conflict terminated, and the vanquished sought security in flight, or in the natural strong-holds of the mountains, some of their conquerors pursued them to their hiding-places, while others repaired to the villages, and destroyed the wives, children, infirm and afflicted relatives, of those who had fled before them in the field. These defenceless wretches seldom made much resistance to the lawless and merciless barbarians, whose conduct betrayed a cowardly delight in torturing their helpless victims. Plunder and revenge were the principal objects in these expeditions. Every thing valuable they destroyed or bore away, while the miserable objects of their vengeance were deliberately murdered. No age or sex was spared. The infant that unconsciously smiled in its mother's arms, and the venerable gray-haired father or mother, experienced unbridled and horrid barbarity. The aged were at once despatched, though embowelling and every horrid torture was practised. The females experienced brutality and murder, and the tenderest infants were perhaps transfixed to the mother's heart by a ruthless weapon—caught up by ruffian hands, and dashed against the rocks or the trees—or wantonly thrown up in the air, and caught on the point of the warrior's spear, where it writhed in agony and died. A spear was sometimes thrust through the infant's head from ear to ear, a line passed through the aperture, and when the horrid carnage has been over, and the kindling brand has been applied to the dwellings, while the flames have crackled, the dense columns of smoke ascended, and the ashes mingled with the blood from the victims, the cruel warriors have retired with fiendish exultation, some bearing the spoils of plunder, some having two or three infants hanging on the spear they bore across their shoulders, and others dragging along the sand those that were strung together by a line through their heads, or a cord round their necks.

When those who had been vanquished in the field did not return to battle, but remained in their strong-holds, another religious ceremony was performed by the conquerors, called the Hora. A large quantity of property, the spoils of victory, was taken to the priests of Oro, partly as an acknowledgment for past success, but chiefly to encourage them to increased intercession that the destruction the god had commenced might not cease till their enemies were annihilated, for their wars were wars of extermination.

One singular result of their dreadful wars is, the existence of a number of wild men inhabiting the fastnesses of the interior mountains of Tahiti. I have not heard of any having been seen in any other island, but they have been more than once met with in the neighbourhood of Atehuru. When I visited this station in 1821, I saw one of these men, who had been some time before taken in the mountains, and was comparatively tame, yet I shall not soon forget his appearance. He was above the middle size, large-boned, but not fleshy. His features and countenance were strongly marked, his complexion was not darker than those of many around, but his aspect was agitated and wild. His beard was unshaven, and his hair had remained uncut for many years. It appeared about a foot and a half in length, in some parts perhaps longer. He wore it parted in the middle of his forehead, but hanging uncombed and dishevelled on the other parts of his head. On the outside it was slightly curled, and hung in loose ringlets. The colour was singular; at the roots, or close to his head, it was dark brown or black, six inches from his head it was of a tawny brown, while the extremities exhibited a light and in some places bright yellow. Many attempts had been made to persuade him to have it cut, but to this he would never consent.

His only clothing was a *maro,* or girdle, with sometimes a light piece of cloth over his shoulder. His nails, for the sake of convenience, he had cut. He said but little, and though he came and looked at us once or twice, he seemed averse to observation, and retired when I attempted to converse with him. He had been driven to the mountains in a time of war, had remained in solitude for years, had been at length discovered by persons travelling in these regions, secured, and brought down, where with great difficulty he had been induced to remain. Mr. Darling said, he was very quiet, but appeared uninterested in most of what was passing around him. It is supposed that, during his solitude, he was under a degree of insanity, probably from the effect of the fright with which he had fled from battle.

Since Mr. Darling's residence at Bunaauïa, others have been seen in the mountains, and one was secured by the people of Burder's Point. They had gone to the mountains for food or timber, and suddenly perceived a man approaching. As soon as he saw them, he fled precipitately; they followed, and, by lying in wait, ultimately succeeded in securing him. When spoken to, he did not reply, and seemed not to understand. They led him down to the shore. He evinced the greatest horror at the sight of men, but they took him to the chief's house, treated him with kindness, and avoided crowding round him. Food and water were brought, which he refused. The first night, they kept watch; the next day, although they placed food before him, he refused it, and maintained the most entire silence. During the second night, he escaped from the house, fled to the mountains, and has never since been heard of. He did not appear to be advanced in years, was without clothing when taken, and, although a finely formed man, exhibited one of the most afflicting spectacles it was possible to behold.

It is supposed that, under the panic which seized those who were defeated in some of the battles that within the last fifty years have been fought in these portions of the island, he had fled to the mountain fastnesses in its more central parts, and perhaps had experienced a degree of mental aberration which had deprived him of memory, and had induced him to wander like a demoniac among the lonely rocks and valleys. It is reported by the natives that others have been seen, and that some of the inhabitants of the lowlands have been in danger of losing their lives from coming in contact with them. After the evidence of the facts above-mentioned, we cannot doubt the existence of such unhappy victims; but at the same time, the circumstance of their being so seldom seen, warrants the hope that they are not numerous.

The captives taken in war, called *ivi* or *titi*, were murdered on the spot, or shortly afterwards, unless reserved for slaves to the victors. The bodies of the slain were treated in a most savage manner. They were pierced with their spears, and at times the conduct of the victors towards their lifeless bodies was inconceivably barbarous.

On the day following the battle, the *bure tuata* was performed. This consisted in collecting the bodies of the slain, and offering them to Oro, as trophies of his prowess, and in acknowledgment of their dependence upon his aid. Prayers were preferred, imploring a continuance of his assistance.

The bodies were usually left exposed to the elements, and to the hogs or wild dogs that preyed upon them.—The victors took away the lower jaw-bones of the most distinguished among the slain, as trophies, and often some of the bones, converting them into tools for building canoes with, or into fish-hooks. Sometimes they piled the bodies in a heap, and built the skulls into a kind of wall around the temple, as at Opoa, but they were commonly laid in rows near the shore, or in front of the camp, their heads all in the same direction. Here the skulls were often so battered with the clubs, that no trace of the countenance or human head remained.

In addition to the preceding indignities, their bodies were sometimes laid in rows along the beach, and used as rollers, over which they dragged their canoes, on landing, or launching them after a battle. We do not know that the Tahitians ever feasted on the bodies of the slain, although this is practised by the Marquesians on the one side, and the New-Zealanders on the other—by the inhabitants of the Dangerous Archipelago in the immediate neighbourhood of the Georgian Islands in the east, and in several of the Harvey Islands in the west—especially Aitutake, where it continued until the abolition of idolatry in 1823.

Here the warriors were animated to the murderous combat by allusions to the inhuman feast it would furnish at the close. In New-Zealand, it is stated that a warrior has been known, when exulting over his fallen antagonist, to sever his head from his body, and, while the life-blood has flowed warm from the dying trunk, to scoop it up in his hands, and, turning to his enemies with fiend-like triumph, drink it before them.

Besides the *atore,* embowelling, which was frequently inflicted, they sometimes practised what they called *tiputa taata.* When a man had slain his enemy, in order fully to satiate his revenge, and intimidate his foes, he sometimes beat the body flat, and then cut a hole with a stone battle-axe through the back and stomach, and passed his own head through the aperture, as he would through the hole of his tiputa or poncho; hence the name of this practice. In this terrific manner, with the head and arms of the slain hanging down before, and the legs behind him, he marched to renew the conflict. A more horrific act and exhibition it is not easy to conceive of, yet I was well acquainted with a man in Fare, named Taiava, who, according to his own confession, and the declaration of his neighbours, was guilty of this deed during one of their recent wars.

Other brutalities were practised towards the slain, which I never could have believed, had they not been told by the individuals who had been engaged in them, but which, though I do not doubt their authenticity, are improper to detail. I should not have dwelt so long on the distressing facts that have been given, but to exhibit in the true, though by no means strongest colours, the savage character and brutal conduct of those, who have been represented as enjoying, in their rude and simple state, a high degree of happiness, cultivating all that is amiable and benevolent.

The bodies of the slain being now abandoned by the victors, they turned their attention to the division of the spoils of the vanquished, and the appropriation of the country, &c. In connexion with this, the *rani arua* was performed, and was indeed considered as a part of the ceremony of devoting the slain to the gods. A human sacrifice was procured, and offered, principally to secure the return of the occupations and amusements of peace; feasting, dancing, &c. The burden of the prayer was— *Tutavae aua i te po, Roonui arena homai te ao*, &c. and which may be rendered, "Let the god of war return to the world of night: Let Roo the god of peace preside in the world, or place of light," &c.

The local situation of the people, and their familiarity with the sea, led them to feel at home upon the water, and on this element many of their bloodiest battles were fought. A description of their *pahis*, or war canoes, has been already given. Their fleets were often large. The Huahinian expedition, according to the account of those still living who were in the battle of Hooroti, amounted to "ninety ships, each twenty fathoms long," on which it is probable a number of smaller canoes were in attendance. When the engagement took place within the reefs, the canoes were often lashed together in a line, the stem of one being fastened to the stern of the other before it. This they called *api*, and adopted it to prevent the breaking of their line, or retreat from the combat. The opposing fleet was, perhaps, lashed or fastened in the same way; and thus the two fleets, presenting one continued line of canoes, with the *revas* or streamers flying, were paddled out to sea, the warriors occupying the platform raised for their defence, and enabling them to command each part of the canoe.

At a distance, stones were slung; on a nearer approach, light spears or javelins were hurled, until they came close alongside of each other, when, under the influence of rage, infatuation, ambition, or despair, they fought with the most obstinate and desperate fury.

It is not easy to imagine a conflict more sanguinary and horrid than theirs must have been. Although the victors, when *faatini'd* or supplicated, sometimes spared the fallen, it was rarely they gave any quarter. Retreat there was none—and, knowing that death or conquest must end the fray, they fought under the power of desperation.

At times, both fleets retired, as was the case at Hooroto; but when victory was evidently in favour of one, the warriors in that fleet sometimes swept through the other, slaughtering all who did not leap into the sea, and swim toward the canoe of some friend in the opposing fleet. I have been informed by some of the chiefs of Huahine, who have been in their battles, that they have seen a fleet towed to the shore by the victors, filled with the wounded and the dead—the few that survived being inadequate to its management.

When the canoes of a fleet were not fastened together, as soon as the combatants perceived they were overpowered, they sought safety in flight, and, if pursued, abandoned their canoes on reaching the shore, and hastened to their fortress in the mountains.

They did not enclose their temporary encampments in the open field, but each party considered a fortification as a security against invasion, and a refuge after defeat in the field.

Their places of defence were rocky fortresses improved by art—narrow defiles or valleys sheltered by projecting eminences—passes among the mountains, difficult of access, yet allowing their inmates a secure and extensive range, and an unobstructed passage to some spring or stream. The celebrated Pare, in Atehuru, was of this kind; the mouth of the valley in which it was situated was built up with a stone wall, and those who fled thither for shelter, were generally able to repel their assailants.

Sometimes they cut down trees, and built a kind of stage or platform called *pafata,* projecting over an avenue leading to the *pare;* upon this they collected piles of stone and fragments of rock, which they hurled down on those by whom they were attacked. In some of the Harvey Islands they planted trees around their places of encampment, and thus rendered them secure against surprise.—These enclosures they called *pa,* the term which is used to designate a fort in the Sandwich Islands.

If those who had been routed on the field of battle were allowed by their pursuers time to wall up the entrances of their places of refuge, they were seldom exposed to assault, though they might be decoyed from them by stratagem, or induced to leave from hunger. The *pari* in Borabora, and some places in Tahiti, are seldom excelled as natural fortresses. Several of these places were very extensive: that at Maeva, in Huahine, bordering on a lake of the same name, and near Mouna-tabu, is probably the best artificial fortification in the islands. It encloses many acres of ground well stocked with bread-fruit, containing several springs, and having within its precincts the principal temples of their tutelar deity. The walls are of solid stone-work, varying in height from six to twelve feet. They are even at the top; in some places ten or twelve feet thick. Openings in the wall appear at intervals for ingress and egress, but during a siege, these were built up with loose stones, when it was considered a *pari haabuea,* an impregnable fortress, or, as the term indicates, place of refuge and life. Such as fled to the rocks or mountains were called *meho.*

If those who had fled were numerous, and the conquering army wished to reduce them, the war often assumed a protracted form. When the assailants had determined on reducing them, they endeavoured to decoy them out; if they failed, they seldom succeeded in scaling or forcing their ramparts. Famine often reduced the besieged to the greatest distress, so that they ate the *pohue,* or wild convolvolus stalks, and other rude kinds of food. They frequently made desperate sallies upon the besiegers, but were often driven back with great slaughter.

In a sally made during one of the wars which occurred in the year 1802, called in the annals of Tahiti, "the war of Rua," this chief, and a number of his fighting-men, were taken, and killed on the spot by the king's order. The next day the king marched to the fortress, but found it well manned, and the greatest determination to resist manifested by the warriors.

An ambassadress, with a flag of truce, passed between the parties, and the besieged manifested an uncommon degree of dauntless obstinacy. When told of the numbers and the persons slain, they appeared as if but little affected by it, pretended not to know them, excepting the chief, who, they said, it was far more likely had been drowned in the river, than that he had fallen into their hands. This they evidently did, to shew that what the king thought would induce them to make an unconditional surrender, had not so subdued them; and the survivor, Taatahee, directed the ambassadress to tell Pomare, that when he had experienced the same fate as Rua, then, and not till then, might he expect peace.

When the reduction of a fortress was a matter of importance, the co-operation of the gods was again invoked, and the Hiamoea performed. This was a religious ceremony, in which the finest mats, cloth, and other valuable spoils, were taken by the victorious party, as near to the fortress as it was safe to approach. Here they took the different articles of property in their hands, and, holding them up, offered them to the gods, who, it was supposed, had hitherto favoured the besieged; the priests frequently exclaiming to the following effect —*Tane* in the interior or fortress, *Oro* in the interior or fortress, &c. come to the sea, here are your offerings, &c. The priests of the besieged, on the contrary, endeavoured to detain the gods, by exhibiting whatever property they possessed, if they considered the god likely to leave them. A warrior would often offer himself, and say, *Eiaha e haere,* "Leave us not, here is your offering, O Oro! even I!" It is hardly possible to avoid admiring the patriotism evinced on such occasions. It was a devotion worthy of a better cause.

Although the besieged might offer their human sacrifices, they must perform what, under these circumstances, would be called *Taaraa-moua,* the fall from the mountain, and which they carried as near the temple of the tutelar deity as their enemies would allow them to approach, when, having deposited their offering, they fled to the fortress, determined to defend it; yet, if the property which the victors had there offered, and devoted as it were to the gods, was valuable and abundant, the besieged became dispirited, believing that the gods had left them, and gone to the party by whom these offerings had been made. They always imagined that the gods were influenced by motives similar to those which governed their own conduct; and when once the vanquished party imbibed the impression that the gods had forsaken them, their defence was comparatively feeble, and they consequently fell a prey to their enemies, who were often indebted more to the superstitious apprehensions of their foes, than to their own skill or power.

If the conquered party surrendered at discretion, their land and property were divided by the conquerors, and the captives either murdered, reduced to slavery, or reserved for sacrifices when the gods required human victims. The bodies of such as were killed in their forts, were treated with the same indignity as those slain in the field; part of the bodies was *eaten* by the priests, the rest piled up in heaps on the sea-coast, where the effects of decomposition have been so offensive, that the people have forborne to fish in the adjacent parts of the sea. On the contrary, when neither party had been subdued, and, by intimation from the gods or any other cause, one party desired peace, an ambassador was sent with a flag of truce, which was usually of native cloth, a bunch of the sacred mero, &c. and proposals of peace. If the other party were favourable, an interview followed between the leaders, attended by the priests and national orators.

They usually sat in council on the ground, either under a shady grove, or on the sandy beach. The orators of those who had sent the proposals made the first harangue; this was followed by a reply from the orator of the other party, who was sitting on the ground opposite, and ten or twenty yards distant. Each held in his hand a bunch of the sacred mero. The king or chiefs sat beside them, while the people stood around, at some distance. When the terms were agreed upon, the wreath of peace was woven with two or three green boughs, furnished by each, as the *manufaiti*, the bond of reconciliation. Two young dogs were then exchanged by the respective parties, and the *apa pia* brought; it was a long strip of *apa*, or cloth, white on one side, and red on the other; the materials were furnished, and the cloth was joined together, by both parties, in token of their union, and imprecations were invoked on those who should *hae*, or rend, the *apa pia*, or band of peace. The apa pia and the green boughs were then offered to the gods, and they were called upon to avenge the treachery of those who should rend the band, or break the wreath. Divinations were also sometimes used, to know whether it would be of a long or a short continuance.

Feasting followed the ceremony, together with the usual native games; besides which, religious rites were performed. The first was the *maioi*, when vast quantities of food were taken to the king, and large offerings to the gods, together with prayers for the establishment and prosperity of the reign. Another was called the *oburoa na te arii*, and consisted also in offerings to the gods, with prayers for their support, and a large present of food to the principal warrior chief, under the king, as an acknowledgment of his important service in the recent struggles, and his influence in establishing the king in his government.

But the most important ceremony, in connexion with the ratification of peace, was the *upoofaatau*, &c. It was commemorative of the establishment of the new government, and designed to secure its perpetuity, and the happiness of the community. A leading raatira was usually the chief proprietor of the entertainment, and master of the ceremonies. The festival was convivial and religious. Food and fruits, in the greatest profusion, were furnished for the altars of the gods, and the banquet of the king.

A *heiva*, or grand dance, formed a part of this ceremony. It was called the dance of peace, and was performed in the presence of the king, who, surrounded by a number of chiefs and warriors, sat at one end of the large house in which it took place. A number of men, and sometimes women, fantastically dressed, danced to the beating of the drum and the warbling of the *vivo*, or flute; and though the king was surrounded by a number of attendants as body-guards, towards the close of the exhibition the men sought to approach the king's person, and kiss his hand, or the females to salute his face; when one or the other succeeded in this, the heiva, or dance, was complete, and the performance discontinued.

This, however, was only part of the ceremony, for while they were thus employed, the priests were engaged in supplicating the gods that these amusements might be continued, and their enjoyments in feasting, dancing, and the pursuits connected with them, might not be again suspended or disturbed by war. Peace was now considered as established, the club and the spear were cleaned, varnished, and hung up in their dwellings; and the festive entertainments, pagan rites, and ordinary avocations of life, resumed, till some fresh quarrel required an appeal to their weapons, and led them to the field of plunder and of death.

I have dwelt longer on this subject than I intended, and perhaps than it required; but the former frequency of war, the motives influencing the parties engaging in the ceremonies connected with it, and the manner in which it was prosecuted, were all adapted to convey, next to their mythology, a correct idea of the national character of the people, who made war, paganism, and vicious amusements, the business of life. In all our converse with them relative to their former state, no subject was so frequently introduced. No event in history, no character in their biography, appeared unconnected with some warlike expedition, or feat of arms; and almost all the illustrations of the most powerful and striking expressions which we sought to investigate, were drawn from the wars.

In connexion with this state of society, how cheering is the contrast exhibited since the introduction of Christianity. Rumours of war have indeed been heard, especially at Tahiti—where, since the death of the late king, very powerful interests, and perhaps some latent feelings of ancient rivalship, have been brought into collision, and where the conduct of some in the highest authority, has not been at all times the most honourable or conciliatory—but no actual hostility has yet existed. In the Leeward Islands also, reports of war, and warlike preparations, have appeared— more particularly in reference to the bold and martial chieftain of Tahaa, and some of the restless spirits among the inhabitants of Borabora, once celebrated for their military prowess, and masters of most of the Leeward group—but here it has been only rumour

The transient affair at Huahine, in connexion with which these remarks have been introduced, and similar occurrences in Raiatea. and Tahaa—between the chiefs, together with a great body of the people, on the one side; and those dissatisfied with the moral restraints the new laws imposed upon their conduct, to which under idolatry they had been unaccustomed, on the other—are the only public disturbances that have occurred. A few disaffected and lawless young fellows in Raiatea, supposing the Missionaries were chiefly instrumental in the adoption and maintenance of the laws, formed a plan for murdering them, and overturning the government. Mr. Williams, who was to have been the first object of their vengeance, averted the threatened danger by what appeared to him, at the time, a circumstance entirely accidental,—but which afterwards proved a remarkable interposition of Providence for the preservation of his life. With these exceptions, the inhabitants have, since their adoption of Christianity, enjoyed uninterrupted peace during a longer period than it was ever before known to exist among them.

Some noble instances of calm determination not to appeal to arms, have been given by Utami, and other governors; the love and the culture of peace, having indeed succeeded their delight in the practice of war, even in the most turbulent and fighting districts. It is well known, Mr. Darling observes, in reference to the district of Atehuru, that the inhabitants of this part of Tahiti were always the first to make war. False reports reached the ear of the king's party, who were told that the people of Atehuru entertained evil designs against the royal family. Rumours of war were spread by the adherents of the king, but instead of rejoicing, as they would for merly have done, every one appeared to dread it as the greatest calamity. They gathered round the house of the Missionary, declaring that if attacked they would not fight, but would willingly become prisoners or slaves, rather than go to war. The threatened war was thus prevented—those with whom the reports had originated were sought out—an appeal was made to the *laws*, instead of the *club* and the *spear*, and the matter submitted to the magistrate rather than to the warrior. The punish ment annexed to the circulation of false and injurious reports was inflicted on the offenders, and the parties united in amity and friendship.

As they feel the blessings of peace increase with its continuance, their desires to perpetuate it appear stronger. Its prevalence and extent are often surprising, even to themselves; and some of the most striking illustrations of the advantages of true religion, and appeals for its support and extension, are drawn from this fact, and expressed in terms like these: Let our hands forget how to *hi te omore*, or *vero ti patia*, lift the club, or throw the spear: Let our guns decay with rust, we want them not; for though we have been pierced with balls or spears, if we pierce each other now, let it be with the word of God: How happy are we now, we sleep not with our cartridges under our heads, our muskets by our side, and our hearts palpitating with alarm: Now we have the Bible, we know the Saviour; and if all knew him, if all bowed the knee to him, there would be no more war on the earth.

It is not in public only that they manifest these sentiments; in ordinary life at home, they act upon them. The most affectionate and friendly intercourse is cultivated between the parties who formerly cherished the most implacable hatred, and often vowed each other's extermination. Offices of kindness and affection are performed with promptitude and cheerfulness; and though by some their weapons are retained as relics of past days, or securities against invasion, by many they are destroyed. Often have I seen a gun-barrel or other iron weapon, that has been carried to the forge, submitted to the fire, laid upon the anvil, and beaten, not exactly into a plough-share or a pruning-hook, (for the vine does not stretch its luxuriant branches along the sides of their sunny hills,) but beaten into an implement of husbandry, and used by its proprietor in the culture of his plantation or his garden. Their weapons of wood also have often been employed as handles for their tools; and their implements of war have been converted with promptitude into the furniture of the earthly sanctuary of Jehovah. The last pulpit I ascended in the South Sea Islands was at Rurutu. I had ministered to a large congregation, in a spacious and well-built chapel, of native architecture, over which the natives conducted me at the close of the service. The floor was boarded, and a considerable portion of the interior space fitted up with seats or forms. The pulpit was well, though rudely constructed; the stairs that led to it were guarded by rails, surmounted by a bannister of mahogany-coloured tamanu wood; the rails were of dark aito wood, and highly polished. I asked my companions where they had procured these rails; and they replied that they had made them with the handles of warriors' spears!

CHAP. XVIII.

Arrival of the deputation at Tahiti—Visit to Huahine—Pomare's death —Notice of his ancestry—Description of his person—His mental character and habits—Perseverance and proficiency in writing—His letters to England, &c.—Fac-simile of his hand-writing, and translation of his letter on the art of drawing—Estimation in which he was held by the people —Pomare, the first convert to Christianity—His commendable endeavours to promote its extension—Declension during the latter part of life—His friendship for the Missionaries uniform—His aid important— Circumstances connected with his death—Accession of his son Pomare III. to the government—Coronation of the infant king—His removal to the South Sea academy—Encouraging progress in learning—Early and lamented death—The extensive use of letters among the islanders— Writing on plantain leaves—Value of writing paper, &c.—The South Sea academy, required by the state of native society—The trials peculiar to Mission families among uncivilized nations—Advantages connected with the visits of Missionaries' children to civilized countries.

We had attended our friends from the Windward Islands to a general meeting at Raiatea, and enjoyed their society a month, when Pomare's vessel called at Huahine, on her way from New South Wales to Tahiti. Circumstances requiring that as many of the Missionaries in the Leeward Islands as could leave their stations should meet those of the Windward group, Mr. and Mrs. Williams, Mrs. Ellis, and myself, accompanied our friends on board the Governor Macquarie, in which they returned to Tahiti.

After five days at sea, finding ourselves near the land, we entered our boat, which had been towed at the stern of the vessel, and, rowing to the shore, landed a few miles to the southward of the settlement at Burder's Point No effort had been wanting on the part of the captain to render our voyage agreeable; but, from the smallness of the cabin, number of the passengers, frequent rains, and contrary winds, it had been tedious and unpleasant, and we were glad to find ourselves on terra firma again. Exhausted by the fatigues of the voyage, we found the walk to the settlement exceedingly laborious, and some of our party were more than once obliged to sit down upon the rocks by the way side. On reaching the dwellings of our friends, the welcome, the refreshment, and the rest, we there received, soon recruited our strength and spirits.

We had accomplished our business, and were at Papeete preparing to return, when, on the 24th of Sept. about three o'clock in the afternoon, a vessel of considerable size was seen approaching Point Venus. By the aid of a glass, we perceived that it was a three-masted vessel, and, in endeavouring to ascertain its signal, we were surprised on beholding a large white triangular banner flying at the top-gallant-mast-head. The ship was too distant to allow of our reading the motto, or perceiving with distinctness the device, and we could only conjecture the character of the vessel, or the object of the visit.

The next morning, a note from Mr. Nott conveyed to us the gratifying intelligence, that the ship was direct from England, and that G. Bennet, Esq. the Rev. D. Tyerman, a deputation from the Society, with three Missionaries, had arrived. The captain had come over in his boat, and, anxious to welcome our newly arrived friends, I accompanied him in his return to the ship. On reaching the Tuscan, we were happy to see Messrs. Jones, Armitage, and Blossom, with their wives, and afterwards proceeding to the shore, had an opportunity of greeting the arrival of the deputation.

The next morning the ship proceeded to Papeete; and, in the forenoon of the same day, Messrs. Williams and Darling, having returned from Eimeo, we met the deputation, read the letters from the Board of Directors, acknowledged the appointment of the deputation as a proof of their attachment, and expressed our sense of their kindness in forwarding supplies.

The letters they had brought, and the accounts of their intercourse with our friends, were cheering; and after spending upwards of a week very pleasantly in their society, I returned to Eimeo in my own boat, Mr. and Mrs. Williams, and Mrs. Ellis, having sailed to Huahine a week before, in the Westmoreland. Contrary winds detained me another week at Eimeo, during which I visited Pomare. On the 12th of October we set sail, and, after passing two nights at sea, reached Fare harbour in safety on the fourteenth.

The year 1821 was an eventful period in the political annals of Huahine, not only in reference to the promulgation of the new code of laws, and the resistance made to their enforcement, but also in regard to the death of Taaroarii, the king's only son, the chief of Sir Charles Sanders' Island, and the heir to the government of Huahine. This event took place very soon after my return from Tahiti.

Soon after our return from Tahiti, the indisposition of Mr. and Mrs. Williams required a suspension of their exertions in Raiatea, and a visit to New South Wales.

On the 8th of December, 1821, the shout of *E pahi, e!* A ship, ho! re-echoed through our valley; we proceeded towards the beach, and, on reaching the sea-side, beheld a large American vessel already within the harbour. The captain soon landed, and informed us that our friends Messrs. Bennet and Tyerman were in the ship. We hastened on board, conducted them to the shore, and welcomed them to our dwellings. Mr. Bennet took up his abode with Mr. Barff, while we were happy to accommodate Mr. Tyerman. The chiefs and people, who had been led to expect a visit from our friends greeted their arrival with demonstrations of joy; these friends remained some time in Fare, and the period they spent with us was one of unusual interest and enjoyment.

In the close of this year, 1821, the Mission and the nation experienced the heaviest bereavement that had occurred since the introduction of Christianity. This was the death of the king, Pomare II. which took place on the seventh of December, the day preceding the deputation's arrival in Huahine. His health had been for some time declining, but his departure at last was sudden. I spent the greater part of a Sabbath afternoon with him at Eimeo, in the beginning of October. He was then unable to leave the house, but was not considered dangerously ill. I was then for some days with him, and had not seen him since. He had long been afflicted with the elephantiasis, a disorder very prevalent among the people; but the principal cause of his dissolution was a dropsical complaint, to which he had been for some time subject.

The conspicuous station Pomare had occupied in the political changes of Tahiti, since the arrival of the Missionaries, the prominent part he had taken in the abolition of idolatry, the zeal he had manifested in the establishment of Christianity, and the assistance he had rendered to the Missionaries, caused a considerable sensation to be experienced among all classes by his death; and as his name is perhaps more familiar to the English reader than that of any other native of the South Sea Islands, some account of his person and character cannot fail to be acceptable.

Pomare, originally called Otoo, was the son of Pomare and Idia: the father was sovereign of the larger peninsula when it was first visited by Cook, and was then called Otoo; subsequently, by the aid of the mutineers of the Bounty, he became king of the whole island, and adopted the name of Pomare, which at his death was assumed by his son, and has since been the hereditary name of the reigning family. Idia, his mother, was a princess of the adjacent island of Eimeo, and sister to Motuaro, one of the principal chiefs at the time of Cook's visit.

Pomare was the second son of Otoo and Idia, the first having been destroyed according to the regulations of the Areois society, of which they were members. He was born about the year 1774, and was consequently about forty-seven years of age at the time of his decease. Tall, and proportionably stout, but not corpulent, his person was commanding, being upwards of six feet high.* His head was generally bent forward, and he seldom walked erect. His complexion was not dark, but rather tawny; his countenance often heavy, though his eyes sometimes appeared to beam with intelligence. The portrait of Pomare, in the frontispiece to the first volume of this work, is from one taken at Tahiti by an

* His father's height was six feet four inches.

artist attached to two Russian ships of discovery, that visited the islands a short time before his death; and, excepting a little undue prominency in the forehead, is a good likeness.

His character was totally different from that of his father—who was a man of enterprise, excessive labour and perseverance, bent on the aggrandisement of his family, and the improvement of his country, clearing waste tracts of land, planting them, and generally occupying the people with some public work. Pomare took no delight in exertions of this kind, but manifested aversion to them; his habits of life were indolent, his disposition sluggish, and his first appearance was by no means adapted to produce a favourable impression on a stranger's mind. Captain Wilson conceived such an idea of his stupidity and incapacity, as to suppose him the last person on whom any favourable impression would be made.

He was, however, though heavy in his appearance and indolent in his habits, inquisitive, attentive, and more thoughtful perhaps than any other native of the islands; —a keen observer of every thing that passed under his notice, although at the time he would not appear to be paying particular regard. He was not only curious and patient in his inquiries, laborious in his researches, but often exhibited a great degree of ingenuity, notwithstanding his dull appearance. I have sometimes been in his company, when he has kept a party of chiefs in constant laughter, as much from the dryness and coolness with which his expressions were uttered, as the humour they contained. He was not, however, fond of conviviality or society, but appeared to be more at ease when alone, or attended only by one or two favourite chiefs.

In mental application to learning, Pomare certainly exceeded every one of his subjects; and, had he been free from practices which so banefully retarded his progress, and enjoyed the advantage of a regular and liberal education, there is every reason to believe the development and culture of his intellect would have shewn that it was of no inferior order.

He had heard much, from the early visitors to his island, of king George, and appeared, on more than one occasion, desirous to make the British sovereign his model. He was walking one day in the district of Pare with great dignity, in the company of the Missionaries, when he suddenly stopped, and said, "Does king George walk in this way?" As soon as he in any degree comprehended the use of letters, he manifested a great desire to be able to read and write, and was one of the first pupils. Looking over the books of the Missionaries one day, he saw a Hebrew Bible: the singularity of the letter attracted his attention; and having been informed that it was the language of the Jews, in which the greater part of the Scriptures was written, he expressed a wish that one of the Missionaries would teach him to read it, inquiring at the same time whether king George understood Hebrew. In this he did not persevere, but he soon made himself master of the English alphabet, and could read in the English Bible, not with fluency, but so as to comprehend the meaning of the plainest parts.

It was, however, in his native language that the Tahitian ruler made the greatest progress, and in writing this, he excelled every other individual. Mr. Nott and Mr. Davies were his principal instructors; the latter has spent many hours with him, sitting on the ground, and teaching him to form letters on the sand, probably before Dr. Bell's system was introduced to general notice in England. The hand-writing of Pomare, during the latter part of his life, was much better than that of any of the Missionaries. His earliest letters or notes, the first ever written by a native, were from Eimeo. In 1805 he wrote a letter to the Missionaries. In 1807 he wrote one to the Missionary Society, which being the first despatch ever forwarded by a native of those islands to Britain, is a great curiosity.

The Directors had written, advising him to banish the national idol, to attend to the instruction of the Missionaries, and to discountenance those sins which were so rapidly depopulating his country. In reply, he wrote a letter in the native language, which the Missionaries translated; he then copied the translation, and both letters, signed by his own hand, were forwarded to London. He expresses a determination to banish Oro to Raiatea, wishes them success in their efforts to instruct the people of Tahiti, which he calls a bad land, a regardless land. He desires them to send a number of men, women, and children, to Tahiti, to send cloth, and then they will adopt the English dress, and tells them, that, should he be killed, they will have no friend in the islands. "Come not here after I am dead," was his expression. He also requested them to send him all the curious things in England, especially those necessary for writing, and, after enumerating pens, ink, &c. concluded his request by stating, "Let no writing utensil be wanting." He signed his name, "Pomare, King of Tahiti," &c. superscribed his letter to "My Friends the Missionary Society, London."

Sedentary occupations and amusements appeared more congenial to Pomare than active pursuits; he found an agreeable occupation in braiding the finest kinds of cinet with the fibre of the cocoa-nut husk; writing, however, was his chief employment and recreation. At first, he had a writing house erected, that he might follow his favourite pursuit, uninterrupted by his domestics or the members of his household; he then sat at a table, but, during the latter part of his life, he usually wrote lying in an horizontal position, leaning his chest on a high cushion, and having a desk before him.

Pomare kept a regular daily journal, and wrote in a book provided for that purpose, every text of scripture that he heard. Sometimes he wrote out the prayers he used in social and private devotion; maintained an extensive correspondence, after the introduction of writing among the people; prepared the first code of laws for his kingdom; copied them out fairly with his own hand, and promulgated them with his voice. He also rendered very important aid to the Missionaries in the translation of the scriptures, and copied out many portions before they were printed.

The king was remarkably pleased with engravings and paintings, and has often called at my house to look at the plates in an Encyclopædia, frequently asking if I thought it possible for him to learn to draw. I always told him it depended on his own industry; that I had no doubt of his capacity, if he would apply. In connexion with these encouragements, I received from him the accompanying note, soon after our settlement at Afareaitu in 1817. I insert it as a specimen of his hand-writing, although it is by no means so carefully written as many of his letters, or his copy of the laws, &c. It will also serve as a specimen of the idiom of the language, as I have affixed a literal translation.

FAC-SIMILE.

Ehoaino e! Friday Morning

Iaorana oe, e to fetii atoa ite ora raa
ia Jesus Christ i te ora mau ra.

Teie tau parau ia oe ehoaino. E papai mai
oe i te huru o te Raau, e te Manu e te fare, e
hapono mai i raronei, ia au e hio eiaha tu
te itea iau, o te hio noa inaeira te hinaaro
nei au ite haapii i te reira ra peu ia ite
au; e ore paha rau e ite. eiaha tu te ite
rahi; ei ite iti ae. E papai mai hoi oe i
ta oe parau ia ite au, Faaite mai hoi oe
iau i to oe hinaaro.

Iaorana oe ia Jesus Christ.

Pomare

TRANSLATION.

DEAR FRIEND.

MAY you and your family be saved by Jesus Christ, the true Saviour. This is my speech to you, my friend, that you draw (or write) the likeness of a tree, and a bird, and a house, and send them down here, that I may see them; they will not be known by me, if I look carelessly at them. I desire to learn that custom or art, that I may know it: perhaps I shall not know it much; I may know it a little. Write also to me, and make known to me your wishes.

May you be saved by Jesus Christ,

POMARE.

Sketched by Capt. R. Elliot. R. N.

Engraved by B. Winkles.

TOMB OF POMARE, AT PAPAOA, IN TAHITI.

His policy as a ruler was deliberative and cautious, rather than prompt and decisive, and most of his measures were pursued more with a view to their ulterior influence upon the people, than to their immediate effect. His views were in many respects contracted, and he was easily imposed upon by bold and heedless advisers. He was more rapacious than tyrannical, but probably would not have been so rigid in his exactions, but for the influence of those constantly round him, who often availed themselves of his authority and influence to advance their own unjust and oppressive proceedings. Though destitute of many essentials in a great prince, the Tahitian ruler was universally respected and beloved by his own family, and by many of the chiefs, who were under great obligations to him, but I do not think he was beloved by the nation at large. It was rather a respectful fear than a fond attachment, that was generally entertained for him. He was exceedingly jealous of any interference with his prerogative or his interest, and was frequently attended by a number of the Paumotuans, or natives of the Palliser Islands, as a kind of body guard. These were considered as in some degree foreigners; and their selection by the king, as the protectors of his person, caused dissatisfaction in the minds of several of the chiefs.

Pomare was not only the first pupil whom the Missionaries taught to read and write, but he was also the first convert to Christianity in the island of which he was king. He made a profession of belief in the true God, and the only Saviour, in 1812; and there is every reason to believe that, according to the knowledge he had of Christianity, and the duties it enjoined, he was sincere. He bore the persecution and ridicule to which he was exposed, on this account, with firmness and temper, mildly entreating those who reviled him, to examine for themselves.

In the year 1813 he proposed to Tamatoa, the king of Raiatea, and Mahine the king of Huahine, to renounce idolatry. They determinately refused; but he still continued firm in his own principles, and persevering in all his endeavours to influence other chiefs in favour of Christianity. It was in consequence of his recommendation, that Taaroarii, the son of the king of Huahine, prohibited the abominations of the Areois, and sent for a preacher to teach him the word of God. Pomare continued the steady disciple of the Missionaries for several years, using all his influence in persuading the people to renounce their dependence on the idols, and to hear about the true God. His conduct in this respect was most commendable, for I never heard that he had recourse to any other means than persuasion, or that he ever held out any other inducements than those which the scriptures present. He had no worldly honours or advantages to bestow, for he was at that time an exile; and the constant reproach of his family and adherents was, that his ruin was inevitable, as he had, by renouncing the national worship, made the gods his enemies.

The conduct of the king in the battle of Atehuru, his treatment of the captives, and his clemency towards the vanquished, have been already detailed, as well as his journeys for the purpose of inducing the people to embrace Christianity. His baptism, and his promulgation of the laws by which the islands of Tahiti and Eimeo are now governed, have been also given.

During the latter part of his life, his conduct was in many respects exceptionable, and his character appeared less amiable than it had been before. He had shewn his weakness in allowing the unfounded representations of a transient visiter to induce him to request that the manufacture of sugar might not be extensively carried on under the management of Mr. Gyles. He was also, as might have been expected from the circumstance of his having been the high-priest of the nation under the system of false religion, and having been identified with all the religious observances of the people, too fond of regulating matters purely connected with Christianity.

A few years before his death he was induced, by the representations of designing and misinformed individuals, to engage in the most injudicious commercial speculations, in connexion with persons in New South Wales. This proved a great source of disquietude to his mind, and probably hastened his death. One or two vessels were purchased for him at a most extravagant price; and the produce of the island was required to pay for them, and expenses connected with their navigation. One of them was seized, a law-suit instituted in consequence at Port Jackson, the rahui or tabu laid upon the island, the rights of property were invaded, and no native was allowed to dispose of any other article of produce, excepting to the agents of the king. He became the chief factor in the island, or rather the instrument of those who were associated with him in these commercial speculations, and who used his authority to deprive the people of the right to sell the produce of the soil, and the fruit of their own labour. The inhabitants were required to bring their pigs, oil, &c. and to receive in return what he chose to give them: the individuals who urged upon him this policy considered all they could obtain by any means as fair emolument. The welfare of the nation, the natural rights of the people, the establishment of commerce upon just and honourable principles, were viewed as inimical to their interest, and consequently beneath their regard. It is needless to add, that these speculations ended in embarrassment and loss.

The habits of intemperance which Pomare was led to indulge, in consequence of these associations, threw a stain upon his character, and cast a gloom over his mind, from which he never recovered, and under the cloud thus induced he ended his days.

He was also reported to be addicted to other vices, but it is not my object to exhibit the dark features of his character—truth and impartiality require what has been said—and it is with far greater pleasure that we contemplate his uniform kindness to the Missionaries, and his steady patronage of them, especially in their seasons of greatest extremity, when civil wars forced them to abandon their home, and seek safety in flight. His unwavering adherence to the profession of Christianity, amidst the greatest ridicule and persecution, and his valuable aid in its introduction, were highly serviceable to the nation. Without presuming to pronounce an opinion on his final state, he certainly was employed by God (who selects his agents from whatever station he chooses, and uses them just so long as he sees fit,) as a principal instrument in subverting idol-worship, introducing the gospel of Jesus Christ, and establishing a code of laws founded on the principles of true religion; he is therefore to be considered, if not a father, undoubtedly as a benefactor to his country. Pomare was not averse to religious conversation and devotional engagements; we conversed very freely together the last time I saw him, which was about two months before his death. He expressed his apprehensions of the increase of his disorder, but did not think it likely to prove fatal; he was shortly afterwards removed to Tahiti, where he died. During his illness, he was attended by Mr. Crook, who reminded him, in their last conversation, of the number and magnitude of his sins, and directed him to Jesus Christ, who alone could save his soul: all the reply he made was, "Jesus Christ alone," and in about an hour afterwards expired.

The lamentations of his friends, and of the people around, were great; a new tomb was erected for his remains, near the large chapel he had built at Papaoa. Messrs. Nott, Davies, and Henry, the senior Missionaries in the island, performed the religious services at his funeral, which was attended by all the Missionaries, and multitudes of the people. Mr. Nott, who had been in habits of closest intimacy with him, and had better opportunities of understanding his character than others, deeply regretted his departure. No one felt the loss of his assistance more than Mr. Nott, who was principally employed in translations of the scripture. For this department Pomare was well qualified, and always ready to render the most important services. He was well acquainted with the language, usages, and ancient institutions of the people, and his corrections were usually made with judgment and care. The compilation of a dictionary of the Tahitian language, would, if completed, have been invaluable; but he had scarcely commenced it systematically, when death arrested his progress, even in the prime of life.

Pomare was succeeded in the government by his son, who being proclaimed king immediately after his father's death, was crowned, with no small ceremony, under the title of Pomare III. on the 21st of April, 1824.

In order that the ceremonies, observed on this occasion, might be performed in the presence of the inhabitants, the greater part of whom were expected to attend, a stone platform was raised, nearly sixty feet square, upon which another smaller platform was erected, where the coronation was to take place.

When the order of the procession was arranged, it advanced towards the place, preceded by two native girls, who strewed the path with flowers. Mahine, the chief of Huahine, and nominally one of the judges of Tahiti, carried a large Bible, and was attended by the deputation from the Missionary Society, who were then at Tahiti, and Messrs. Nott and Henry; the rest of the Missionaries followed. Then came the supreme judges, three abreast; Utami, the chief of Atehuru, bearing a copy of the Tahitian code of laws. Three other judges followed; and Tati, the chief of Papara, walking in the centre, carried the crown. The young king, seated on a chair, was next borne in the procession, by four young chieftains, an equal number of chiefs' sons supporting the canopy over his head; his mother and his sister walking on one side, and his aunts on the other. His brother-in-law walked immediately behind, and was followed by Tamatoa, the king of Raiatea, and the members of the royal family. The governors, judges of districts, and magistrates, walking four abreast, closed the procession.

When they reached the place of coronation, the wives, children, and friends of the Missionaries, who had also walked in the procession, sat on the platform. The king was seated in his chair; in the centre before him, on small tables, the crown, the Bible, and the code of laws, were placed. Those who were to take part in the transactions of the day were seated around and behind the king.

The youthful Pomare, being only four years of age, was necessarily passive in the important business. Mr. Davies, one of the senior Missionaries, spoke for him; and as all were requested to take a part in the ceremonies, when the king had been asked if he promised to govern the people with justice and mercy, agreeably to the laws and the word of God, Mr. Nott placed the crown on his head, and pronounced a benediction upon the young ruler; Mr. Darling then presented him with a Bible, accompanying the presentation with a suitable address.

As soon as the coronation ceremony was closed, a herald proclaimed pardon to all who were under the sentence of the law. Every exile was directed to return, and all were exhorted to become good members of society. The assembly afterwards repaired to the Royal Mission Chapel, where Divine service was performed, and thus the first Christian coronation in the South Sea Islands closed.

The kings of Tahiti were not formerly invested with any regal dignity by receiving a crown, but by being girded with the *maro ura*, or sacred girdle, of which ceremony an account has been already given. On that occasion they bathed the king in the sea, before girding him with the sacred maro. On the present occasion they anointed his person with oil; a part of the ceremony which, I think, might have been as well dispensed with.

Shortly after his coronation, the young Pomare III. was placed at the South Sea Academy, in the island of Eimeo, under the care of Mr. and Mrs. Orsmond, for the purpose of receiving, with the children of the Missionaries, a systematic English education. His disposition was affectionate, his progress encouraging, and he promised fair to gain a correct acquaintance with the English language, which, had he lived, by giving him the key to all the stores of knowledge contained in it, would have conferred on him a most commanding influence among the people, over whom the providence of God had made him king. So far as his faculties were developed, they were not inferior to those of European children at the same age, but he was soon removed by death.

Being attacked with a complaint that passed through the islands about the middle of December, 1826, he was immediately conveyed to his mother's residence in Pare, where he lingered till the eleventh of January, 1827, when he died in Mr. Orsmond's arms. His mother and other friends standing by, when they saw him actually in the agonies of death, were so affected that they could not bear to look upon his struggles, but cast a cloth over Mr. Orsmond and the dying child he held in his arms; they removed it in a few minutes, and found his spirit had fled.

He was Pomare's only son, and the sole child of his surviving widow. A daughter of Pomare II. by a second wife, whose name is Aimata, and who is about sixteen years of age, being his only surviving child, has succeeded to the government: she was married some years ago to a young chief of Tahaa, to whom her father had given his own name, so that Pomare is still the regal name. Her character, perhaps, is yet scarcely formed, but the Missionaries speak very favourably of her principles and conduct, and hope she will prove a blessing to the nation, and a nursing-mother to the cause of Christianity.

Although Pomare II. was the first pupil whom the Missionaries taught to write, and although he excelled all others; his example induced many to make an attempt, while his success encouraged them to proceed; and now it is probable that as great a proportion of the population of the Georgian and Society Islands can write, as would be found capable of doing so in many portions of the United Kingdom. Some progress had been made, by several of the most intelligent of the converts, before the abolition of idolatry in 1815, but it is only since that period that writing has become general.

Various methods of instruction have been adopted: some of the natives have been taught altogether by writing on the sea-beach, or on sand in the schools; others have learned to write on the broad smooth leaves of the plantain-tree, using a rather bluntly pointed stick, instead of a pen or pencil. The delicate fibres of the leaf being bruised by the pen, become brown, while the other parts remain fresh and green. If it was necessary to read it immediately after being written, when held up to the light, the letters were easily distinguished. These plantain-leaf letters answer very well for short notes to pass among the natives themselves, but are liable to injury if conveyed to any great distance, or kept any length of time. They are always rolled up like a sheet of parchment, and have a remarkably rustic appearance, being usually fastened with a piece of bark, tied round the roll, the length of which being formed by the breadth of the leaf, is about twelve or fifteen inches. I have often seen the chief's messenger hastening along the road with two or three plantain-leaf rolls under his arm, or in his hand, which were the despatches of which he was the bearer.

Some of the chiefs learned to write on a slate, but these have always been articles too scarce and valuable to be in common use; they were very highly prized, and preserved with the greatest care. The greatest favour a chief could shew his son, has sometimes been to allow him to practise on his slate. We have often regretted that the supply was not more abundant, and though several hundreds of the thick slates, without frames, such as are used in the national schools, have been sent out by the Society, and others by the liberality of friends, they have not been sufficient to supply the different schools; so that many of the natives, who desire to possess them as their own, are still destitute. Framed slates are sometimes taken by traders, as articles of barter; but they are so liable to break, that the people greatly prefer the kind above alluded to.

A copy-book has never been used for the purpose of learning to write; paper has always been too scarce and valuable amongst them, to admit of such an appropriation. And a copy-book, although highly prized, is used rather as a journal, common-place-book, or depository of something more valuable than mere copies. Writing paper is still a very valuable article, and would prove one of the most acceptable presents that could be sent them.

I have often been amused on beholding a native, who had several letters to write, sitting down to look over his paper, and finding perhaps that he possessed but one sheet, has been obliged to cut it into three, four, or five pieces, and regulate the size of his letter, not by the quantity of information he had to communicate, but by the size of the paper he had to fill. I have recently received upwards of twenty letters from the natives, some of them, although they were to travel fifteen thousand miles, written on very small scraps of paper, and that often of a very inferior kind: part of the small space for writing being occupied by apologies for the small paper, and urgent requests, that if I do not return soon, I will send them some paper; and that if I return, I will take them a supply.

The art of writing is of the greatest service to the people in their commercial, civil, and domestic transactions, as well as in the pursuit of knowledge. They are not so far advanced in civilization as to have a regular post; but a native seldom makes a journey across the island, and scarcely a canoe passes from one island to another, without conveying a number of letters. Writing is an art perfectly congenial with the habits of the people, and hence they have acquired it with uncommon facility; not only have the children readily learned, but many adults, who never took pen or pencil in their hands until they were thirty or even forty years of age, have by patient perseverance learned, in the space of twelvemonths or two years, to write a fair and legible hand. Their comparatively small alphabet, and the simple structure of their language, has probably been advantageous; their letters are bold and well formed, and their ideas are always expressed with perspicuity, precision, and remarkable simplicity.*

The South Sea Academy, in which the young king was a pupil, is a most important institution, in connexion

* Writing apparatus and materials of every kind are in great demand among them; most of the letters I have received contain a request that, if possible, I will send them out a writing-desk, or an inkstand, penknife, pens, a blank paper book, &c. The youthful widow of Taaroarii has begged me to bring her a writing-desk.

with the Missionary establishments in this part of Polynesia. It had long been required by the circumstances of the European families, and the peculiar state of Tahitian Society; and the establishment of the Academy was designed to meet their peculiar necessities in this respect.

There are many trials and privations inseparable from the situation of a Christian Missionary among a heathen people. The latent enmity of the mind familiar with vice, to the moral influence of the gospel; the prejudices against his message, the infatuation of the pagan in favour of idolatry, and the pollutions connected therewith—originate trials common to every Missionary; but there are others peculiar to particular spheres of labour. The situation of a European in India, where, although surrounded by pagans, he yet can mingle with civilized and occasionally with Christian society, is very different from that of one pursuing his solitary labours, year after year, in the deserts of Africa, or the isolated islands of the South Sea, where they have been five years without hearing from England—where there is but one European family in many of the islands—and where I have been twelve or fifteen months without seeing a ship, or hearing a word of the English language, except what has been spoken by our own families.

There are disadvantages, even where the Missionary is in what is called civilized society, but they are of a different kind from those experienced from a residence among a rude uncultivated race. In either barbarous or civilized countries, the greatest trials the Missionaries experience are those connected with the bringing up of a family in the midst of a heathen population; and it probably causes more anxious days, and sleepless nights, than any other source of distress to which they are exposed. This was the case in the South Sea Mission. There were at one time nearly sixty children or orphans of Missionaries; and there are now, perhaps, forty rising up in the different islands, under circumstances adapted to awaken in their parents the most painful anxiety.

In the Sandwich Islands, during our residence there, although our hearts were cheered, and our hands strengthened, by the great change daily advancing among the people, yet the situation of our children was such as constantly to excite the most intense and painful interest. It is impossible for an individual, who has never mingled in pagan society, and who does not understand the language employed in their most familiar intercourse with each other, to form any adequate idea of the awfully polluting character of their most common communications. Their appearance is often such, as the eye, accustomed only to scenes of civilized life, turns away in pain from beholding. Their actions are often most repulsive, and their language is still worse. Ideas are exchanged, with painful insensibility, which cannot be repeated, and whose most rapid passage through the mind must leave pollution. So strongly did we feel this in the Sandwich Islands, that the only play-ground to which our children were allowed access, was enclosed with a high fence; and the room they occupied was one from which the natives, who were in the habit of coming to our dwelling, was strictly interdicted.

We were always glad to inspire the natives with confidence, and admit them to our houses, but when any of the chiefs came, they were attended by a large train of followers, whose conversation with our own servants we could not restrain, and which we should have trembled at our children listening to. The disadvantages under which they must have laboured, are too apparent to need enumeration. Idolatry had indeed been renounced, but, during the earlier part of the time we spent there, nothing better had been substituted in its place, and the great mass of the people were living without any moral or religious restraint.

Our companions, the American Missionaries, felt deeply and tenderly the circumstances of their rising families, and made very full representations to their patrons; they have also sent some of their children to their friends in their native country. The children of the Missionaries in the South Sea Islands were not in a situation exactly similar to those in the northern islands. The great moral and religious change that has taken place since the subversion of idolatry, had very materially improved the condition of the people, and elevated the tone of moral feeling among them; still it must be remembered, that though many are under the controlling influence of Christian principle and moral purity, these are not the majority, and there is not that fine sense of decency, which is a powerful safeguard to virtue; and, besides this, the circumstances of the families are far from being the most pleasing.

In only two of the islands is there more than one Missionary; and only at the Academy, where Mr. Blossom is associated with Mr. Orsmond, is there more than one family at a station. The duties of each station, from the partially organized state of society, and the multitude of objects demanding his attention, are such, that the Missionary cannot devote the necessary time to the education of his own children, without neglecting the public duties of his station; hence he experiences a constant and painful struggle between the dictates of parental affection and the claims of pastoral care. To relieve, as far as possible, from this embarrassment, the South Sea Academy was established by the deputation from the Society, and the Missionaries in the islands, in the month of March, 1824.

In compliance with the earnest recommendation of the deputation, and the solicitation of his brethren, Mr. Orsmond removed from Borabora, to take charge of the institution, and has continued to preside over it, to the satisfaction of the parents, and the benefit of the pupils. The first annual meeting was held in March, 1825; the children had not only been taught to read the Scriptures, and to commit the most approved catechisms to memory, but had also been instructed in writing, grammar, history, and the arts and sciences. During the examination, portions of scripture were read and recited, copy-books examined, problems in geometry worked, and parts of catechisms on geography, astronomy, and chronology, repeated. The whole of the proceedings gave satisfaction to all present, and left an impression on each mind, that great attention must have been paid by Mr. and Mrs. Orsmond to the scholars, during the short period they had been in the school. Subsequent examinations have been equally satisfactory.

The institution is under the management of a committee, and its primary design was to furnish a suitable, and, so far as circumstances would admit, a liberal education to the children of the Missionaries, "such an education as is calculated to prepare them to fill useful situations in future life." Native children of piety and talent have access to its advantages, and it is designed as preparatory to a seminary, or college, for the purpose of training up native pastors for the different stations in the South Sea Islands. It is certainly a most important institution, and will probably exert no ordinary influence on the future character of the nation at large, as well as prove highly advantageous to the individuals who become its inmates. It merits the countenance of the friends of Missions. Several individuals have kindly enriched its library with suitable elementary books, philosophical apparatus, &c. but these are still very inadequate to the accomplishment of the design contemplated.

While the establishment of this institution is a just occasion of gratitude to the Missionaries, it by no means removes all anxiety from their minds with regard to the future prospects of their families. The nature of their station, and the spirit and principles of their office as ministers of Christ, prevent the parents from making any provision for their families. The comfortable settlement of their children is an object of most anxious solicitude to Christian parents at home—to foreign Missionaries it is peculiarly so. Their remote and isolated situation precludes their embracing those openings in Divine Providence for placing their children in comfortable circumstances, of which they might avail themselves in Christian and civilized society. The prospect of filling comfortable stations there, are all uncertain; professions there are none; commerce is in its infancy, as will appear from the fact of its being still carried on by exchange or barter. The circulation of money is very limited, and its use known to but few.

The fondest hope of every Missionary is, that his children may grow up in the fear of God, be made partakers of his grace, and, under the constraining influence of the love of Christ in their hearts, imbibe their parent's spirit, select his office, spend their lives in supplying his lack of service, and carry on that work which he has been honoured to commence. In prosecuting this, they will have advantages their parents never possessed, they will have been identified with the people among whom they labour, and will not appear in language and idiom as foreigners; but they will labour under more than counteracting disadvantages, if they never visit the land of their fathers, and must necessarily be far less efficient teachers of the truths of Christianity than their predecessors in the work.

There are a thousand things known to an individual who has received or finished his education and passed his early days in England, which can only be known under corresponding circumstances, and which a Missionary can never, in such situations as the South Sea Islands, teach his child. Those born there may indeed have access to English literature; but many books, however familiar and perspicuous to an ordinary English reader, will, in many perhaps important parts, appear almost enigmatical to those who have never seen any other society than such as that now under consideration. It has always appeared to me, in reference to a rude, uncivilized, illiterate people, who are to be raised from ignorance, barbarism, and idolatry, to a state of intelligence, enjoyment, and piety—where their character, habits, taste, and opinions, have to be formed principally, if not entirely, by the Missionary—that for some generations, at least, every Missionary's child, trained for the Missionary work even by a father's hand, and blessed with the grace of God, ought to finish his education in the land of his parents, prior to entering upon the work to which his life is devoted.

Many a Missionary spends the greater part of his life without being able to produce any powerful or favourable impression upon the people among whom he has laboured; others expire in a field, on which they have bestowed fervent prayer, tears, and toil, but from which no fruit has been gathered; the second generation have to commence under circumstances corresponding with those under which their predecessors began. When success attends their efforts, and a change takes place decisive and extensive as that which has occurred in the South Sea Islands; yet so mighty is the work, so deep the prejudices, so difficult to be overcome are evil habits, and so slow the process of improvement upon a broad scale, even under the most favourable circumstances, that the ordinary period of a Missionary's life in actual service, would appear too short to raise them from their wretchedness, and elevate them to a standard in morals, habits, intelligence, and stability in religion, at which those who were instrumental in originating their emancipation, would like to leave them. They never can be expected to advance beyond those who are their models, their preceptors, and their guides; and if the successors of the first Missionaries be in any respect inferior to their predecessors, the progress of the nation must, in regard to improvement, be retrograde—unless this deficiency be supplied from some other source.

On this account, it does appear exceedingly desirable that the successors to the first Missionaries among an uncivilized people, who may even renounce idolatry, should be in every respect equally qualified for this office with those by whom they were preceded, and that even the children of the Missionaries should be able to carry on, to a greater degree of perfection, that work which their parents were privileged to commence.

I am aware that the expense attending a measure of this kind will probably prevent its adoption in those Institutions by whom the first Missionaries are sent out; but this does not render the measure less desirable or important in its immediate or more remote and permanent influence upon the nations converted from idolatry. The same difficulties occur with regard to the promotion of civilization, and the culture of the mechanic arts, among the barbarous nations. The primary design of all Missionary contributions is the communication of Christianity to the heathen; and it is to be regretted that the smallest portion of the pecuniary means furnished by Christian liberality for this purpose, should be appropriated to any other purpose than the direct promulgation of the gospel. It has been already stated, that a sort of Civilization Society, an institution for the purpose of promoting intellectual culture, scientific pursuits, agricultural and mechanical arts, and advancing all that we understand by civilization, among the barbarous and unenlightened nations of the earth, would be highly advantageous. The agents of such an institution would merit the sanction and support, not only of every Christian, but of every friend to improvement, virtue, and humanity. They might remain distinct, and yet co-operate harmoniously with the Christian Missionary: the one directing his attention to present circumstances, the other principally to the future destinies of those to whom they were sent.

As it is, however, the Missionary Societies, in reference to unenlightened nations, where any measure of success attends their efforts, have no alternative, but are, from circumstances, under the necessity of attending to these departments of effort. They must either supply the apparatus and sustain the heavy expense of carrying on the work of civilization, or leave those on whom they have been instrumental in bestowing the light of revelation a prey to indolence, or to unprincipled individuals, whose influence will be exerted to neutralize the advantage Christian instruction may have conferred.

Christianity and civilization ought never to be separated, and although we rejoice in the temporal advantages which follow the former, by the introduction of the arts and comforts of society, it is to be regretted that the Missionary Societies should be prevented from sending the gospel to waiting nations, by the drafts made upon their resources for the establishing and maintaining agencies for the purpose of attending only to temporal concerns. Civilization never precedes, but invariably follows Christianity, and until some other means of facilitating its progress be supplied, it will not be neglected by those who are employed in the propagation of the gospel throughout the world. The difficulties already alluded to, connected with the Missionary stations, are not the only ones that exist. They would operate powerfully, supposing the children were all that the parents could wish; supposing they were qualified by talent, disposed by deliberate choice, and prepared by Divine grace, for the work of a Christian Missionary; but these indispensable requisites, it is unnecessary to remark, a parent, with all his solicitude and care, cannot always secure. God may see fit to withhold those decisive evidences of genuine piety, without which the fondest parent would tremble at the idea of introducing even his own child into the sacred office of an evangelist. However Missionary pursuits may have been accounted the honour, or have proved the happiness, of the parent, the child, as he grows up, may not even possess a desire to engage in the same: that desire the parent cannot give; and without it, it would be both cruel and injurious to every party to urge it.

The alternative is most distressing to contemplate. There are at present no situations of comfort to fill, no trade or business that can be followed. Productive plantations, regular labour, mercantile establishments, warehouses, and shops, it is to be expected, will ultimately exist and flourish in these islands, but they cannot be looked for in the short period of fifteen years from the time when the people emerged from the grossest ignorance, the most inveterate vices, and the most enervating and dissipating idleness. The circumstances of the female branches of the Mission families is, perhaps, still more discouraging.

I have extended these remarks much beyond what I intended, when speaking of the South Sea Academy; and although they may be less interesting to the general reader than other matters, they will serve to shew what are some of the heaviest trials of a Missionary life among an uncivilized people; and may not only awaken the sympathies of the friends of Missionaries, but lead to such a consideration of the subject, as may result in the suggestion or application of a remedy, which, if it shall not altogether remove them, shall, at least, alleviate their pressure; which is, perhaps, felt more heavily by the present generation, than it will be by their successors.

CHAP. XIX.

Voyage to Borabora—Appearance of the settlement—Description of the island—Geological peculiarities of Borabora, Maurua, &c.—New settlement in Raiatea—Arrival of the Dauntless—Designation of native Missionaries—Voyage to the Sandwich Islands—Marriage of Pomare and Aimata—Former usages observed in marriage contracts—Betrothment—Ancient manner of celebrating marriage—Resort to the temple—Address of the priest—Proceedings of the relatives—Prevalence of polygamy—Discontinued with the abolition of idolatry—Christian marriage—Advantageous results—Female occupations—Embarkation for England—Visit to Fare—Improvement of the settlement—Visit to Rurutu and Raivavai—Propagation of Christianity by native converts—Final departure from the South Sea Islands.

MR. ORSMOND, who removed to Raiatea in the close of the year 1818, was accompanied by Mrs Orsmond, who, in the communication of useful instruction to her own sex, and in every other department of female Missionary labour, was indefatigable, until her decease, which took place very soon after her removal from Huahine.

In November 1820, nearly two years after this, Mr. Orsmond, in compliance with the urgent request of the chiefs and people, removed to the island of Borabora, where he established a christian mission, and continued his valuable labours till required by the united voice of the Missionaries, in the Windward and Leeward Islands, to take charge of the South Sea Academy, founded at Eimeo, in 1824. During the year 1821, the inhabitants of Borabora erected a substantial place of worship; and in the month of January 1822, according to a previous engagement with Mr. Orsmond, I visited the island, for the purpose of preaching at the opening of the new Chapel. Indisposition detained Mr. Bennet at Huahine, but the late Rev. D. Tyerman, his colleague, kindly accompanied me.

On reaching the settlement, to which the people had given the appellation of Beulah, we were gratified no less with the warm reception we experienced, than with the evident improvement among the inhabitants. The school was regularly attended, and many were well informed in the great truths of revelation; the observance of the sabbath was strictly regarded; four or five neat plastered houses were finished, others were in progress. Three causeways, upwards of six feet wide, and elevated two or three feet above the water, extended about three hundred and sixty feet into the sea, and united at the extremity. The chapel, which was one of the best that had been erected in the islands, was part of a large building one hundred and sixty feet by forty-eight, comprising a place of worship, school, and court-house.

Contrary winds detained us some days in the pleasant settlement at the head of Vaitape bay, on the west side of the island, which is situated in 16° 32 S. Lat. and nearly 152° W. Long. Borabora, as well as the other islands of the group, is surrounded by a reef rising to the water's edge, at unequal distances from the shore. On this reef there are three low coral islands covered with trees and verdure, equal to that which adorns those around Raiatea and Tahaa. There are also four other islands separated from the main land, which is about sixteen miles in circumference. These islands, like Papeorea in Huahine, are not of coral formation, but resemble in structure the promontories on the adjacent shore. Tobua, the principal, forming the south or west side of Vaitape bay, is not less than three or four hundred feet above the sea.

In the geology of Borabora, the only peculiarity is the existence of a species of feldspar and quartz, but the appearance and shape of the island is singular and imposing. The high land in the interior is not broken into a number of small mountain ridges, but, uniting in one stupendous mass, rears its magnificent form, which resembles a double-peaked mountain, to an elevation perhaps little below 3000 feet above the water. The lower hills and small islands are not seen at a distance, so that when viewed from the sea or the other islands, especially Huahine, (from the north and western parts of which it is generally visible,) it appears like one solitary and gigantic obelisk or pyramid rising from the ocean and reaching to the clouds.

The settlement at the head of Vaitape bay commands a view of every diversity in scenery. The lofty interior mountain clothed with verdure, and the deep glens that indent its sides, stand in pleasing contrast with the hilly or coralline islands that appear in the west, while the uniformity and nakedness of the distant horizon is broken by the appearance of the conical or circular summits of the mountains of Maupiti or Maurua, upwards of thirty miles distant. This island was frequently visible from Borabora, during our visit at this time.

Maupiti is but circumscribed in extent, and its mountains are less broken and romantic than those of others in the group; it has, however, some peculiarities. It is the only place in the Georgian or Society Islands in which primitive formations are found to any extent. Besides the cellular volcanic rock and the several kinds of basalt, common to all the islands, a species of granite is found here in considerable abundance, which presents an anomaly as striking in the geology of these islands, as that furnished by the existence of carbonate of lime in the island of Rurutu, where garnets are also obtained. Hornblende and feldspar are found in Huahine, as well as in some of the other islands. Ancient lava, containing olivine, augite, and zeolite, are also met with, together with pumice and cellular lava, some kinds of which, found in Sir Charles Sanders' Island, are of a dark blue colour, and so light as to float on the water, though apparently containing a portion of iron. A large specimen of the latter kind, which I have from this island, is more porous than any I ever met with among the volcanoes of the Sandwich Islands, and is so completely honeycomb in its structure, that it is difficult to account for its formation.

After remaining some time at Borabora, we took leave of our friends, and sailed for Huahine.

On our way we touched at Raiatea, and were gratified with the prosperous appearance of the station. It was then at Vaoaara, but since that period Mr. Williams, the only remaining Missionary, has removed to *Utumaoro*, a fine extensive district near the northern extremity of the island, and adjacent to the opening in the reef called the *Avapiti*, or double entrance. This station was commenced in 1823; and, in consequence of the extent of land by which it is surrounded, and the proximity of the harbour, has been found much more convenient than that formerly occupied. The improvement has been rapid, and the transformation so astonishing, that in a short period, three hundred enclosures for the culture of sugar, coffee, and tobacco, with other kinds of produce, were completed; a substantial place of worship, schools, and a house for the Missionaries, had been finished, and the neat plastered dwellings of the natives extended for two miles along the beach. The scenery of this district of the island is much less picturesque than in many other parts; yet it is impossible to behold the neat and extensive settlement, with its gardens, quays, schools, capacious chapel, and cottages, stretching along the shore, which but a few years before was covered with brush-wood and trees, without astonishment and delight. The accompanying plate, from a drawing taken on the spot, although it exhibits the general outline of the station, will convey but a faint representation of the interesting features of the Missionary's settlement, or the surrounding country.

On the twentieth of January, shortly after our return from Borabora, his Majesty's ship Dauntless, commanded by Capt. G. C. Gambier, touched at Huahine. We were happy to introduce the commander of the Dauntless, Capt. R. Elliot, and the officers of the vessel, to the governor and chiefs of the island, and to welcome them to our humble dwellings, as well as to experience their hospitality on board. The recollection of the polite and kind attentions of Captain Gambier, Captain Elliot, and other gentlemen of the ship, is still grateful to the Missionaries and the inhabitants of Huahine.

In a week or two after the departure of the Dauntless, the colonial government cutter Mermaid arrived in Fare harbour, on her way to the Sandwich Islands, with a small schooner, the Prince Regent, as a present from the British government to the king of those islands. The captain intimated his intention of touching at the Marquesas on his return from Hawaii, and politely

Drawn by F. Finch. Engraved by J. Davies.

SETTLEMENT AT UTUMAORO, IN RAIATEA.

offered a passage to any of us who might be desirous of visiting these islands. We had long been anxious to introduce Christianity among the inhabitants of the former, and as the present appeared a favourable opportunity, we communicated the same to the deputation. It appeared to them desirable to visit these places, and the captain's offer was accepted. Our views were then made known to the members of the church, and we proposed to them to send some of their own number, to introduce the gospel to the Marquesians. They approved of the proposal, and a public meeting was held soon afterwards. The duty or desirableness of communicating Christian instruction to the heathens around, had been discussed before, both in public and private, and some had expressed their desires to engage in this work; two pious and intelligent men, with their wives, now offered to go, and, after being approved, they were appointed by the church to this important enterprise.

The arrangements for the voyage being completed, we assembled at the chapel about ten o'clock on the forenoon of the 24th of February; the native Christians were animated by kind and appropriate addresses from the church, and were affectionately encouraged by Mr. Barff and Mr. Orsmond, the latter being on a visit with us. The native Missionaries then took leave of their fellow Christians in a most solemn and impressive manner; and, as it had been arranged by Mr. Barff and myself that I should accompany them, to aid in the commencement of their labours, I addressed the people, and, recommending Mrs. Ellis and our dear children to their kind attentions under God, I took leave of them. The meeting was peculiarly impressive and affecting; and after mutually committing each other, under deep intensity of feeling, to the guidance and the keeping of the God of all our mercies, the whole congregation walked from the chapel to the sea-shore, where we exchanged our last salutations; the deputation, the two native Missionaries and their wives, five other natives and myself, now embarked, and the Mermaid stood out to sea.

The weather was on the whole pleasant, and we reached the Sandwich Islands in about a month after our departure from Huahine.

While supping at our table, on the night previous to our embarkation, the captain had, in answer to Mrs. Ellis's inquiries, assured her that he expected to return in three months; but seven months passed without any appearance of our vessel. In the mean time, a piratical ship touched at Huahine; some of the pirates absconded, and remained on shore. It was found that they knew something of our vessels; but as they refused to say what they knew, surmises arose, and reports were circulated that they had met us at sea, and either sunk our vessel or murdered the passengers. Such was the influence of this report when first circulated, that it was necessary to protect the deserters from the indignation of the populace. The whole of their statements was invested with a degree of mystery, which, together with the very protracted period of our absence, augmented the distress of Mrs. Ellis and our friends in Huahine. From this painful state of anxious uncertainty, they were however relieved by the appearance of the Mermaid off Fare harbour early in the month of October, and by our landing in health and safety in the evening of the same day. The pirates had fallen in with the schooner, which had been separated from us during the early part of the voyage; they by this means heard of our destination, &c. and this partial information accounted for the vagueness of their reports. In the close of the same month, the invitation I had received from the chiefs in the Sandwich Islands, and the American Missionaries, to remove thither, was submitted to the consideration of the Missionaries in the Leeward Islands, and they, with the deputation, were unanimous in opinion, that we ought to proceed to that important station by the earliest opportunity.

The Active, a small schooner, commanded by Captain Charlton, arriving at Huahine soon after, was engaged to convey us to the Sandwich Islands. While we were preparing for our departure, viz. in the month of December, 1822, a marriage took place between Pomare, the young chief of Tahaa, and Aimata, the only daughter of the late king of Tahiti. The parties met at Huahine, which was midway between the residence of the families to which they respectively belonged. Young Pomare had received his name, as a mark of special favour from the king of Tahiti.

More than a week before his intended bride arrived from Tahiti, Pomare sailed from Tahaa, and landed at Fare, where he was entertained with the attention and respect suited to his rank and prospects, by the chiefs of Huahine. It was not, however, at that time supposed that his consort would become the queen of Tahiti, as her brother, with whom her father had left the government, was then living.

In the month of December, Aimata, accompanied by her mother and aunt, arrived at Huahine, on board the Queen Charlotte, a brig belonging to the king. The afternoon of the day on which the vessel anchored was fixed on for her landing, and introduction to her future husband. We walked down to the settlement, to witness the meeting of the youthful pair. A small open house, belonging to the governor, was the place appointed for their first interview. When we reached the spot, we beheld the young chieftain, who, for his age, was remarkably stout, dressed in full native costume, with a large purau, and a flowing tiputa; he wore, also, an English beaver hat. He was seated at one end of the building on an *iri*, or native seat, waiting with gravity of appearance the arrival of Aimata.

About a quarter of an hour after we had reached the place, two or three boats from the vessel rowed towards the shore. Several of the attendants of the young princess arrived in the first; and the queen and her sister, with the youthful Aimata, landed from the second. The visitors were met on the beach by the governor of the island, and a number of chief women, who conducted Aimata to the house where Pomare and his friends were waiting. They entered, and, after greeting the friends present, took their seats near where the young chief was sitting.

Pomare continued motionless, neither rising to welcome his guests, nor uncovering his head. Aimata sat close by her mother's side, occasionally glancing at the individual who was to be her husband, and who sat like a statue before her.

This was the first time either Pomare or Aimata had seen each other, and the interview was certainly a singular one; for, after sitting together for about twenty minutes, the queen and her companions rose, and repaired to the house provided for their accommodation, and Pomare and his friends returned to their encampment. During the whole of the time they had been in each other's company, they had not exchanged a single word.

Shortly after this meeting, they were publicly married, and afterwards removed to the island of Tahiti, which has ever since been their principal residence. Pomare was about sixteen years of age, and his consort but little, if any, younger. Since the death of her brother, which took place in 1827, she has been considered queen of Tahiti, Eimeo, &c., though the regency, appointed to govern the islands during the minority of the late king, still manages the political affairs, acting, however, in the name of Aimata, instead of that of her brother.

Pomare was very young when the inhabitants of his native island embraced Christianity; the first time we saw him was in 1819, when he appeared nine or ten years of age. His establishment, however, was at that time nearly as large as it has been since. He possessed a number of houses in different parts of Raiatea and Tahaa, and was surrounded by a numerous train of attendants; one or two chiefs of rank and influence, acting as his guardians, usually accompanied his movements. During the early parts of his life, he was frequently carried about on men's shoulders, according to the ancient custom of the kings of the Society Islands. When the king of Tahiti embraced Christianity, this, with other practices connected with idolatry, was laid aside in the Windward Isles. It was occasionally adopted by the young chief of Tahaa, more, perhaps, to gratify the pride of some of his attendants, than to afford any satisfaction to his own mind. By him it has now been discontinued for a number of years, and young Pomare is probably the last Tahitian chieftain that will ever ride in state on the necks of his people.

Aimata, the only surviving child of the king of Tahiti, although about the same age, appeared in perfect contrast to her husband. Her form was neither athletic nor corpulent, her countenance open and lively, her jet-black eye sparkling and intelligent, her manners and address engaging, her disposition volatile, and her conversation cheerful. In these respects she was the very opposite to Pomare, who was taciturn and reserved.

She gave early indications of superior intellectual endowments; and, had her mental faculties been properly cultivated, she would probably have excelled most of her own sex in the society in which she was destined to exert the highest influence. The restraint and application, however, which this required, were ill suited to her lively disposition, and uncontrolled habits of life. She has, nevertheless, been a frequent, and, while she continued, a promising pupil of the Missionaries, having, in a short time, made a pleasing progress in the acquisition of knowledge. She has for some time made a profession of Christianity, and her conduct has generally been in accordance with the same. To the Missionaries she has invariably proved friendly; and, since she has been the queen of Tahiti, has patronized and encouraged their efforts.

Pomare and Aimata had been by their respective families betrothed to each other for some time prior to their meeting in Huahine. Considerable preparations had been made for the celebration of the marriage, and as the parties were nearly related to the reigning families in the Windward and Leeward Islands, arrangements were made for entertainments corresponding with the rank and dignity of Pomare and his bride.

About noon on the day appointed, the young chieftain

with his guardian and friends reached the chapel, where we were waiting to receive them. Aimata, attended by her mother-in-law, the queen of Tahiti, her sister, and the wife of Mahine, chief of Huahine, arrived shortly after. The royal party were attended by the dependants of Hautia, the governor of the island. In honour of the distinguished guest, these dependants or guards were not only arrayed in their best apparel, which was certainly any thing rather than uniform, but they also marched under arms. Many of the raatiras of Huahine attended, out of respect to the reigning family.

When the ceremony commenced, Mr. Barff and myself took our station near the communion table in front of the pulpit; Pomare and his friends standing on our right, and Aimata with her relatives on the left. The raatiras formed a semicircle three or four deep immediately behind the bride and bridegroom, while the body of the chapel was filled with spectators. Most of the chiefs appeared in European dresses, some of which being large loose gowns of highly glazed chintz of a brilliant red and yellow colour, intermixed with dresses of black and blue broad cloth, presented a novel spectacle.

The principal part of Pomare's dress was manufactured in the islands, and worn after the ancient fashion. Aimata wore a white English gown, a light pink scarf, and a finely platted hibiscus bonnet trimmed with white ribands. The queen, Pomare-vahine, and all the females of the royal party, appeared in white dresses of foreign manufacture. The raatiras wore the native costume peculiar to their rank and station, while the dress of the multitude behind them presented almost every variety of European and native clothing.

The rich and showy colours exhibited in the apparel of the chiefs, the uniform white raiment of the queen and her companions, in striking contrast with their olive-coloured complexions and dark glossy curling hair, presented an unusual appearance. The picturesque dress of the raatiras, who wore the purau or beautifully fine white matting tiputa, bordered round the neck and the edges with a most elegant fringe, and bore in the right hand a highly polished staff, or kind of halbert, of black iron-wood, together with the diversified appearance of the spectators, greatly increased the novel and imposing effect of the whole.

During the ceremony, I observed a tear moistening the eye of the youthful bride. Agitation of feeling, perhaps, produced it, as I have every reason to believe no cloud of anticipated evil overshadowed her prospects; and she is reported to have said, that had she not been betrothed, but free to choose her future partner, she should have selected the individual her friends had chosen for her.

When the service was over, the registry made, and the necessary signatures affixed, the parties returned, to partake of the entertainment provided. We were invited to join them, but declined the honour; yet walked down to see the preparation, and, among other articles of dessert, noticed two barrels full of pine-apples. As soon as the ceremony was concluded, the governor's guards, who were drawn up on the outside of the chapel, fired several volleys of musketry, and a British, vessel lying in the harbour, saluted them with twenty-one guns.

Betrothment, as in the instance of Pomare and Aimata, was the frequent method by which marriage-contracts were made among the chiefs, or higher ranks in society. The parties themselves were not often sufficiently advanced in years to form any judgment of their own, yet, on arriving at maturity, they rarely objected to the engagements their friends had previously made.

The period of courtship was seldom protracted among any class of the people; yet all the incident and romantic adventure that was to be expected in a community in which a high degree of sentimentality prevailed, was occasionally exhibited, and the disappointed, perhaps, led to the commission of suicide, under the influence of revenge and despair. Unaccustomed to disguise either their motives or their wishes, they generally spoke and acted without hesitation; hence, whatever barriers might oppose the union of the parties, whether it was the reluctance of either of the individuals, or of their respective families, the means used for their removal were adopted with much less ceremony than is usually observed in more civilized society. Several instances of this kind occurred during our residence in Huahine: one regarded a chief of Eimeo, attached to Taaroarii the king's son. His figure was tall and gigantic, his countenance and manners not unpleasing, and his disposition mild and humane. He was upwards of twenty years of age. Some time after our arrival in Huahine, he became attached to the niece of the principal raatira in the island, and tendered proposals of marriage. Her family admitted his visits, and favoured his design, but the object of his choice declined every proposal he made. No means to gain her consent were left untried, but all proved unavailing. He discontinued his ordinary occupations, left the establishment of the young chief who had selected him for his friend, and repaired to the habitation of the individual whose favour he was so anxious to obtain. Here he appeared the subject of deepest melancholy, and, leaving the other members of the family to follow their regular pursuits, from morning to night, day after day he attended his mistress, performing, with apparent satisfaction, offices of humiliating servitude, and constantly following in her train whenever she appeared abroad.

His friends interested themselves in his behalf, and the disappointment, of which he was the subject, became for a time the topic of general conversation in the settlement. At length she was induced to accept his offer; they were publicly married, and lived very comfortably together. Their happiness, however, was but of short duration, for his wife, for whom he appeared to cherish the most ardent affection, died a very few months afterwards.

Another instance of a rather different kind, subsequently occurred. A party of five or six persons arrived in a canoe from Tahiti, on a visit to their friends in the Leeward Islands. Though Borabora was their destination, they remained several weeks at Huahine, the guests of Taraimano. During this period, a young woman, one of the belles of the island, belonging to the household of their hostess, became exceedingly fond of the society of one of the young men, and it was soon intimated to him that she wished to become his companion for life. The intimation, however, was disregarded by the young man, who expressed his intention to prosecute his voyage. The young woman became unhappy, and made no secret of the cause of her distress. She was assiduous in redoubling her efforts to please the individual whose affection she was desirous to obtain. At this period I never saw him either in the house of his friend, or walking abroad, without the young woman by his side.

Finding the object of her attachment, who was probably about eighteen years of age, unmoved by her attentions, she not only became exceedingly unhappy, but declared, that if she continued to receive the same indifference and neglect, she would either strangle or drown herself. Her friends endeavoured to dissuade her from her purpose; but, as she expressed her determination to be unaltered, they used their endeavours with the stranger, who afterwards returned the attentions he had received, and the parties were married at Huahine. His companions pursued their voyage to Borabora, and afterwards returned to Tahiti, while the new-married couple continued to reside with Taraimano. Their happiness was of short duration; not that death dissolved their union, but that attachment, which had been so ardent in the bosom of the young woman before marriage, was superseded by a dislike as powerful; and although I never heard the slightest charge of unkindness preferred against the husband, his wife not only treated him with insult, but finally left him. Instances of such unhappy marriages, though not unusual formerly, are now of rare occurrence.

It is only among the middle and lower ranks of society, that the contract is made by the parties themselves. I am not aware that the husband received any dowry with his wife, unless the rank of her family was inferior to that of his own. The suitor often made presents to the parents of the individual whom he wished to marry, in order to gain their consent.

Among the higher ranks, the individuals themselves were usually passive, and the arrangements were made by their respective friends. They were usually betrothed to each other during their childhood, and the female thus betrothed was called a *vahine pahio*. As she grew up, a small platform, of considerable elevation, was erected for her abode, within the dwelling of her parents. Here she slept, and spent the whole of the time she passed within doors. Her parents, or some member of the family, attended her by night and by day, supplied her with every necessary, and accompanied her whenever she left the house.

When the time fixed for the marriage arrived, and the parties themselves agreed to the union, great preparations were made for the dances, amusements, and festive entertainment, usual on such occasions. A company of Areois generally attended, and, on the day preceding the nuptials, commenced their *upaupa*, or dance, and pantomimic exhibitions.

On the morning of the marriage-day, a temporary altar was erected in the house of the bride. The relics of her ancestors, perhaps their skulls or bones, were placed upon it, and covered with fine white native cloth; presents of white cloth were also given by her parents, and those relatives of the family who attended.

The sanction of the gods they considered essential to the marriage contract. and these preliminaries being adjusted, the parties repaired to the *marae*, or temple. The ceremony was generally performed in the family marae, excepting when the parties were connected with the reigning family, which rendered it necessary that it should be solemnized in the temple of Oro or of Tane, the two principal national idols. On entering the temple, the bride and bridegroom changed their dresses, and arrayed themselves in their wedding garments, which were afterwards considered sacred; they took their stations in the place appointed for them, the bride on one side of the area, and the bridegroom on the other, five or six yards apart.

The priest now came forward, clad in the habiliments of his office, and, standing before them, addressed the bridegroom usually in the following terms: *Eita anei oe a faarue i ta oe vahine?* "Will you not cast away your wife?" to which the bridegroom answered, *Eita;* "No." Turning to the bride, he proposed to her the same question, and received a similar answer. The priest then addressed them both, saying, "Happy will it be, if thus with ye two." He then offered a prayer to the gods in their behalf, imploring for them that they might live in affection, and realize all the happiness marriage was designed to secure.

The relatives now brought a large piece of white cloth, which they call *ahu vauvau*, spreading cloth: it was spread out on the pavement of the marae. The bridegroom and bride took their station upon this cloth, and clasped each other by the hand. The skulls of their ancestors, which were kept carefully preserved by the family, who considered the spirits of the proprietors of these skulls as the guardian spirits of the family, were sometimes brought out and placed before them.

The relatives of the bride then took a piece of sugar-cane, and, wrapping it in a branch of the sacred mero, placed it on the head of the bridegroom, while the new-married pair stood holding each other's hands. Having placed the sacred branch on the bridegroom's head, they laid it down between them. The husband's relatives then performed the same ceremony towards the bride. On some occasions, the female relatives cut their faces and brows with the instrument set with shark's teeth, received the flowing blood on a piece of native cloth, and deposited the cloth, sprinkled with the mingled blood of the mothers of the married pair, at the feet of the bride.

By the latter parts of the ceremony, any inferiority of rank that might have existed was removed, and they were considered as equal. The two families, also, to which they respectively belonged, were ever afterwards regarded as one. Another large piece of cloth, called the *tapoi*, covering, was now brought, and the ceremony concluded by the relatives throwing it over the bridegroom and bride.

The cloth used on these occasions, as well as the dress, was considered sacred, and was taken to the king, or appropriated to the use of the Areois. The parties returned to their habitation, where sumptuous feasting followed, the duration of which was according to the rank or means of the families thus united.

Such were the marriage ceremonies formerly observed among the inhabitants of the South Sea Islands. There was much in them curious and affecting, especially in the blood of their parents, and the skulls of their ancestors, presented before the parties. The one, perhaps, as the emblem of their union, and the other as witnesses of the agreement. Considering these, and the other significant usages, it is truly surprising how a people, so uncivilized and rude as in many respects they certainly were, should ever have instituted observances so singular and impressive, in connexion with the marriage contract.

Notwithstanding all this ceremony and form in entertering into the engagement, the marriage tie was probably one of the weakest and most brittle that existed among them; neither party felt themselves bound to abide by it any longer than it suited their inclinations and their convenience. The slightest cause was often sufficient to occasion or to justify their separation, though among the higher classes the relation was nominally continued long after it had actually ceased.

Polygamy was practised more extensively by the Tahitians than by the inhabitants of the Sandwich Islands, and probably prevailed to as great an extent among them as among any of the Polynesian tribes. Many of the raatiras, or inferior chiefs, had two or three wives, who appeared to receive an equal degree of respect and support. With the higher chiefs, however, it was different; although they might, like Hamanemane, keep a number of females, it was rather a system of concubinage, than a plurality of wives, that prevailed among them. The individual to whom the chief was first united in marriage, or whose rank was nearest his own, was generally considered as his wife, and, so long as she lived with her husband, the other females were considered as inferior. When the rank of the parties was equal, they often separated; the husband took other wives, and the wife other husbands; and if the rank of the wife was in any degree superior to that of her husband, she was at liberty to take as many other husbands as she pleased, although still nominally regarded as the wife of the individual to whom she had been first married.

With the abolition of idolatry all the ceremonies originally performed at the temple were discontinued, and, shortly after the reception of Christianity by the nation, Christian marriage was instituted, and it is now universally observed. From this moral revolution some perplexing questions relative to polygamy have naturally arisen, but for the principal difficulties, the code of laws inserted in a preceding chapter has made suitable provisions.

In the marriage ceremony, the use of the ring has not been introduced, and the only distinction that prevails in society, in reference to married and unmarried females, is, that the wife ceases to be called by her original name, and is designated by that of her husband: excepting where the name of the wife was also an hereditary title of rank or honour, in which case it is retained.

No change in their customs or usages has taken place in connexion with the introduction of the religion of the Bible, more extensive or beneficial in its influence on every class in society, than the institution of Christian marriage. Instances of unfaithfulness are not indeed unknown, but, considering their former habits of life, the partial influence of regard to character, and the slight inconvenience in reference to the means of support, by which they would probably be followed, they have but seldom occurred. The solemn and indissoluble obligations of the marriage vow are recognized by all who profess to be Christians; and the domestic, social, and elevated happiness it has imparted, is readily acknowledged. It has entirely altered the tone of feelings, and imparted new principles of conduct in regard to the conjugal relation.

Originating from the institution of marriage, and nurtured by its influence, domestic happiness, though formerly unknown even in name, is now sedulously cultivated, and spreads around their abodes of order and comfort, its choicest blessings. The husband and the wife, instead of promiscuously mingling with the multitude, or dwelling in the houses of their chiefs, live together in the neat little cottages reared by their own industry, and find satisfaction and comfort in each other's society. Every household virtue adorns their families; the children grow up the objects of their mutual affection, and call into exercise new solicitudes and unwonted emotions of delight. Often they appear sitting together reading the Scriptures, walking in company to the house of God, or surrounding, not indeed the family hearth, or the domestic fireside, which in their warm climate would be no addition to their comfort, but the family board, spread with the liberal gifts of divine bounty. The father at times may also be seen nursing his little child at the door of his cottage, and the mother sitting at needlework by his side, or engaged in other domestic employments. These are the delights it has imparted to the present race—while the rising generation are trained under the influence of the principles of Christianity, and these examples of social and domestic virtue.

Marriages frequently take place at an early age among the people; they do not, however, appear to be less happy than those celebrated when the parties are further advanced in life. In former times the men were often cruel in their treatment of the women, and considered them as their slaves; but the husbands now treat their wives with respect, and often cherish for them the most sincere affection. The female character is elevated in society; the husbands perform the labours of the plantation or the fishery, recognizing it as their duty to provide the means of subsistence for the family; while the preparation of their food, (especially where the European mode of living has been adopted by them,) together with attention to the children, and the making of clothing, native or foreign, for themselves and the other members of the family, is now considered the proper department of the females. They occasionally accompany their husbands and elder children to work in the plantation or garden, at particular seasons of the year; but it is a matter of choice, and not from fear of cruel treatment, as formerly. They go to assist their husbands in planting or gathering in the crops, instead of undertaking alone these labours, while the men were idling away the noon-day hours in heedless slumbers, or spending them in songs or other amusements.

The establishment of schools has in a great degree overcome their love of wandering, and habituated them to regularity and perseverance in their occupations, although at first found irksome and difficult. Desire of mental improvement, general acquaintance with writing, and fondness for epistolary correspondence, furnish new and agreeable occupation for their leisure hours. The introduction of needlework, the universal desire for European clothing, together with the preservation of these articles of dress, having increased their domestic duties, occupies a great portion of their time.

With the close of the year 1822, we terminated our regular labours in the South Sea Islands; and on the 31st of December, soon after the marriage of Pomare and Aimata, accompanied by two native teachers, Taua and his family, and Taamotu, a female who had been a member of the church, a teacher in the school, and an affectionate and valuable companion and assistant to Mrs. Ellis during my voyage to Huahine, we embarked in the Active, and reached Oahu on the 5th of the following February. The result of my observations, and the detail of a part of my labours, prosecuted in affection and harmony with the American Missionaries, have been already published. Towards the close of 1824, an afflictive dispensation of Divine providence removed us from these islands. This was, the severe and protracted illness of Mrs. Ellis; the only hope of whose life was derived from the effects of a voyage to England. On our return we visited Huahine, anchored in Fare harbour, and had the high satisfaction of spending a fortnight in delightful intercourse with our Missionary friends, and the kind people of the settlement.

Early in the month of November we again took leave of our friends and fellow-labourers, hoping to revisit them when we should return to the Pacific; feeling, at the same time, that, with regard to some, perhaps many, we should not meet again in this world; but cheered with the anticipation of meeting in a region where parting would be unknown. When our anchor was raised, and our sails spread, the vessel moved slowly out of the harbour. The day was remarkably fine, and the wind light, and both these afforded opportunities of leisurely surveying the receding shore. As the different sections of the bay opened and receded from my view, I could not forbear contrasting the appearance of the district at this time with that presented on my first arrival in 1818.

There was the same rich and diversified scenery, but, instead of a few rustic huts, a fine town, two miles in length, now spread itself along the margin of the bay; a good road extended through the settlement; nearly four hundred white, plastered, native cottages appeared, some on the margin of the sea, others enclosed in neat and well-cultivated gardens. A number of quays were erected along the shore; the schools were conspicuous; and prominent above the rest was seen their spacious chapel, since rebuilt, and now capable of accommodating 2000 worshippers. The same individuals, who on the former occasion had appeared uncivilized and almost unclothed islanders, now stood in crowds upon the beach, arrayed in decent apparel, wearing hats and bonnets of their own manufacture; while, beyond the settlement their plantations and their gardens adorned the mountain's side. These were but indications of a greater change among the people. All were professing Christians. Most of them could read the Bible, and between four and five hundred had been united in church-fellowship. This number has been increased to five hundred, who are walking in the ordinances and commandments of the Lord blameless. Agriculture has since increased, and some acres are now planted, or preparing for the culture of coffee.

After sailing from Huahine, we touched at Rurutu, and subsequently at High Island, two branch Missionary stations, and, on parting from the latter, took our final leave of the Polynesian Islands, and the interesting people by whom they are inhabited.

THE END.

London: Henry Fisher, Son, and P. Jackson, Printers.